Polymer Science and Engineering

PRENTICE-HALL INTERNATIONAL SERIES
IN THE PHYSICAL AND CHEMICAL ENGINEERING SCIENCES

NEAL R. AMUNDSON, EDITOR, *University of Minnesota*

ADVISORY EDITORS

ANDREAS ACRIVOS, *Stanford University*
JOHN DAHLER, *University of Minnesota*
THOMAS J. HANRATTY, *University of Illinois*
JOHN M. PRAUSNITZ, *University of California*
L. E. SCRIVEN, *University of Minnesota*

AMUNDSON *Mathematical Methods in Chemical Engineering*
ARIS *Chemical Reactor Analysis*
ARIS *Introduction to the Analysis of Chemical Reactors*
ARIS *Vectors, Tensors, and the Basic Equations of Fluid Mechanics*
BALZHIZER, SAMUELS AND ELIASSEN *Chemical Engineering Thermodynamics*
BERAN AND PARRENT *Theory of Partial Coherence*
BOUDART *Kinetics of Chemical Processes*
BRIAN *Staged Cascades in Chemical Processes*
CROWE ET AL. *Chemical Plant Simulation*
DOUGLAS *Process Dynamics and Control: Volume 1, Analysis of Dynamic Systems*
DOUGLAS *Process Dynamics and Control: Volume 2, Control System Synthesis*
FREDRICKSON *Principles and Applications of Rheology*
HAPPEL AND BRENNER *Low Reynolds Number Hydrodynamics*
HIMMELBLAU *Basic Principles and Calculations in Chemical Engineering*, 2nd ed.
HOLLAND *Multicomponent Distillation*
HOLLAND *Unsteady State Processes with Applications in Multicomponent Distillation*
KOPPEL *Introduction to Control Theory with Applications to Process Control*
LEVICH *Physiochemical Hydrodynamics*
MEISSNER *Processes and Systems in Industrial Chemistry*
PERLMUTTER *Stability of Chemical Reactors*
PATERSEN *Chemical Reaction Analysis*
PRAUSNITZ *Molecular Thermodynamics of Fluid-Phase Equilibria*
PRAUSNITZ AND CHUEH *Computer Calculations for High-Pressure Vapor-Liquid Equilibria*
PRAUSNITZ, ECKERT, ORYE AND O'CONNELL *Computer Calculations for Multicomponent Vapor-Liquid Equilibria*
WILDE *Optimum-Seeking Methods*
WHITAKER *Introduction to Fluid Mechanics*
WILLIAMS *Polymer Science and Engineering*
WU AND OHMURA *The Theory of Scattering*

PRENTICE-HALL, INC.
PRENTICE-HALL INTERNATIONAL, UNITED KINGDOM AND EIRE
PRENTICE-HALL OF CANADA, LTD. CANADA

Polymer Science and Engineering

DAVID J. WILLIAMS

*Associate Professor
of Chemical Engineering
The City College of the
City University of New York*

PRENTICE-HALL, INC.

Englewood Cliffs, New Jersey

© 1971 by Prentice-Hall, Inc.
Englewood Cliffs, N. J.

All rights reserved. No part of this book may be reproduced, by mimeograph or any other means without permission in writing from the publishers.

Current printing (last digit):
10 9 8 7 6 5 4 3 2 1

13 – 685636 – 5

Library of Congress Catalog Card Number: 75-160255
Printed in the United States of America

PRENTICE-HALL INTERNATIONAL, INC., *London*
PRENTICE-HALL OF AUSTRALIA, PTY. LTD., *Sydney*
PRENTICE-HALL OF CANADA LTD., *Toronto*
PRENTICE-HALL OF INDIA PRIVATE LIMITED, *New Delhi*
PRENTICE-HALL OF JAPAN, INC., *Tokyo*

Preface

Polymers are ubiquitous and essential. Biological polymers are of unparelleled importance as constituents of our bodies and of the food we eat. Wood was certainly one of the first engineering materials, and its use is still widespread. Synthetic polymers—the subject of this book—enter virtually every aspect of our lives; most of the objects that surround us are, if not entirely, at least partially polymeric in constitution.

In spite of our longstanding dependence on polymers, their intensive study as a class of materials distinct from their low molecular weight counterparts is a recent phenomenon. One can fix the birth of modern polymer science in the 1920's with the pioneering work of Hermann Staudinger. In dilute solution, polymers exhibit properties analogous to suspended colloids. For this reason it was first thought that they were colloid-like aggregates of low molecular weight materials bound together by physical forces. Staudinger showed that polymers were actually giant molecules comprised of low molecular weight materials bound together by chemical bonds rather than physical forces. For the championship of this viewpoint, as well as other elements of his work, he was awarded the Nobel Prize in 1953.

Polymer science and engineering is a rapidly advancing discipline. The development of nylon in the 1930's and of synthetic rubber during World War II marked the beginning of polymers as important commodities. The polymer industry has grown at nearly four times the annual growth rate of the national economy. Today, thousands of polymer products are manufactured, and over 50% of all chemists and chemical engineers are associated with the polymer industry. Finally, when the American Institute of Chemical Engineers posed the question to a panel of chemical engineering authorities: "What are chemical engineers' ten biggest all-time feats?"; three of the choices were: establishment of the plastics, synthetic fiber, and synthetic rubber industries.*

Polymers are complex materials that exhibit correspondingly complex behavioral patterns. These complexities are reflected in the aura of excitement

---
* Chemical Engineering – December 4, 1967, page 81.

surrounding their study. Many aspects of polymer behavior are centers of considerable debate, and numerous fundamental problems remain to be solved. Nevertheless, there is an underlying body of knowledge that is on a firm basis and that forms the cornerstone of current thought and research. Although many fine texts discuss specific areas of technology or research, or summarize the state-of-the-art, none serve, in any satisfactory way, to bridge the gap between the fundamentals of nonmacromolecular disciplines and the underlying concepts of polymer science and engineering. My primary objective has been to bridge this gap. A firm grasp of the principles herein developed will enable the student to read this advanced literature and to establish his own position in this exciting discipline.

Emphasis is placed on discussing general classes of polymers and their general patterns of behavior. Specific polymers are referred to in citing examples, but the stress is always on the underlying concepts. In selecting and organizing the subject matter, I have been careful to balance scope and depth so as to foster interest and maintain readability. In order to add perspective and a unifying theme, I have taken the viewpoint of the polymer engineer; that is, I have stressed the relation of composition and structure to a polymer's physical and mechanical properties as well as the relevance of the synthesis process to the design of a desired polymer product.

The text centers on a discussion of synthetic organic polymers. It is most suitable for physical chemists, chemical engineers, and materials engineers at the senior or first or second year graduate school level. It is hoped that the subject matter will prove valuable and stimulating to the practicing scientist and engineer as well.

The book has been developed over a six year period through the use of class notes, with presentation to both undergraduate and graduate chemical engineers. There is ample material for a two-semester program. Problem and discussion sets have been included. References are cited in an appended bibliography. The book is divided into four parts: I Introduction, II Polymer Synthesis, III Physics of the Solid State, and IV Polymer Rheology. The nature of the subject matter contained within the four parts is discussed in an introduction preceding each part. Part I is must reading, but the others may be read independently as individual needs and interests dictate. Cross references are supplied where appropriate.

In an undertaking of this sort it is impossible to acknowledge all of one's debts. Nonetheless, special thanks are due to Professor Alois X. Schmidt for his guidance and inspiration during my early years at The City College; to the DuPont Company for their financial support during the summers of 1965 and 1966; to Michael Grancio and David Blum who made contributions to the material for the manuscript; and to Mrs. Norma Cohen who typed the bulk of the manuscript.

David J. Williams

Contents

Part I INTRODUCTION   1

1 Introductory Definitions and Concepts   3

101 *The Nature of High Polymers*, 3.
102 *Seat of Special Properties of High Polymers*, 8.
103 *Examples of Special Polymer Behavior*, 9.
104 *Functionality and Polymerization*, 10.
105 *Two Fundamental Types of Polymerization*, 11.
106 *Copolymerization*, 12.
107 *Polymer Blends and Composites*, 13.
108 *Chemical Modification of High Polymers*, 14.
109 *Degradation and Depolymerization*, 15.
110 *Molecular Weight and Degree of Polymerization of High Polymers*, 16.

2 Intermolecular Forces of Attraction in High Polymers   19

201 *The Nature of Secondary Intermolecular Bonding Forces*, 20.

MOLECULAR ARCHITECTURE

202 *Molecular Weight*, 24.
203 *Molecular-Weight Distributions*, 27.
204 *Development of the Nylons*, 27.

STERIC FACTORS

205 *Structural Regularity*, 29.
206 *Repeat Unit Structure: Composition of the Backbone*, 32.

207 *Repeat Unit Structure: Pendant Groups*, 34.
208 *The Effects of Crosslinking*, 36.

MOLECULAR PACKING

209 *Conformation*, 39.
210 *Morphology and Order*, 42.

Part II POLYMER SYNTHESIS   43

3 Step-Reaction Polymerization   46

301 *Chemistry of Step-Reaction Polymerization*, 46.
302 *Ring versus Chain Formation*, 49.

THEORETICAL ANALYSIS OF POLYCONDENSATION REACTIONS

303 *Mechanisms of Polycondensation Reactions*, 50.
304 *Kinetics of Polycondensation Reactions*, 52.
305 *Stoichiometry of Linear Systems*, 55.
306 *Size Distribution in Linear Condensation Systems*, 57.
307 *Crosslinking in Step-Reaction Systems*, 60.
308 *Criteria for Incipient Formation of Infinitive Networks*, 62.
309 *The Probability* α, 64.
310 *Gel Point Observations*, 65.
311 X_n *in Polyfunctional Systems*, 67.
312 *Size Distributions in Polyfunctional Systems*, 68.
313 *Post-Gel Relations in Polyfunctional Systems*, 72.
314 *Low Temperature Polymerization*, 74.
315 *Copolymerization*, 77.

PRODUCTION OF CONDENSATION POLYMERS

316 *Bulk Polymerization of Condensation Polymers*, 78.
317 *Polymerization of Polyfunctional Systems*, 79.

4 Chain Polymerization   80

401 *Kinetic Processes of Chain Polymerization*, 81.
402 *Effect of Substituents*, 84.
403 *Free Radical Polymerization*, 85.

IONIC POLYMERIZATION

404 *Cationic Polymerization*, 87.
405 *Anionic Polymerization*, 89.

406 *Coordination Polymerization*, 92.
407 *Branching Mechanisms*, 94.
408 *Steady State Kinetic Analysis*, 96.
409 *Kinetics of Homogeneous Free Radical Polymerization*, 97.
410 *Instantaneous Number of Average Degree of Polymerization*, 102.
411 *Thermal Dependence of Kinetic Expressions*, 104.
412 *Molecular-Weight Distributions*, 105.

PRODUCTION OF ADDITION POLYMERS

413 *Homogeneous and Heterogeneous Polymerization*, 110.
414 *Bulk or Mass Polymerization*, 112.
415 *Solution Polymerization*, 112.
416 *Free Radical Suspension Polymerization*, 113.
417 *Suspension Polymerization*, 114.
418 *Emulsion Polymerization*, 114.
419 *Kinetics of Constant-Rate Styrene Emulsion Polymerization*, 120.
420 *A Core-Shell Morphology*, 123.
421 *Molecular Weight Development*, 126.

5 Additional Copolymerization   128

RANDOM AND ALTERNATING COPOLYMERIZATION

501 *The Copolymer Equation*, 129.
502 *The Monomer Reactivity Ratios*, 131.
503 *The Distribution of Sequence Lengths*, 133.
504 *Instantaneous Feed and Polymer Composition*, 134.
505 *Integration of the Copolymer Equation*, 137.
506 *Rate of Copolymerization*, 141.

CHEMISTRY OF ADDITION COPOLYMERIZATION: RANDOM AND ALTERNATING

507 *Radical Copolymerization*, 144.
508 *Ionic Copolymerization*, 147.
509 *Copolymerization of α-Olefins*, 148.
510 *The Q-e Scheme*, 149.

BLOCK AND GRAFT COPOLYMERIZATION

511 *Generalized Concepts*, 151.
512 *Block Copolymerization by the Living Polymer Technique*, 152.
513 *Block Copolymerization by Free Radical Processes*, 152.
514 *Graft Copolymerization by Free Radical Processes*, 155.
515 *Graft Copolymer Formation by Ionic Mechanisms*, 159.

Part III PHYSICS OF THE SOLID STATE 161

6 Morphology and Order in Crystalline Polymers 163

601 *Polymeric Crystal Structures*, 164.
602 *Some Representative Crystal Structures*, 167.
603 *Polymer Single Crystals Grown From Solution*, 172.
604 *Hollow Pyramids*, 176.
605 *More Complex Morphologies*, 178.
606 *The Morphology of Polymers Crystallized from the Melt*, 182.
607 *Interlamellar Ties*, 185.
608 *Polyethylene Crystallized Under Pressure*, 187.
609 *Annealing Polymer Crystals*, 188.
610 *Orientation and Drawing*, 190.

7 Transitional Phenomena 193

701 *First-Order Transitions*, 193.
702 *The Glass Transition*, 196.
703 *Mechanical Properties*, 198.
704 *The Principle of Corresponding Temperatures*, 200.
705 *Transition Temperatures and Engineering-Use Temperatures*, 201.
706 *Molecular Motion and Transitional Phenomena*, 202.
707 *Secondary Transitions*, 205.

THE EFFECT OF COMPOSITION ON POLYMERIC TRANSITIONAL PHENOMENA

708 *The Boyer-Beamen Rule*, 206.
709 *Intermolecular Bonding*, 207.
710 *External Plasticization*, 210.
711 *Steric Factors*, 212.
712 *Molecular Weight*, 216.

COPOLYMERS AND POLYBLEND SYSTEMS

713 *Copolymers with Structural Regularity*, 218.
714 *Homogeneous Copolymers and Compatible Polyblends*, 219.
715 *Semicompatible Copolymer and Polyblend Systems*, 222.

8 Polymer Chain Conformation in Random Systems 225

801 *The Randomly Chain Coil*, 225.
802 *Average Chain Dimensions*, 227.

803 *The Freely Orienting Chain Model*, 230.
804 *The Polymethylene Chain Model*, 231.
805 *The Effect of Restricted Rotation*, 233.
806 *General Ideal Expression for Average Chain Dimensions*, 235.
807 *Random Flight Analysis*, 237.
808 *Some Fundamental Probability Concepts*, 239.
809 *Distribution Function for End-to-End Distances*, 241.
810 *The Distribution of Segments Relative to The Center of Mass*, 246.

9 Rubber Elasticity 248

901 *Definition and Properties of Elastomers*, 248.

THEORY OF RUBBER ELASTICITY

902 *Thermodynamics of Rubber Elasticity*, 253.
903 *Equilibrium Stress-Elongation Experiments*, 256.
904 *Statistical Rubber Elasticity Theory*, 260.
905 *Nonideal Networks*, 265.
906 *Elastically Effective Chain-Sections*, 266.

VERIFICATION OF RUBBER ELASTICITY THEORY

907 *Testing the Stress-Strain Relation*, 267.
908 *The Effect of Network Structure*, 271.

Part IV POLYMER RHEOLOGY 273

10 Introduction to Polymer Rheology 275

1001 *Types of Mechanical Deformation*, 276.
1002 *Simple Rheological Responses*, 277.
1003 *Introduction to the Viscoelastic Properties of Polymers*, 279.

SOME SIMPLE LINEAR VISCOELASTIC MODELS

1004 *The Maxwell Model*, 281.
1005 *The Voigt Model*, 283.
1006 *Series Combination of Maxwell and Voigt Models*, 284.
1007 *The Material Response Time*, λ, 286.
1008 *The Deborah Number*, 287.

GENERALIZED LINEAR VISCOELASTICITY

1009 *The Boltzman Principle*, 288.

RETARDATION AND RELAXATION SPECTRA

1010 *The Maxwell-Weichert (Relaxation) Model,* 290.
1011 *The Generalized Voigt (Creep) Model,* 292.

11 The Linear Viscoelastic Behavior of Polymeric Solids   293

1101 *Creep Experiments,* 294.
1102 *Stress-Relaxation Experiments,* 296.
1103 *Stress-Strain Experiments,* 296.
1104 *Oscillatory Experiments,* 298.

THE ELASTIC MODULUS

1105 *Time-Temperature Equivalence,* 299.
1106 *The Time-Temperature Superposition Principle,* 299.
1107 *Five Basic Properties,* 302.

OSCILLATING EXPERIMENTS

1108 *Rheologically Simple Bodies,* 303.
1109 *Introduction to Viscoelastic Responses,* 304.
1110 *The Voigt Element,* 307.
1111 *The Maxwell Element,* 310.
1112 *The Maxwell Element with Mass,* 311.

PHENOMONOLOGICAL ASPECTS OF DYNAMIC MECHANICAL TESTING

1113 *Time-Temperature Relations,* 314.
1114 *Damping Peaks,* 320.
1115 *The Effect of Crosslinking,* 320.
1116 *The Effects of Plasticization,* 321.
1117 *The Effect of Copolymerization,* 322.
1118 *Multiple Relaxation (Transition) Peaks in Polymers,* 325.
1119 *Pendant Group Motion and Secondary Relaxation,* 328.
1120 *In-Chain Motion and Secondary Relaxation,* 330.
1121 *Relaxation and Retardation Spectra,* 334.

12 Introduction to Polymer Melt Rheology   340

1201 *The Viscous Character of Non-Newtonian Fluids,* 341.
1202 *The Flow Characteristics of Polymer Melts,* 342.
1203 *The Capillary Rheometer,* 345.
1204 *Aspects of Flow Behavior in Capillaries,* 346.

THE VISCOUS CHARACTER OF POLYMER MELT FLOW

1205 *The Laminar Flow of Newtonian Fluids*, 348.
1206 *The Consistency Variables*, 352.
1207 *The Power Law*, 353.
1208 *A General Treatment of Isothermal Viscous Flow in Tubes*, 355.
1209 *The Effect of Temperature*, 357.
1210 *Entrance and Exit Effects*, 360.
1211 *The Relaxation Region and Jet Swelling*, 364.
1212 *Viscometric Flow*, 365.
1213 *Unstable Flow*, 366.
1214 *Inlet Borne Instability*, 370.

Part V APPENDICES   373

A. Polymer Nomenclature,   375

B. Problems,   377

C. Bibliography,   389

    Index,   393

PART I

Introduction

In Chapters 1 and 2, which comprise Part I, numerous aspects of polymeric materials are introduced at an elementary level in order to establish a broad overview of the subject. Many of the topics touched on here will be developed in detail in later chapters. Others must remain undeveloped, but references are provided in the bibliography to allow the reader to pursue these subjects in more detail if he so desires.

In addition to the obvious need for establishing working definitions of what comprises a polymer and how they are formed, a primary objective will be to see how and why the behaviors of polymeric materials differ from those of low molecular weight materials. We shall see that these differences are profound, and that polymers often exhibit unique properties directly attributable to their polymeric constitution.

Chapter 2 begins the detailed development of an underlying theme: to relate polymer composition and structure to physical and mechanical properties.

1

Introductory Definitions and Concepts

High polymers are large molecules, with molecular weights on the order of 10^4 to 10^6. They may be synthetic or natural in origin. Polymeric materials are among the most common of all materials. They include such useful and essential materials as the films that package the foods we eat, the fibers that make up the clothes we wear, the rubber that goes into all our truck and auto tires, and the plastic that makes so many of the articles we use in our daily living. Polymeric materials also find their way into more exotic applications, such as in medicine and space. Indeed, our own bodies and the food we eat are largely composed of high polymers.

101 The Nature of High Polymers

These high-polymer molecules are built up by the repetitive chemical linking of small, simple units into long, chainlike structures much as a chain is constructed of links. Thus polyethylene, one of the simplest high-polymer molecules, is composed of many —CH_2—CH_2— units linked together by covalent bonding. The word *polymer* is derived from the Greek and means "many parts." The starting materials from which polymers are derived are known as *monomers*. For example, polyethylene is derived from ethylene.

The units that are repeated throughout the polymer chain and that characterize the chemical composition of the polymer are known as *repeat units*. The repeat unit is generally composed of one or two monomer units and is well defined for simple polymer systems. A few examples of such systems are given in Table 1-1. Rules of polymer nomenclature are discussed in Appendix A. In polystyrene, the repeat unit is composed of one monomer unit and the chemical compositions are nearly identical. On the other hand, the repeat unit in poly(hexamethylene adipamide) is composed of two monomer units and the compositions differ by two molecules of water.

TABLE 1-1

EXAMPLES OF SOME SIMPLE POLYMER MATERIALS

Polymer	Repeat Unit	Monomer Unit(s)
Polyethylene	$-CH_2-CH_2-$	$CH_2=CH_2$
Polystyrene	$-CH_2-CH(C_6H_5)-$	$CH_2=CH(C_6H_5)$
Polyisoprene	$-CH_2-C(CH_3)=CH-CH_2-$	$CH_2=C(CH_3)-CH=CH_2$
Poly(hexamethylene adipamide)	$-NH(CH_2)_6-NH-C(=O)-(CH_2)_4-C(=O)-$	$NH_2-(CH_3)_6-NH_2$ and $HOOC-(CH_2)_4-COOH$
Polycaprolactam	$-NH-(CH_2)_5-C(=O)-$	$\overset{\frown}{NH-CH_2-(CH_2)_4-C=O}$
Poly(dimethyl siloxane)	$-Si(CH_3)_2-O-$	$(CH_3)_2-SiCl_2$ and H_2O

Let us depict a single repeat unit in a polymer molecule as a bead and the entire molecule as a strand of these beads. If the repeat units are arranged in a single-stranded structure as shown in Figure 1-1(a), the molecule is said to be linear, no matter how the strand may be snaked or flexed, and we speak of a *linear polymer*. If the units are joined in a three-dimensional array as depicted in Figure 1-1(c) (in two dimensions), the polymer is said to be *crosslinked*. Such a polymer may be visualized as an array of linear molecules joined by chemical crosslinks into a network of macroscopic proportions. If an otherwise linear chain has side-chain appendages, as in Figure 1-1(b), where each molecule is still separate and discrete from its neighbors, the molecule is said to be nonlinear, and it is now referred to as a *branched polymer.* As will be explained later, fundamental differences exist between these three types, and they should not be confused. We will see that structural variations on the linear chain also occur.

In addition to the molecular shape fixed by chemical bonding, as in the linear, branched, and crosslinked varieties of molecules, variations in the overall shape and size of molecules arise through rotation of the chain atoms about primary valence bonds. Just as a chain of beads exhibits some degree of flexibility, so does a polymer molecule. For example, some molecules like polyethylene exist in the crystalline state in a planar zigzag form (bond length = 1.54Å, bond angle = 109° 28′), as illustrated in Figure 1–2(a).

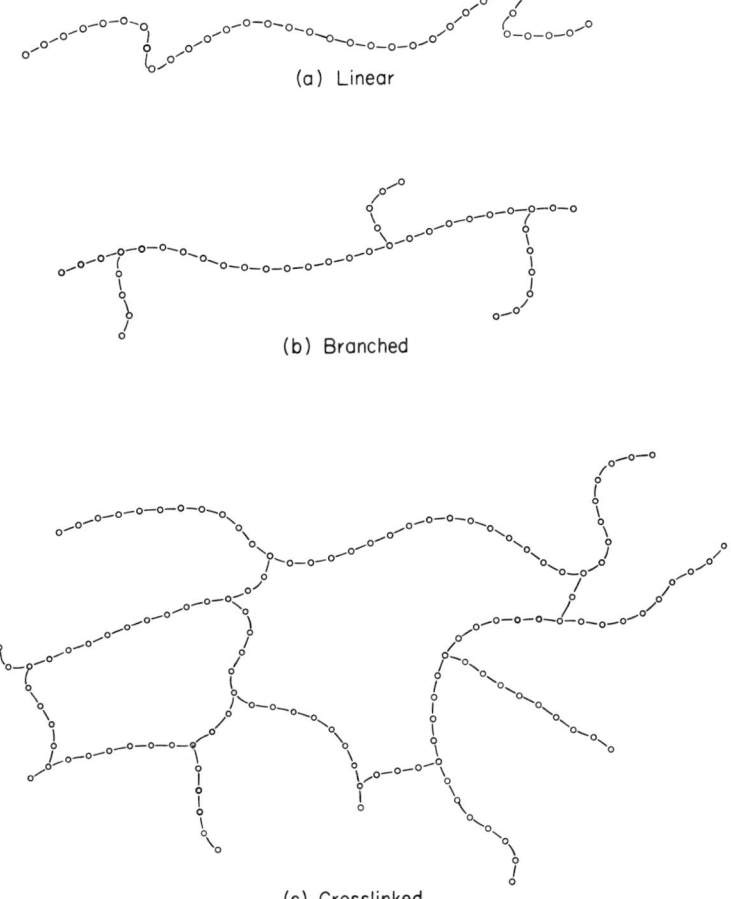

Figure 1–1. Representation of linear, branched, and crosslinked polymer systems.

We visualize that in the molten state, the same molecules may exist in the form of random coils, as illustrated in Figure 1–2(b). For any particular polymer, the degree of flexibility will be determined by such structural parameters as bond lengths and bond angles, as well as by various other steric factors.

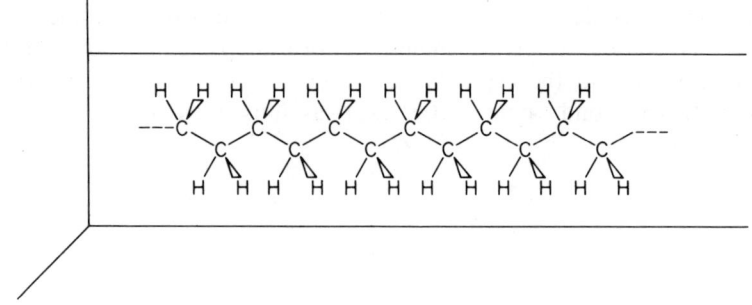

(a) Planar zigzag form

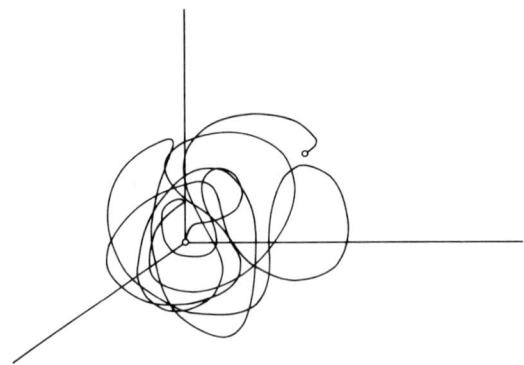

(b) Random coil*

*Restrictions which would be imposed by the individual atoms and structural details are not shown.

Figure 1–2. Examples of polymer shape established by rotation about covalent bonds.

Thus polymer shape must be considered in two contexts. Polymer shape established by primary valence bonding is herein referred to as polymeric *configuration*, whereas polymer shape established by rotation about primary valence bonds is herein referred to as polymeric *conformation*. Both the chains of Figure 1–2 have linear configurations, but they are shown in different conformational arrangements. The distinction between configuration and conformation will be amplified in Chapter 2. Here it is only necessary to realize that such considerations are especially important in polymer systems compared to nonpolymer systems because of the wide variety of shapes in which even a polymer consisting of a single, simple repeating unit may be found. For example, polyethylene is produced in linear, branched, and crosslinked configurational forms; and as we have already indicated, the linear form can exist in both planar zigzag and random conformational forms.

The vast majority of commercially important polymers are organic in nature, and these polymers will be of primary concern in this book. Many polymers, however, may be purely inorganic in nature, or they may be composed of a mixture of organic and inorganic components, in which case they are referred to as *organometallic polymers*. Asbestos $[Mg(OH)_6Si_4O_{11}]$ is a naturally occurring inorganic polymer. The polyphosphate chain in the polynucleotide deoxyribonucleic acid (DNA) is a biological polymer of unparalled importance. This material is a mixture of inorganic and organic components. The composition of the phosphate chain is

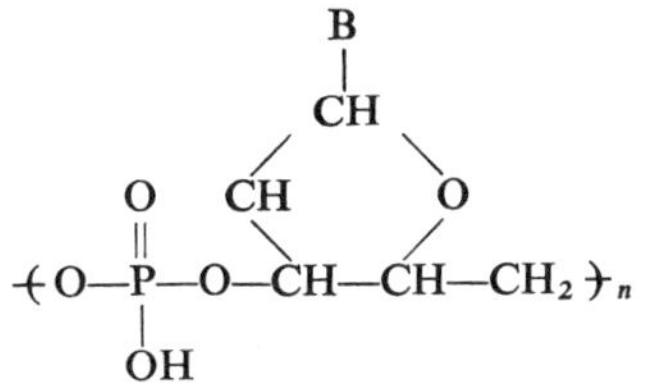

$$\left(O-\underset{\underset{OH}{|}}{\overset{\overset{O}{\|}}{P}}-O-CH-\overset{\overset{B}{|}}{\underset{\underset{O}{\diagdown}\diagup}{CH}}-CH-CH_2 \right)_n$$

where B represents one of the purine or pyrimidene bases.

PURINES

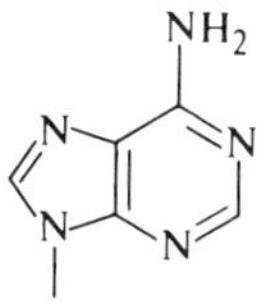

Adenine 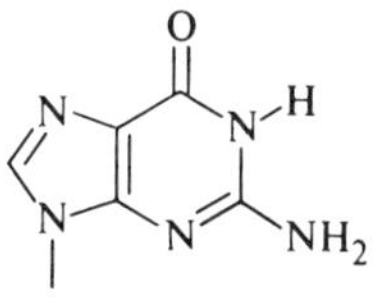   Guanine

PYRIMIDENES

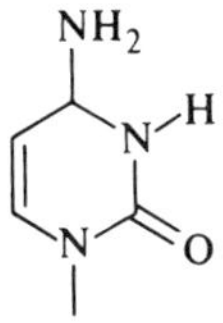

Cytosine 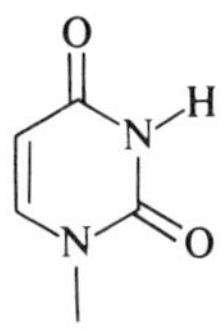   Uracil 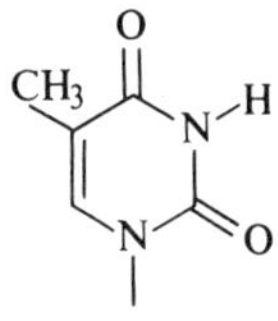   Thymine

Poly(dimethyl siloxane) is the only significant synthetic, organometallic polymer. A wide variety, generically known as silicone polymers, are available.

102 Seat of Special Properties of High Polymers

It is well known that high-polymeric materials exhibit behaviors and properties that are different from those of ordinary, low-molecular-weight compounds. Consider, for example, the paraffins, which are the low-molecular-weight homologs of polyethylene. Paraffins are waxy solids with little mechanical strength, whereas polyethylene is tough with considerable mechanical strength. The forces accounting for these differences, however, are a matter of degree rather than kind. In general, the bulk behavior of a collection of molecules depends on the nature and magnitude of the intermolecular forces existing between molecules. *The remarkable and sometimes unique behavior of a collection of polymer molecules resides primarily in their large size and chain-like structure.* Their large size and chain-like structure, in turn, result in strong fields of attraction between these molecules as well as in a high degree of physical entanglement or interaction.

There are also a number of additional fundamental factors that must be considered. It is convenient to discuss all factors relating to the determination of polymeric properties, including molecular size and shape, under the following headings: (1) the nature of the secondary intermolecular bonding forces, (2) molecular architecture, and (3) molecular packing. A large part of Chapter 2 will be devoted to elucidating the nature of these factors and citing pertinent examples. In addition, the development of this text will tend to center around them. Briefly, the properties of a polymer are governed to a large extent by the nature of the secondary *intermolecular bonding forces* associated with the molecules. For example, given all other factors equal, we would expect poly(vinyl alcohol)

$$-(CH_2-CH)_n-$$
$$\quad\quad\quad |$$
$$\quad\quad\quad OH$$

to have a higher-cohesive energy density (because of hydrogen bonding and polar forces) and, therefore, a higher-intermolecular force field associated with it than polyethylene $-(CH_2-CH_2)_n-$. *The molecular architecture* refers to the size and shape of the molecule imparted through primary valence bonding. Molecular weight or length is of primary concern here, for the high intermolecular forces of attraction in polymers are directly attributable to their large size. Branching and crosslinking are also shape factors imparted through chemical bonding. *Molecular packing* refers to the manner in which aggregates of molecules are arranged in the bulk. In fibrous materials, for instance, intermolecular forces exert a maximum effect because the molecules are packed in a highly ordered, closely packed, parallel arrangement. On the other hand, these forces are relatively weaker in amorphous materials, for the molecules are packed in a random fashion.

103 Examples of Special Polymer Behavior

While on the subject of the factors accounting for the unusual behavior of polymeric systems, it would seem appropriate to cite a few such examples. Many examples will, of course, be cited throughout the text.

One of the most outstanding examples is to be found in the range of their rheological behavior—that is, in their deformation and flow properties. For low-molecular-weight fluids, the flow behavior is virtually always Newtonian, and the behavior of structural solids is often strictly elastic. Polymers in the molten state, or in anything but dilute solution, are not only non-Newtonian but also elastic. The behavior of polymeric solids is characterized in many instances by considerable viscous flow as well as by elastic behavior. Materials that simultaneously exhibit both viscous flow and elastic deformation under a wide variety of rheological conditions are classified as *viscoelastic*. Polymers are highly viscoelastic and often dramatically so.

Another example of the remarkable behavior of polymeric materials is found in the variation of their mechanical properties with temperature, as illustrated in Figure 1–3. At sufficiently low temperatures, amorphous polymers are hard and glasslike, and they have a relatively high mechanical strength. This state of affairs is maintained as the temperature is raised until a critical temperature region is reached. Within this region, the polymer will change from a hard, glasslike, inflexible material to a softer, rubbery, flexible

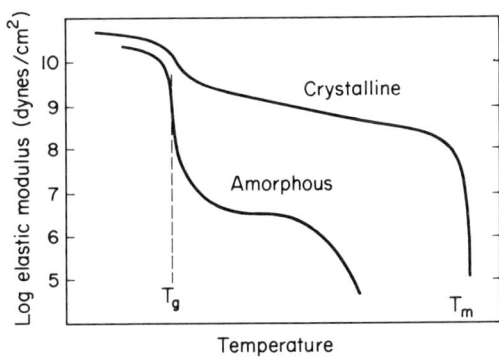

Figure 1–3. Log elastic modulus vs. temperature for amorphous and crystalline polymers.

material. Indeed, the ability to exhibit rubber elasticity is a property unique to polymers. This drastic transition in mechanical properties is referred to, quite descriptively, as the *glass-transition*, and a critical temperature, known as the *glass-transition temperature*, T_g* is thereby defined. In addition, crystalline polymers exhibit a *crystalline melting transition* analogous to that encountered in low-molecular-weight organic compounds. Again, as shown in Figure 1-3, there is a dramatic loss in mechanical strength at a critical transition temperature, which is *near* the equilibrium crystalline melting point, T_m. Unlike low-molecular-weight crystalline materials, crystalline polymers melt over a temperature range, instead of at a well-defined single temperature, and the phenomenon is subject to a strong hystersis effect.

104 Functionality and Polymerization

Natural high polymers are formed in geological processes and the life processes of plants and animals by mechanisms that are now only beginning to be elucidated. The well-read student cannot help but notice the wide interest among biologists and biology-oriented scientists in the chemistry of life processes, especially with what are known as the biological polymers or biopolymers. Perhaps the most publicized advance in this field has been the elucidation of the structures and roles of DNA and RNA in the genetic processes. Needless to say, the results of other studies are also of importance and interest. As interesting as this field of biological polymers might be, we will not discuss them here, for they are outside the scope of this text.

As has already been noted, a synthetic polymer is made by causing a great number of monomer units to join chemically by covalent bonding to form large molecules. In order for this process to occur, it is essential that the chemical nature of these monomers be such that each can "hook up" by covalent bonding with at least two other units. We define *functionality* as the ability to form covalent bonds, and we note that monomer units must have a functionality of at least two. If this latter condition is met, processes leading to molecular growth and the generation of high-molecular-weight polymers can take place. The chemical and physical processes required to generate high polymer may be quite complex. Collectively, we describe these processes as *polymerization*.

As a simplified illustration of these principles, consider a hydroxycarboxylic acid whose structure is HO—R—COOH, where R represents an alkyl chain. This molecule has two functional groups, the —OH and the —COOH, both of which can participate in esterification. Under the proper conditions, two of these monomers can be made to react.

*Although this transaction actually occurs over a modest temperature range, for convenience we specify a single value that is in the middle of the range. See Figure 1-3.

$$2 \text{ HORCOOH} \longrightarrow \text{HORCO—ORCOOH} + H_2O$$

The resulting dimer is an ester roughly twice the molecular weight of the original monomer, and at one end it is still an alcohol and at the other end a carboxylic acid. It can react with another monomer to form a trimer, another to form a tetramer, and so forth. This stepwise process may be repeated many times to form high polymer. The reaction can be represented as follows:

$$n \text{ HORCOOH} \longrightarrow \text{H}(\text{ORCO})_n \text{OH} + (n-1)H_2O$$

The number and kinds of functional groups present in any set of reactants bear a simple and fundamental relationship to the product obtainable from the reaction. When the functionality of all reactants is two, as was the case in the preceding example, the resulting polymeric configuration is linear. Monomer systems that form linear polymers may be of two types. A *bifunctional system* consists of one monomer unit capable of reacting with itself—for example, a hydroxcarboxylic acid. A *bi-bifunctional system* consists of two monomer units that do not react with themselves but with each other—for example, glycol and a dicarboxylic acid.

A *polyfunctional system* is one in which one of the reactants is at least trifunctional and the rest of the system is bifunctional or bi-bifunctional; for instance, a glycerol can be combined with a glycol and a dicarboxylic acid. Obviously, new configurational possibilities present themselves. Each polyfunctional unit that becomes incorporated in a polymer molecule serves as a *branch unit* by introducing sites for the growth of a branch chain. A branch may grow via bifunctional units for a period of time and then react with a branch unit from another chain, in which case a *chemical crosslink* forms. This process may continue until an infinite, three-dimensional network forms, or it may never proceed past the stage of adding a few branches to form a *branched polymer*. The nature of the products formed always depends on the functionality of the reactants, their stoichiometric proportions, and the extent to which they have reacted. Generally, in such a polyfunctional system, linear, branched, and crosslinked structures are all formed, the relative amounts depending on the factors listed.

105 Two Fundamental Types of Polymerization

There are two fundamental polymerization mechanisms. Classically, they have been differentiated as *condensation polymerization* and *addition polymerization*. In the condensation process, a low-molecular-weight by-product is evolved, as illustrated in Section 104; whereas in the addition process, no by-product is evolved, as in the polymerization of ethylene.

$$n(CH_2=CH_2) \longrightarrow (CH_2-CH_2)_n$$

More exactly, in that the actual mechanism of polymerization is reflected in the terminology, these types are currently referred to as step-reaction polymerization and chain-reaction polymerization, respectively.

Polymers formed by *step-reaction* (*condensation*) *polymerization* are usually referred to as condensation polymers. They form by the stepwise, intermolecular condensation of reactive groups. Each step of the process—that is, reaction between each pair of functional groups—requires individual activation. For each reaction pair, the mechanism is analogous to that for the condensation of their low-molecular-weight, monofunctional counterparts.

Polymers formed by *chain-reaction* (*addition*) *polymerization* do so by the successive addition of unsaturated monomer units in a chain reaction promoted by an active center. The establishment of an active center requires the initial activation of an initiator or the presence of a catalyst. Monomer units may then add to the center at fantastically rapid rates—for example, on the order of 10^3 to 10^4 units per second. These active centers may be free radicals, anions, or cations; or they may consist of a coordination complex between a catalyst surface, the growing polymer chain, and the adding monomer unit. (In passing, we should note that the last type of active center has been the object of much recent interest in that a whole new class of polymers has resulted, namely, polymers with a stereo-regular architecture.) Monovinyl monomers are bifunctional, and as such they should yield linear polymers. In certain instances, however, branching may occur as a side reaction. Polyfunctional systems can be made by the use of diene and divinyl monomers: for instance, butadiene and divinyl benzene, respectively.

106 Copolymerization

The simultaneous polymerization of two or more bifunctional or bi-bifunctional monomer systems, each of which is capable of polymerizing by itself (*homopolymerization*), is a process called *copolymerization*. For example, styrene and butadiene can individually undergo homopolymerization; combined, they readily copolymerize in such a way that their units are randomly distributed along the chain structure. When the weight ratio of butadiene to styrene is 75 to 25, a copolymer is obtained that is the source of much of today's tire rubber. It was developed during World War II to meet the needs caused by the shortage of natural rubber.

By varying both the kinds and relative amounts of two monomers used in the copolymerization, a series of products may be manufactured with a considerable spread in chemical, physical, and mechanical properties. The range of useful products may be further extended by the copolymerization of three, four, or even more monomeric species, often of radically different structures and chemical composition. From a relatively small number of monomeric raw materials, it is thus possible to make a tremendous number of polymeric products. In a sense, copolymerization does for high polymers what alloying does for metals. *Of the synthetic polymers in use today, by far the greater number are copolymers.*

For the copolymerization of two bifunctional or two bi-bifunctional monomer systems, we can define four types of copolymers. These represent ideal models that are not always obtained in practice. They are defined and illustrated schematically below, using the symbols A and B for the repeating structures:

1. *Random copolymers.* The units A and B recur in random-length sequences along the polymer chain.

—A—B—B—B—A—A—A—A—A—B—A—A—B—B—B—A—B—B—B—B—B—A—

2. *Alternating copolymers.* The units appear in the chain in alternating positions.

—A—B—A—B—A—B—A—B—

3. *Block copolymers.* The units appear in long, alternating segments of random length.

—A—A—//—A—A—B—B—//—B—B—A—A—//—A—A—

4. *Graft copolymers.* One of the units makes up the main chain, whereas the other units appear in short branches grafted to this chain.

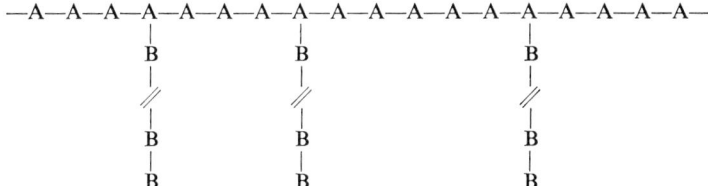

Generally, as one can see, the repeat unit is not well defined for copolymers. Nomenclature is discussed in Appendix A.

107 Polymer Blends and Composites

Further modification in properties can be achieved by the physical combination of two polymers or by the physical combination of a polymer with a non-polymeric component. Two types of systems are encountered, depending on the nature of the components and the way in which they are combined.

Polymer blends or polyblends are formed by intimately blending two polymers. This process may be done by mixing the polymers while they are in solution or in a molten, fluid state. Because of the inherent incompatibility of polymer chains of differing composition, polymer blends rarely form true solid solutions; and in the solid state, one polymer component is found dispersed in a continuous matrix of the other polymer component. Moreover, because of the incompatibility of polymer chains of differing composition, block and graft copolymers have much in common with polymer blends. Judicious blending leads to materials with improved impact resistance and yield strength.

Composites are formed by combining a polymer of some type with a nonpolymeric solid or by combining a polymer of one form with a polymer of another form. Composites are among the most important types of engineering materials. A few examples will illustrate this point. Automobile tires are composed of a rubbery polymer, a fibrous polymer (tire cord), carbon black, and steel wire to reinforce the bead. Paints primarily contain a polymeric binder, filler, pigment particles, and a polymeric viscosity modifier. Reinforced fiber glass consists of laminated layers of fiber-glass cloth impregnated and coated with a three-dimensional network polymer. Plywood is formed from alternating layers of thin sheets of wood (a natural polymer) and a polymeric adhesive. Foam products result from dispersing cells of air in a continuous matrix of polymer.

Novel composites using graphite and saphire whiskers as the reinforcing agent have raised the modulus of elasticity beyond that of any engineering material. These so-called *whisker composites* have modulus values of up to 50×10^6 psi compared to 29×10^6 psi for stainless steel. When combined with temperature-resistant polymers, the use temperatures of polymer systems can be raised to as high as 650 to 900 °C compared to 100 °C for many common linear, noncrystalline polymers or 300 °C for common crosslinked systems. Whisker composites find their most exotic engineering role in the aerospace applications for which they were specifically developed.

As important and exciting as this area of composite materials may be, further discussion is beyond the scope of this book. The subject of polymer blends and the closely associated block and graft copolymers will be discussed in subsequent chapters.

108 Chemical Modification of High Polymers

Chemical transformations are sometimes performed on synthetic polymers after they are formed to improve their properties. Natural polymers that come ready-made from various biological and geological processes seldom have properties that suit us; therefore we modify them, usually by chemical transformation. We cite a few important examples here to indicate the scope of modifications that are made.

Natural rubber and synthetic rubber contain double bonds in the main chain, enabling us to crosslink them with sulfur. These materials would be useless as rubbers without this light crosslinking, and the discovery of this process by Goodyear in 1839 marked a significant event in the development of polymer materials for useful artifacts. He named the process *vulcanization* after Vulcan, the Roman God of Fire.

The double bond, however, does create a source of weakness in that the rubber is open to the deteriorating actions of heat, light, oxygen, and ozone. By the addition of chlorine, chemically stable polymer derivatives that are excellent for certain applications can be made.

One synthetic polymer that is not obtained by polymerizing the monomer —because the monomer is nonexistent—is poly(vinyl alcohol). It is made by the hydrolysis of poly(vinyl acetate).

Cellulose was probably the first natural polymer to be modified, and it still is one of the most important sources of what one might call derived polymer materials. In spite of its linear configuration, cellulose in its native form (cotton and wood fiber) is insoluble, infusible, and nonmoldable because each cellulose repeat unit

$$\begin{array}{c} \text{OH} \quad \text{OH} \\ | \quad\quad | \\ \text{CH}-\text{CH} \\ \diagup \quad\quad\quad \diagdown \\ -\text{CH} \quad\quad\quad \text{CH}- \\ \diagdown \quad\quad\quad \diagup \\ \text{CH}-\text{O} \\ | \\ \text{CH}_2\text{OH} \end{array}$$

has three hydroxy groups; therefore the material is strongly polar and has high intermolecular cohesive forces. Soluble, moldable, derivatives may be made by substituting ester, ether, or other groups for part or all of the original groups. Important examples are cellulose acetate and cellulose nitrate.

109 Degradation and Depolymerization

Just as polymers may be built up, they may also be broken down. This degradation might be caused by heat, ultraviolet light, oxygen, ozone, or hydrolysis. Most common polymers thermally decompose in an inert atmosphere, on the order of 250 to 300 °C. Two types of thermal degradation processes are step reactions and chain reactions.

1. *Step reaction degradation* is a random, stepwise process where the

breaks in the original chain occur at random along the backbone. The result is a dispersed mixture of products.

2. *Chain reaction depolymerization* literally involves an "unzippering" of the original chain. The products consist largely of the original monomers.

These decremental processes are highly undesirable in most polymer applications, for they considerably shorten the useful life of a material. In such instances, every precaution is taken to guard against them. They frequently occur in polymer-processing equipment where considerable heat may be generated through mechanical working. One of the severest limitations imposed on the use of polymeric materials is the relatively low temperatures at which they decompose. Much research is being devoted to devise means of improving their thermal resistance. In space reentry vehicles, however, thermal degradation is used to advantage in a process known as ablation; in this case, the degrading polymer serves as a heat sink, absorbing heat to activate the degradation. In addition, the escaping gases carry away heat as they flow away from the vehicle.

110 Molecular Weight and Degree of Polymerization of High Polymers

The size of a single polymer molecule may be expressed in terms of its molecular weight or its degree of polymerization. The *degree of polymerization* of a polymer molecule is determined by the number of monomer units required to form that molecule. In bifunctional systems, the number of repeat units equals the degree of polymerization, whereas in bi–bifunctional systems, the degree of polymerization is twice the number of repeat units. The molecular weight is simply determined by the molecular weight of the repeat unit and the number of these units in the polymer chain.

During the synthesis process, polymer molecules are subjected to a series of random events such that they do not all grow to the same size. Instead, in a collection of polymer molecules, there is usually a distribution of sizes. Under certain conditions (e.g., linear growth), this distribution may be fairly narrow, while under others (e.g., chain branching) it can be quite broad. Biopolymers may be remarkably uniform in size. In systems where there is a distribution of sizes, it is necessary to express a characteristic molecular weight or degree of polymerization as an average value. The average degree of polymerization is defined and measured as the total number of monomer units converted to polymer divided by the total number of polymer molecules. This is a number average, and the average degree of polymerization is represented by $\overline{DP}_n$ or $\bar{X}_n$. Mathematically, it is expressed as

$$\bar{X}_n = \frac{N(0)-N(t)}{P(t)} \tag{1-1}$$

where $N(0)$ = number of monomer units present initially
$N(t)$ = number of monomer units present at time t
$P(t)$ = number of polymer molecules present at time t

Four molecular-weight averages are generally defined; these are defined mathematically below in equivalent forms.

$$\text{Number average} = \bar{M}_n = \frac{\Sigma N_i M_i}{\Sigma N_i} = \frac{1}{\Sigma(w_i/M_i)} \tag{1-2}$$

$$\text{Weight average} = \bar{M}_w = \frac{\Sigma W_i M_i}{\Sigma W_i} = \frac{\Sigma N_i M_i^2}{\Sigma N_i M_i} = \Sigma w_i M_i \tag{1-3}$$

$$z \text{ average} = \bar{M}_z = \frac{\Sigma N_i M_i^3}{\Sigma N_i M_i^2} = \frac{\Sigma W_i M_i^2}{\Sigma W_i M_i} \tag{1-4}$$

$$\text{Viscosity average} = \bar{M}_v = \left(\frac{\Sigma N_i M_i^{1+a}}{\Sigma N_i M_i}\right)^{1/a} = (\Sigma w_i M_i^a)^{1/a} \tag{1-5}$$

where a = constant between 0.6 and 0.8
M_i = molecular weight of the ith species
N_i = moles of molecules with molecular weight M_i
W_i = weight of material with molecular weight M_i
w_i = weight fraction of molecules of the ith species

The weight fraction w_i is computed from the following relationship:

$$w_i = \frac{N_i M_i}{W_t} = \frac{W_i}{W_t} \tag{1-6}$$

where W_t = total weight of polymer

The particular average obtained depends on the analytic device used in its determination.* The number, weight, and z averages can be determined directly from experiments, whereas the viscosity average is an empirical technique utilizing standards for calibration in the measurement of dilute-solution viscosities. The manner in which these averages are related to

---

*A discussion of how the molecular weights are determined analytically will not be taken up here although it is a very important topic. Reference should be made to the literature cited.

polymeric properties will be discussed later. The number average corresponds to our usual concept of the average, and the number-averaging process is the one encountered in most of our everyday averaging procedures. The significance of $\bar{M}_w$ is more abstruse, but some insight into its nature can be gained by way of an analogous physical situation.

Consider a cantilever beam that is loaded, as illustrated in Figure 1-4, with a plate of, say, sheet steel. The shape of the plate, and hence the distribution of weight on the beam, is meant to simulate a distribution of molecular weights. If we designate distance from the fulcrum point as M and then make a moment balance by dividing the plate into many vertical slices, each slice weighing W_i and being M_i distance from the zero point, we obtain for our resultant moment arm $\Sigma W_i M_i / \Sigma W_i$, which is exactly equal to $\bar{M}_w$.

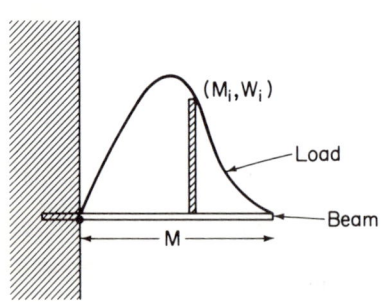
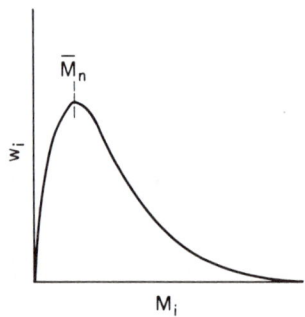

Figure 1-4. Cantilever beam loaded with plate steel.

Figure 1-5. Typical molecular weight distribution.

Figure 1-5 shows a typical molecular-weight-distribution curve, where w_i is plotted against M_i. It can be shown for a number of systems that the peak of the distribution curve nearly corresponds to $\bar{M}_n$ as shown. From Eq. (1-2) through (1-5), we note, that for a system with a distribution, the following relationships hold: $\bar{M}_n < \bar{M}_v < \bar{M}_w < \bar{M}_z$. The ratios $\bar{M}_z/\bar{M}_n$ and $\bar{M}_w/\bar{M}_n$ are often used to indicate the breadth of the distribution. For many linear systems, $\bar{M}_w/\bar{M}_n$ is about 2. For a homogenous molecular-size system, $\bar{M}_n = \bar{M}_v = \bar{M}_w = \bar{M}_z$. In highly nonlinear (branched) systems, $\bar{M}_w/\bar{M}_n$ may range as high as 20 to 50. Considerably more insight will be gained into the nature of these averages during subsequent discussion.

2

Intermolecular Forces of Attraction in High Polymers

The unusual properties of high polymers are not due to bonding forces that are unique to polymer systems, but rather they are the same bonding forces—both intramolecular and intermolecular—that exist in their low-molecular-weight counterparts. To reiterate: the remarkable behavior of an aggregation of polymer molecules results primarily from their large size and flexible, chain-like structure. These two factors in turn, result in large total force fields of attraction between the individual molecules plus a high degree of physical entanglement and physical interaction. The magnitude of these intermolecular interactions depends on at least three fundamental factors. These factors were presented in Chapter 1 and are shown in expanded outline form in Table 2-1. They are discussed in the remainder of the chapter in enough detail to give the

TABLE 2-1

FUNDAMENTAL FACTORS AFFECTING THE MAGNITUDE
OF INTERMOLECULAR FORCE OF ATTRACTION IN HIGH POLYMERS

I		Nature of secondary intermolecular bonding forces
II		Molecular architecture
		Molecular weight
		Steric factors
		Crosslinking
III		Molecular packing
		Conformation
		Morphology and order

reader an idea of their fundamental importance and to outline the scope of the terms. Further amplification of these factors and how they relate to the properties of polymers will be a recurring process throughout this text.

Although Table 2-1 should serve as a valuable study guide, the reader should also be advised that it is difficult to reverse the process and use the concepts embodied in the table as a basis for the design of polymer systems. The reason is because very often it is difficult to control the factors independently. Several factors may be operating simultaneously at cross purposes in a particular system. One of the most desirable goals in polymer science is the design of polymer systems based on the quantification or analytic representation of such factors, but even the qualitative use of these ideas is in its infancy. This approach, however, is certain to receive considerable attention in the future development of this discipline.

201 The Nature of Secondary Intermolecular Bonding Forces

The potential of a high-polymer system to exert strong intermolecular bonding, and thus have high mechanical strength, resides fundamentally in the nature of the *secondary molecular bonding forces*. The strength of secondary valence forces, as reflected in their bonding energies, ranges from 0.5 to 10 kcal/mole. Energies of such a magnitude seem low when compared to the energies of primary valance forces, which range from 50 to 200 kcal/mole. As we shall see in the discussion of the role of molecular weight in Section 202, it is the fact that the secondary forces are associated with extremely large molecules that is of paramount importance.

Most organic polymer molecules contain $-CH_2-$ sequences, which may range in length from a very few to tens of thousands. For example, in polystyrene, the sequence is only one unit long; in poly(1,4-ethylene terephthalate) it is two units long; and in poly(hexamethylene adipamide) there are two $-CH_2-$ sequences, one four and the other six units long. Polyethylene represents the extreme case where the sequence lengths are as long as 20,000 units. These $-CH_2-$ units have only weak van der Waals forces associated with them, in which bonding energies range from 0.5 to 5 kcal/mole.

Many polymers contain such chain unit sequences as the following:

$$-\underset{\text{amide}}{NH\overset{\overset{\displaystyle O}{\|}}{C}-} \;,\; -\underset{\text{urethane}}{N\overset{\overset{\displaystyle O}{\|}}{C}HO-} \;,\text{ or }\; -\underset{\text{urea}}{NH\overset{\overset{\displaystyle O}{\|}}{C}NH-}$$

These groups provide sites for hydrogen bonding in which the energy ranges from 5 to 10 kcal/mole. Materials containing appreciable amounts of these units will be stronger and have higher crystalline melting points than, say, polyethylene. This fact is indeed true as can be seen in comparing the melting points of polyethylene and poly(hexamethylene adipamide)—135°C versus 265°C.*

Molecules that contain —O— units or that have side-chain units of —CN, —Cl, —F, or —NO$_2$ exhibit polar bonding with energies generally between those of hydrogen bonding and van der Waals bonding.

It is also possible to bond chains by ionic bonding. Such bonding is properly considered as resulting from primary valance forces, but we consider it here because it is reversible. (In Section 207 we shall discuss the properties of irreversibly bonded or crosslinked systems.) Certain monomers may be copolymerized with acrylic acid, CH_2=C(COOH)H, or with methacrylic acid, CH_2=C(COOH)CH$_3$. The Zn and Cd salts of such copolymers form ionic bond as

$$\begin{array}{c} -CH_2-CH- \\ | \\ C \\ O^{\nearrow} \quad \searrow O \\ \cdot\cdot Zn^{++} \\ ^{\ominus}O \quad \quad O \\ \searrow C^{\nearrow} \\ | \\ -CH_2-CH- \end{array}$$

with bonding energies on the order of 100 kcal/mole. Bonds formed with monovalent salts show correspondingly lower bonding energies.

In order to describe the energetics of attraction between polymer molecules quantitatively, we borrow a concept from the theory of low-molecular-weight solutions. An important property of ordinary liquids is the *cohesive energy density*, CED, which is defined as the molar energy of vaporization divided by the molar volume. Thus we are considering cohesive energy per unit volume. This quantity is determined from easily measured properties of liquids through the relationship

$$\text{CED} \ \frac{\Delta E_v}{V} = \frac{\Delta H_v - RT}{V} \qquad (2\text{-}1)$$

---

* For reasons that will be discussed later, it is not possible to specify unique values for the crystalline-melting and glass-transition temperatures of polymers. Their determination depends on the conditions under which they are measured; consequently, the literature contains narrow ranges of values. The values of T_m and T_g cited in this text will be largely taken from *The Polymer Handbook*, J. Bandrup and E. H. Immergut, John Wiley and Sons, 1966, and they will be representative or approximate values.

where ΔE_v is the molar energy of vaporization, ΔH_v is the molar heat of vaporization, RT is the perfect gas value of the molar work of expansion during vaporization, and V is the molar volume of the liquid. The *solubility parameter* δ is defined as the square root of the cohesive energy density; it is an interaction parameter very often associated with polymer theory and the one to which we will now refer. Numerous values of solubility parameters for simple liquids are listed in Table 2–2.

TABLE 2–2

SOLUBILITY PARAMETERS FOR SEVERAL LIQUIDS[a]

n-pentane	7.0	Chlorobenzene	9.5
n-hexane	7.3	Methyl chloride	9.7
diethyl ether	7.4	Acetone	9.9
n-heptane	7.4	Carbon disulfide	10.0
n-octane	7.6	Ethyl amine	10.0
Diisobutylene	7.7	Nitrobenzene	10.0
Vinyl chloride	7.8	Acetic acid	10.1
Diethyl amine	8.0	Aniline	10.3
Cyclohexane	8.2	Dichloro acetic acid	11.0
Methyl isobutyl ketone	8.4	Ethylene oxide	11.1
Carbon tetrachloride	8.6	Methyl amine	11.2
Xylene	8.8	Cyclohexanol	11.4
Toluene	8.9	Hydrogen cyanide	12.1
Ethyl acetate	9.1	Ethyl alcohol	12.7
Benzene	9.2	Methyl alcohol	14.5
Ethyl chloride	9.2	Ammonia	16.3
Methyl ethyl ketone	9.3	Water	23.4

[a] J. Bandrup and E. H. Immergut, *The Polymer Handbook* (New York: Wiley, 1966).

Because polymers do not boil, their solubility parameters cannot be obtained directly. Instead a lightly crosslinked polymer is prepared, and several samples of this polymer are placed in a series of solvents with known solubility parameters. The polymer will not dissolve because of the crosslinks but merely swell. The amount of swelling for each sample is plotted as cubic centimeters of solvent imbibed per gram of polymer against δ for each of the solvents. The result, as shown in Figure 2–1, is typically a bell-shaped curve centering around a maximum in the swelling. The solubility parameter corresponding to the maximum point is taken as the solubility parameter, δ_p for the polymer. Some selected values of δ_p are shown in Table 2–3. Note that the trend is for increasing δ_p as the strength of the intermolecular bonding forces increases.

Although the development and application of the solubility parameter concept to polymers marked a significant step forward, it still does not completely characterize molecular interaction. Inconsistencies are sometimes noted when polymer properties are correlated in terms of the solubility parameter. From the data in Table 2–3, one sees that polyethylene terephthalate is below poly(hexamethylene adipamide) in the solubility parameter scale and yet their crystalline melting points are virtually identical—265 °C. Furthermore, observe that polyethylene is below cis–polyisoprene and that it too has a higher melting point—135 °C versus 25 °C. One difficulty

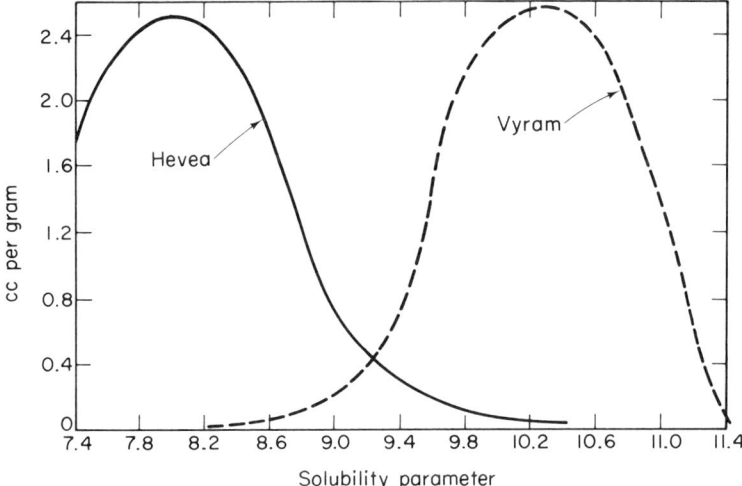

Figure 2–1. Swelling for cis-polyisoprene (Havea rubber) and N–5400 (Vyram rubber) vs. solubility parameter of the solvent. [After M. H. Wilt, Monsanto Co., Tech. Bull., April 1, 1954.]

in the solubility parameter concept is that it involves the resolution of the composite influences of van der Walls, polar, and hydrogen bonding forces. Advanced treatments would require the consideration of separate parameters for each of these influences. Inconsistencies are also observed because of differences in other fundamental factors, such as conformation, steric makeup, or the crystalline morphology.

MOLECULAR ARCHITECTURE

Here we are concerned with the size and shape of the polymer molecule as established by covalent bonding. The molecular architecture of a polymer system depends on the nature of the repeat unit and the manner in which these units are linked together. The term *configuration* will be used to dis-

TABLE 2-3

SOME REPRESENTATIVE SOLUBILITY PARAMETERS FOR
SEVERAL POLYMERS[a,b]

Polymer	Solubility Parameter
Polytetrafluoroethylene	6.2
Poly(dimethyl siloxane)	7.55 (s)
Polyisobutene	7.8 (s)
Polyethylene	7.9
Cis-polyisoprene	8.1 (s)
Polybutadiene	8.1 (s)
Polychloroprene	8.19 (s)
Poly(butadiene–co–styrene) (75/25)	8.45
Polystyrene	9.1 (s)
Poly(vinyl acetate)	9.35
Poly(methyl methacrylate)	9.5 (s)
Poly(vinyl chloride)	9.55
Ethyl cellulose	10.3
Polymethacrylonitrile	10.7
Poly(ethylene terephthalate)	10.7
Cellulose diacetate	10.9
Poly(vinylidene chloride)	12.2
Poly(hexamethylene adipamide)	13.6
Polyacrylonitrile	15.4

[a] *Polymer Handbook, op. cit.*
[b] Just as with glass transition and crystalline melting transitions (footnote page 21), it is not possible to specify unique values for polymer solubility parameters. The values actually determined depend on the experimental means of measurement; consequently, ranges of values are reported in the literature. Those determined by swelling experiments are cited where appropriate and are so indicated by (s). The others are representative values selected from the middle of the range cited in *Polymer Handbook*.

tinguish those structural features of the polymer molecule that can be altered only through alteration of primary valence bonds. Control can be exerted over certain molecular architectural features by proper choice of the monomer(s) and control of the polymerization process. Others are subject to the whims of random molecular processes during polymerization. The reaction mechanisms, kinetics, and process conditions that influence the development of molecular architecture will be a subject of primary concern in later chapters.

202 Molecular Weight

This is the most important property of any polymer system. First and foremost, polymeric materials exhibit unique properties because they do have high molecular weights. Consider the normal paraffins. Since the molecular densities for all members of the series are identical, a higher-molecular-weight member will possess a larger surface area than a lower-molecular-weight member. If it is assumed that the intensity of the residual force fields per unit length for the series is the same, then the larger molecules will exert the larger total attractive forces on the surrounding molecules. If both large and small molecules are in a liquid state, they will exhibit a certain tendency to volatalize due to their translational energy. In order to leave the liquid and enter the vapor, however, they must overcome intermolecular attractive forces. Because of the larger intermolecular attractive forces associated with them, the escaping tendency of the larger molecules will be less, the vapor pressure will be less, and at constant pressure the normal boiling point will be higher. Normal pentane, for example, boils at 36 °C, whereas cetane, n-$C_{16}H_{34}$, boils at 288 °C. On the other hand, heptacontane, $C_{70}H_{142}$, with a molecular weight of 982 decomposes before it reaches its boiling point. The bonding energies involved in the primary valences holding the atoms of a molecule together are on the order of 50 to 200 kcal/gram mole, whereas secondary forces are much weaker, ranging from 0.5 to 10 kcal. But when the molecules are large enough and have a large enough surface, the secondary forces bridging them to one another may build up to the point where, in the aggregate, they are greater than the primary valence forces holding two atoms in the molecule together. When this is the case, primary valence bonds will be ruptured before the molecules can be separated against the extensive bridging forces, and thermal decomposition results before a boiling point is reached.

In general, as any homologous series is ascended, properties that might logically be attributed to intermolecular forces of attraction should increase. Now, the molecules of any particular species of high polymer at various degrees of polymerization are members of a homologous series just as are the paraffins. Among the important polymer properties that will vary with the degree of polymerization are the glass-transition and melt-flow temperatures, the melt and solution viscosities, plus the tensile and impact strengths.

A logical question to ask at this point is: How high must the molecular weight be for the polymer to be a high polymer? If one plots the degree of polymerization of a polymer versus a property depending on the intermolecular forces of attraction, one generally observes a situation where the property changes rapidly with increasing degree of polymerization until a critical point is reached. At this point, the relative change in the property decreases substantially. This critical point is taken as the threshold for high-polymer behavior. For many of the important properties, this change occurs as shown schematically in Figure 2–2. The critical point differs according to the strength of the intermolecular bonding prevalent in the polymer. For polymers with strong bonding forces, it is lowest and the degree of polymerization is on the order 200. For polyhydrocarbons, where bonding is weakest, degrees of polymerization on the order of 500 must be attained. For most polymers, the threshold for high-polymer behavior lies between these two values.

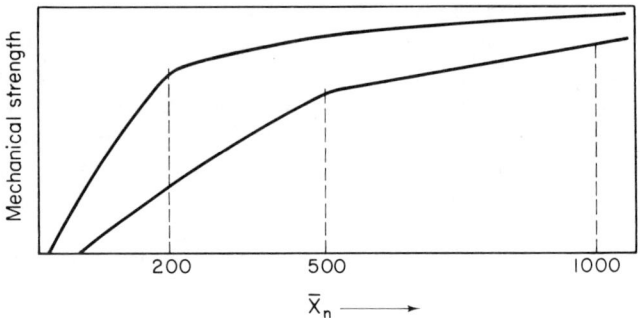

Figure 2–2. Variation of mechanical strength with degree of polymerization.

Another interesting manifestation of the effect of molecular size can be observed in the solution or melt viscosities of polymers. In this instance, the rate of increase in the size-dependent property—that is, the viscosity—is low, below the critical point. But, as shown in Figure 2–3, the viscosity increases rapidly beyond the critical point. This behavior is observed because, below the critical point, the polymer molecules are free to flow as single entities, but at larger sizes the molecules begin to entangle so that the size of hydrodynamic flow units increases markedly. *Network flow* is said to occur; and as the molecules become larger, the degree of entanglement increases rapidly. The flow networks become larger and the viscosity also continues to increase rapidly.

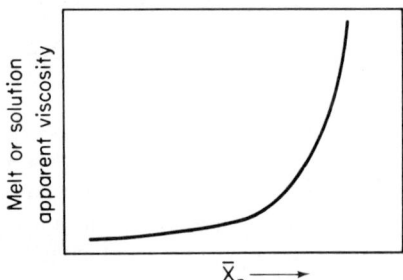

Figure 2–3. Variation of viscosity with degree of polymerization.

It is also interesting to consider what might constitute an upper practical limit to molecular weight. Eventually, even with linear polymers, the molecular weight can become so high that the polymers cannot be dissolved, worked mechanically, or made to flow in a molten state. This intractability usually sets in at a molecular weight somewhat above 10^7. Thus the molecular-weight range of utility for most linear polymers is on the order of 10^4 to 10^7. An important exception to this upper limit occurs in the case of what is referred to as ultrahigh-molecular-weight polyethylene. This material has several interesting properties that make it useful in engineering applications—namely, a remarkable abrasion resistance, toughness, and a low coefficient of friction. It is more resistant to abrasion than hardened steel. Perhaps its most exciting use is for coating the bottom of the modern snow ski, to produce a product that is both fast and durable.

Among those properties that are essentially independent of, or change very little with, the degree of polymerization are refractive index, hardness, electrical properties, color, and density.

203 Molecular–Weight Distributions

As previously indicated, the size of most polymer systems is characterized by a molecular-weight distribution. This distribution has an important effect on a polymer's mechanical behavior. For example, it is well known that with most elastomers a narrow distribution means superior mechanical properties but relatively poor processing properties.

204 Development of the Nylons

To illustrate direct application of the foregoing concepts, it would seem appropriate to trace one of the most important advances in modern polymer technology—the development of the nylons. The term *nylon* is the accepted generic term for synthetic polyamides. Polyamides are a class of polymers formed from the condensation of an acid and an amine and that consequently contain the amide linkage in their repeat units.

The research leading to the production of synthetic fibers from nylon was initiated in the duPont Company laboratories in 1928 by W. H. Carothers.* In his early work he studied polyesters formed by reacting glycols and dibasic acids. By observing the properties of his first products, which

---

* *Collected Papers of Wallace H. Carothers on Polymerization*, Edited by H. Mark and G. S. Whitby (New York: Interscience, 1940).

ranged in molecular weight from 2500 to 5000, he recognized the importance of achieving higher molecular weights. These early polymers were hard, insoluble, waxy materials, but they did not yet possess the attributes of high polymers. By removing the water liberated in the reaction, Carothers succeeded in shifting the reacting equilibrium sufficiently to obtain products with molecular weights on the order of 25,000. The properties of these materials were so remarkably different from their low-molecular-weight homologs that Carothers termed them "superpolymers."

These superpolyesters were tough, opaque solids that melted at moderately elevated temperatures, 75–80 °C, to form clear, viscous liquids. Carothers was able to form filaments from these materials by touching the surface of the molten polymer with a glass rod and withdrawing it. On cooling, these filaments could be drawn to several times their original length, to produce a fiber that was tough, strong, and elastic. X-ray diffraction studies showed them to be crystalline and highly oriented—characteristic properties of all fibrous materials. The commercial utilization of these materials was never realized, however, because of their low melting points (too low to allow for laundering or ironing) and their low strength compared to silk fibers.

It had long been known, however, that silk—a natural fiber with outstanding qualities—was a high polymer composed of repeat units derived from the α-amino acids of the form $H_2NCHRCOOH$. The properties of silk are of course, largely attributable to the presence of the amide linkage which provides sites for hydrogen bonding. Upon recognizing this additional factor, Carothers began the phase of his research that ultimately lead to the development in 1935 of the polyamide, poly(hexamethylene adipamide), otherwise known as Nylon 66. It produced fibers as the earlier polymers had done, but the new material had superior mechanical properties and a melting point of about 265°C. In addition, also because of strong intermolecular bonding, the material was insoluble in all common solvents (a prerequisite for dry cleaning). The material was selected for commercialization, and even today it is the most commercially significant nylon. It is also interesting to note that at the time of these developments, adipic acid was made only in Germany and hexamethylene diamine was little more than a laboratory curiosity. The need for nylon during World War II gave considerable impetus to the rapid growth of large-scale production facilities.

In retrospect, it is easy to explain what required Carothers years to discover. His brilliant pioneering studies, however, represented an outstanding achievement for their time. He is rightly regarded as the father of modern polymer chemistry, and it is remarkable how often his lessons are overlooked even today, more than 30 years after his death in 1937.

STERIC FACTORS

Once molecular-weight properties and intermolecular bonding forces for a polymer are established, it is necessary to consider the effect of those steric factors that control chain flexibility and molecular packing. *Chain flexibility* is controlled primarily by the structure and composition of the groups that make up the polymer backbone as well as that of the groups that are attached to the main chain. *Molecular packing* is largely determined by the way the monomer units are linked together in the polymerization process— that is, by the structural regularity—as well as by the nature of the repeat unit structure.

Chain flexibility is a major determinant in establishing such properties as bulk stiffness or elastic modulus, the glass and crystalline-melting transitions, and thermal stability. It is not hard to visualize that a bulk polymer composed of relatively inflexible or stiff chains will possess a relatively high bulk stiffness as reflected in its elastic modulus. Melting of crystalline regions involves the disruption of uniform chain packing as the chains take on a quantity of translational energy, sufficient to overcome intermolecular bonding. The more flexible the polymer chains are, the more readily segments of the chain will acquire translational motion, and, therefore, the lower will be their melting points. The threshold of thermal degradation is marked as that point where the chain vibrations become so strong and violent that they literally tear the molecule apart. Stiffer chains will resist such strong vibrations, and a higher temperature will be required for the thermal decomposition of a stiff chain than a more flexible one.

The manner in which *chain packing* is accomplished will determine the extent to which the intermolecular bonding forces can exert an effect. If the polymer chains have a regular steric structure, the chains will be able to pack in a well-aligned, parallel fashion; the material will be capable of crystallizing, and intermolecular attraction will be maximized. If the structure is not regular, the chains will not be well packed, crystallization probably will not occur, and intermolecular attraction will be relatively weak.

205 Structural Regularity

Here we are concerned with the shape of a molecule as viewed along the contour length of the chain. In order to view the molecule in its proper perspective, we place the chain, at least hypothetically, in the planar, zigzag form. If the molecule in this form has a chemically and sterically regular structure, it is said to possess *structural regularity*. As we have stated, the latter is a prerequisite for well-developed crystallinity and orientation and, subsequently, high mechanical strength. In simple linear molecules, such as polyethylene, the polyamides, the polyesters, and poly(vinyl alcohol), the side groups are too small to disrupt the ability of these molecules to pack closely, and these polymers are noted for their high-crystallizing tendencies. On the other hand, chain branching and copolymerization tend to disrupt structural regularity.

Consideration of structural regularity is further divided into a discussion of recurrence regularity and steroregularity. *Recurrence regularity* refers to the regularity with which the repeat unit recurs along the polymer chain. To amplify by illustration: a type of recurrence regularity that is inherent in all monosubstituted or 1,1–disubstituted vinyl homopolymers is the so-called *head-to-tail configuration*. That is, a substituted vinyl monomer, $CH_2=CHX$, will polymerize such that the repeat units appears in the backbone as

$$-CH_2-\underset{X}{CH}-CH_2-\underset{X}{CH}-CH_2-\underset{X}{CH}-$$

It is conceivable that a *head-to-head, tail-to-tail configuration* could result.

$$-CH_2-\underset{X}{CH}-\underset{X}{CH}-CH_2-CH_2-\underset{X}{CH}-$$

However, no known polymer contains this configuration to any measurable extent.

The presence of a high degree of recurrence regularity is *not* generally sufficient to ensure crystallizability in many polymers. One must also investigate the spatial properties of a polymer molecule. Again, to illustrate, consider the isomeric forms of natural rubber; Hevea rubber is cis-polyisoprene

$$\begin{array}{ccc} CH_3 & H & CH_3 & H \\ \diagdown & \diagup & \diagdown & \diagup \\ C=C & & C=C \\ \diagup & \diagdown & \diagup & \diagdown \\ -CH_2 & CH_2-CH_2 & CH_2- \end{array}$$

whereas gutta-percha is trans-polyisoprene.

$$\begin{array}{ccc} CH_3 & CH_2-CH_2 & H \\ \diagdown & \diagup & \diagdown & \diagup \\ C=C & & C=C \\ \diagup & \diagdown & \diagup & \diagdown \\ -CH_2 & H & CH_3 & CH_2- \end{array}$$

Materials possessing spatial regularity are said to be *stereoregular* or *stereospecific*, and they inherently possess recurrence regularity. Hevea rubber is soft and pliable with a crystalline melting point (T_m) of 25 °C. Gutta-percha is tough and hard, as well as polymorphous, with T_m for the β form equal to 56 °C and for the α form equal to 65 °C. Hevea rubber is used for making vehicle tires, whereas gutta-percha finds use in golf-ball covers. Early attempts to polymerize isoprene led only to polymers that contained cis and trans isomers mixed in the same chain. Even though such materials may have had recurrence regularity, they did not resemble the natural rubbers, and they possessed no usable properties because spatial regularity was lacking.

The concept of stereoregularity is most important in considering the monosubstituted vinyl polymers of the α-olefin type. When such a polymer is exhibited in the planar zigzag form, one can visualize three possible configurational arrangements for the substituted groups, as illustrated in Figure 2–4: the *isotactic* form, where the substituent appears always on the same

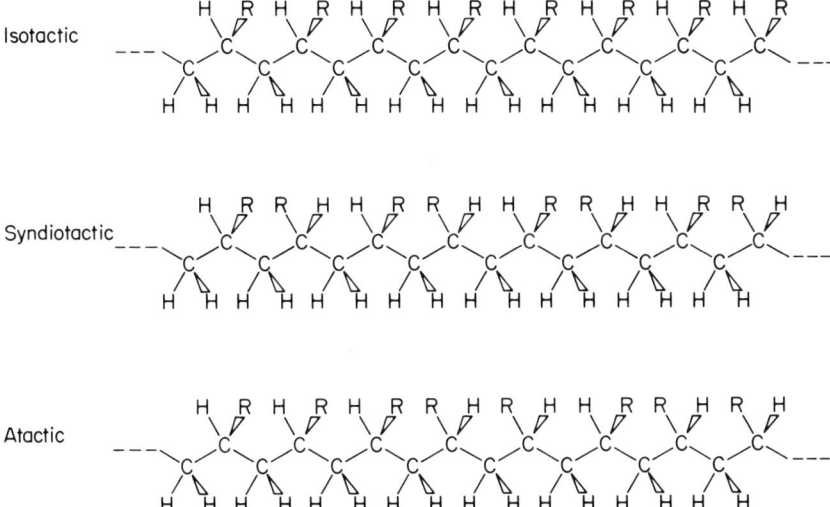

Figure 2–4. The steric forms of poly (α–olefins).

side of the main chain; the *syndiotactic* form, where the substituent is located on alternate sides of the main chain; and the *atactic**, where substitution is completely random. The first two tactic forms are stereoregular structures, and as such they exhibit strong tendencies to crystallize. Except for poly-(vinyl alcohol), the atactic forms of poly(α-olefins) are universally amorphous.

* The names for the various steric or tactic forms were derived from the Greek words *tatto*—"to put in order," *iso*—"the same," and *syndio*—"every two."

The discovery in the early 1950s and the subsequent development of the processes by which stereoregular polymers could be made marked an extremely important period in modern polymer science. Just to indicate the importance of this development, we should note that the two men associated with the original work, Karl Ziegler and Giulio Natta, were awarded the 1963 Nobel Prize in chemistry for their efforts. The synthesis of both the cis and trans forms of polyisoprene was the first time man was able to duplicate one of nature's stereoregular polymers, albeit a very simple example. The well-read student is certainly aware of and can certainly appreciate the stereocomplexity of such natural polymers as DNA and RNA. To the engineer, the process introduced an entirely new class of polymers with new properties and applications.

206 Repeat-Unit Structure: Composition of the Backbone

In discussing the effect of repeat-unit structure on chain stiffness and polymer properties, we look first at the composition imparted to the main chain and then at the structure of side-chain appendages. If we use the straight hydrocarbon chain, composed only of $—CH_2—$ units linked together, as a basis for discussion, we can consider variations on this unit that will increase or decrease the freedom of adjacent units to rotate about one another. Such variations, in turn, will increase or decrease the overall chain flexibility. The flexibility of a straight carbon chain polymer can be considerably increased if some of the carbon atoms are replaced by those of oxygen and considerably reduced if some of the carbon atoms are replaced by phenyl groups. Such a variation is caused primarily by the variation of freedom of rotation about $—O—$, $—C—$, and $—\bigcirc—$ groups. Note, as shown by the melting points in Table 2–4, the opposing effects of increasing or decreasing flexibility where such substitutions were made.

Another interesting manifestation of the stiffness factor is found in comparing the properties of poly(vinyl alcohol) and cellulose. Both have the same ratio of carbon atoms to hydroxyl groups, but poly(vinyl alcohol) is soluble, moldable, fusible, whereas cellulose, in its native form, is virtually intractible. The difference must be attributed to the stiffness imparted by the ring structure of the cellulose repeat unit.

One of the primary factors limiting the utility of ordinary polymeric materials is their relatively low thermal stabilities. This low stability may be evidenced as a loss of strength because of crystalline melting and/or thermal decomposition. In order to create heat-resistant polymers, chemists have

TABLE 2-4

CRYSTALLINE MELTING TEMPERATURES AS A FUNCTION OF BACKBONE COMPOSITION[a]

Polymer	Repeat Unit	Melting Point, °C
Poly(p-xylene)	—CH$_2$—C$_6$H$_4$—CH$_2$—	~400
Polyethylene	—CH$_2$—CH$_2$—	~135
Polyoxyethylene[b]	—CH$_2$—CH$_2$—O—	~65

[a] *Polymer Handbook, op. cit.*
[b] To illustrate how countering factors can influence polymer properties, polyoxymethylene —CH$_2$—O—, has a melting point of about 180°C. In this case, the induced flexibility is more than offset by the increased bonding forces resulting from polarity.

sought and found ingenious ways to stiffen the polymer chain. Some measure of heat stability is imparted by incorporating the phenyl group into the polymer backbone.* Some of the most successful aromatics are shown in Table 2-5, but even these fail at temperatures in excess of 300 °C.

The next generation of heat-stable polymers were the heterocyclic aromatics, the most successful of which were the aromatic polyimides. The aromatic polyimides have melting points of about 600 °C, and films of these materials have maintained their mechanical integrity for over a year at a continuous temperature of 275 °C. When used in glass-cloth composites or as metal-to-metal adhesives, they can withstand temperatures of up to 650 °C. Aromatic polyimides contain

in their backbone.

The third generation of heat-stable polymers, able to withstand even higher temperatures, will probably come from the so-called *ladder polymers*. There are several varieties, but essentially they consist of two distinct chain-

---

* The addition of an aromatic ring offers an additional mechanism for thermal stability. It is more difficult to disrupt a double bond with four shared electrons than a single bond with two electrons. The aromatic ring with its double bond character is therefore inherently more stable than a straight hydrocarbon chain.

TABLE 2–5

SOME AROMATIC HIGH-TEMPERATURE POLYMERS

Class of Polymer	
Polyamide	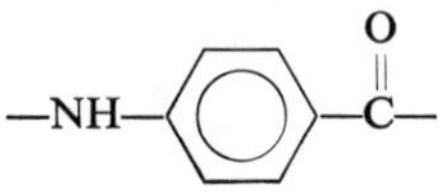 —NH—⟨◯⟩—C(=O)—
Polycarbonate	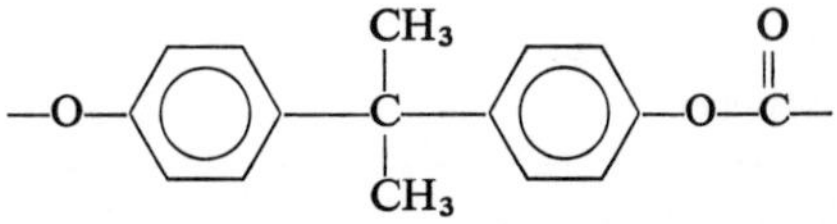 —O—⟨◯⟩—C(CH₃)(CH₃)—⟨◯⟩—O—C(=O)—
Polyphenoxy	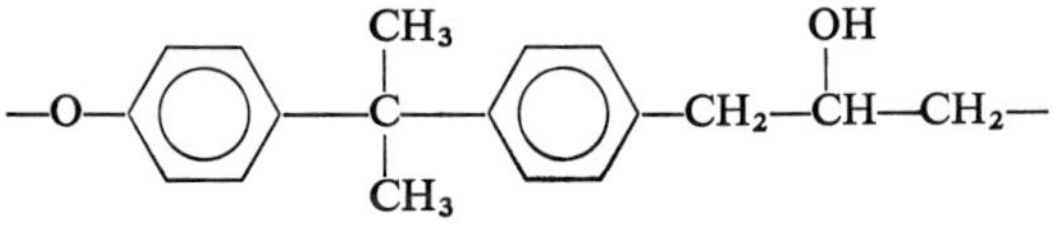 —O—⟨◯⟩—C(CH₃)(CH₃)—⟨◯⟩—CH₂—CH(OH)—CH₂—
Polysulfanone	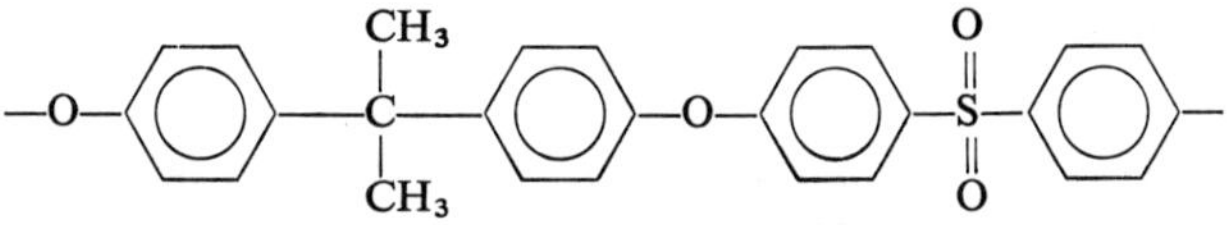 —O—⟨◯⟩—C(CH₃)(CH₃)—⟨◯⟩—O—⟨◯⟩—S(=O)(=O)—⟨◯⟩—

like structures bound together at regular intervals by "rungs." The aromatic heterocyclic character of the previous generation is maintained. An example of a ladder structure is the following:

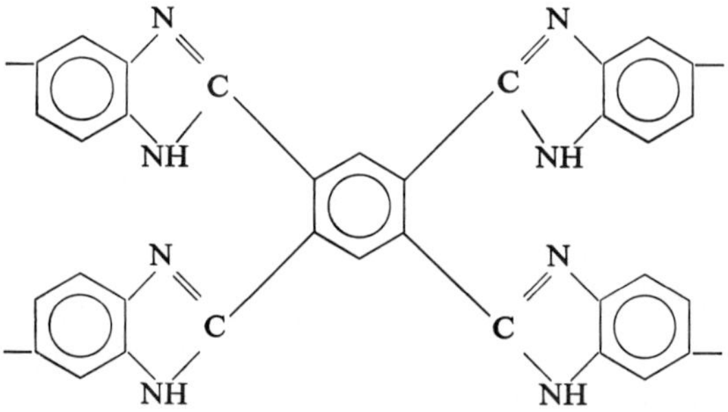

207 Repeat Unit Structure: Composition of Pendant Groups

Pendant Groups are those parts of a polymer molecule that are attached as appendages to the main backbone of the chain. They are introduced into the repeat-unit structure during the polymerization process as an integral part of the monomer structure. We have already briefly discussed monosubstituted vinyl monomers in introducing the concept of stereoregularity, but here we will superpose the effect of the size and structure of these side groups on chain packing and chain stiffness. Generally we will be interested in those groups that are larger than hydroxyl groups or the hydrogen and oxygen atoms. In addition to a steric influence, some substituents account for a major part of the intermolecular forces of attraction.

TABLE 2-6

CRYSTALLINE MELTING POINTS OF SOME ISOTACTIC POLY(α-OLEFINS)[a]

Pendant Groups	Melting Point, °C	No. of Monomer Units per No. of Turns[b]
—CH_3	165[c]	3/1
—CH_2—CH_3	125[c]	3/1
—CH_2—CH_2—CH_3	75[c]	3/1
—CH_2—CH_2—CH_2—CH_3	−55[d]	7/2
—CH_2—$CH(CH_3)$—CH_2—CH_3	196[d]	3/1
—CH_2—$C(CH_3)_2$—CH_2—CH_3	350[c]	(Not Available)

[a] After F. W. Billmeyer, *Textbook of Polymer Science*.
[b] *Polymer Handbook*, op. cit. Section III.
[c] T. W. Campbell and A. C. Haven Jr., *J. Appl. Polymer Sci.*, **1**, 73, (1959).
[d] C. E. H. Bawn, *Chem. & Ind.*, 1960, p. 388.

As pointed out previously, the effect of chain packing and chain stiffness on polymer behavior is reflected in the crystallizing tendencies and crystalline melting point. Consider Table 2-6, which shows melting points for a series of isotactic, α-olefin polymers. When the linear alkyl portion increases in length, the melting point decreases because the volume of the side chains prevents the molecules from packing closely which leaves the backbone more room for vibrational motion. However, when the bulk of even the longest side chain is increased close to the backbone, just by the addition of one methyl group, to form a branched side chain, a marked increase in the melting point is noted. The addition of another methyl group in the same position brings another drastic increase in the melting point. This second trend is caused by an increase in chain stiffness. In atactic (amorphous) polymers, this same trend can be noted in the glass-transition temperature, which is, in general, influenced by the same factors as the melting point in crystalline polymers.

To summarize, the structure of pendant groups *may* cause either of two effects. If the substituent is linear, loose packing results, and the transition temperature will be commensurately low. If, on the other hand, the substituent is bulky and *close* to the main chain, the loose packing will be more than compensated for by increased overall chain stiffness.

208 The Effects of Crosslinking

The preceding discussion of intermolecular forces acting in high-polymer systems has been restricted to systems where the molecules were discrete entities and bonded to one another only by secondary valence forces. In a crosslinked polymer system, the chains are connected to one another in a manner that utilizes primary valence bonding. Crosslinking may be imparted to a system during polymerization with polyfunctional reactants or in the postpolymerization treatment of a linear system that contains functional groups capable of further reaction.

The physical properties of crosslinked systems will vary considerably from linear and branched systems. Continuous flow can only occur in masses in which the individual molecules are free to slide past one another. Similarly, a polymer will dissolve only if the molecules of the solid can be separated from one another. Thus linear and branched polymers, with rare exceptions, can be induced to flow with the application of heat and/or stress, and they can be dissolved in suitable solvents. Because of their ability to flow with the application of heat and/or stress, such polymers can be molded; consequently, uncrosslinked polymers are frequently referred to, as a class, as *thermoplastics*.* In crosslinked systems, on the other hand, continuous free flow of the molecules and their separation by solution are prevented by strong primary valence forces. Chemical decomposition must occur before separation or flow can be induced. Crosslinked polymers will not flow freely with the application of heat and/or stress. Thus they cannot be molded; for this reason, crosslinked polymers are said to be *thermosetting*. Also as a result of the crosslinking that takes place, the molecular weight of crosslinked polymers is considered to be infinite; and in spatial terms one thinks of an infinitely large network system of interconnected chains.

Just as crosslinked polymers differ markedly from uncrosslinked polymers, they can, as a class, differ markedly from one another. For example,

---

* The term "plastics" should not be construed as a synonym for polymers. Its use as such should be avoided. *Plastics* are a distinct class of polymers that are thermoplastic in character.

the vulcanized natural rubber (see Section 108) used in tire manufacture is soft, extensible, and pliable, but the phenol-formaldehyde resins* produce a product that is hard and rigid. In addition, although vulcanized rubber will not dissolve in solvents, it will swell quite readily and extensively in suitable solvents, but the phenol-formaldehyde products are completely impervious to solvents. Such physical property differences can generally be attributed to differences in (1) the character of the main chains, (2) the character of the crosslinks, and (3) the so-called *degree of crosslinking*, measured in terms of the number of crosslinks per unit mass of material. As shown in Figure 2–5, for our previous examples, the main chains of vulcanized natural rubber are inherently flexible, the crosslinks are fairly flexible, and the crosslink density is relatively low. In phenol-formaldehyde systems, the chains are stiff, the crosslinks (which are indistinguishable in this case from "main" chains) are short and stiff, and the number of crosslinks is very high.

MOLECULAR PACKING

The way in which the molecules are packed together in a polymer system has a profound effect on the properties of the material. Recall that the molecular forces of attraction fall off very rapidly as intermolecular distances increase. The secondary bonding forces and the high molecular weight of polymers create a high-residual force field per molecule; but if these forces are to exert a maximum effect, it is essential that the molecules be able to pack closely in perfect parallel alignment. We have already observed that the molecules will tend to pack closely and align if they exhibit a high degree of structural regularity. Complete parallel alignment is never achieved in polymeric systems because their large size, and structural defects tend to complicate the packing and aligning process. The highest level of perfection is found in those polymers that exhibit fibrous behavior—that is, in those possessing a high degree of crystallinity and crystal orientation. For example, polyethylene exhibits fibrous behavior, because of its structural regularity, even though the magnitude of its intermolecular bonding forces is low. The total effect of packing and attraction is the important factor, and in polyethylene this total effect is high. The synthetic polyamides have both structural regularity and high molecular attractive forces, so that these materials are even stronger than polyethylene.

Diametrically opposed in characteristics to the fibrous materials are the

---

* The latter material is familiarly known to us as "Bakelite". Its synthesis in 1906 marked the first fully synthetic polymer yet produced by man. This class of resins is still one of the most important today. They were named after their discoverer, Leo H. Baekeland, by the discoverer himself.

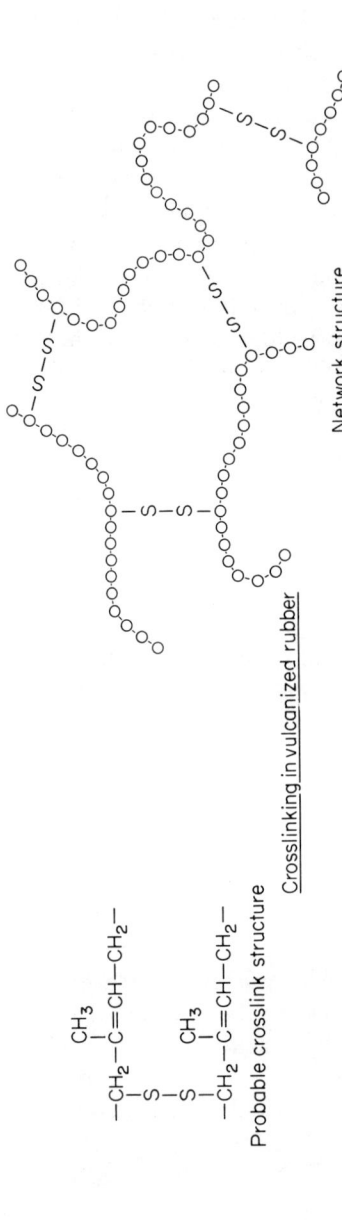

Figure 2-5. Structures in crosslinked systems.

weakest amorphous polymers—the rubbery materials. Here the molecules may have neither high intermolecular attractive forces nor structural regularity. In such materials, the molecules are neither closely packed nor aligned, but the molecules are relatively distant and packed in a random manner. The net result is weak intermolecular bonding. This factor, in turn, is reflected in their lack of crystallinity and their relatively weak, rubberlike behavior.

These two cases represent two extremes in polymer types; in between there is a wide spectrum of materials with an equally wide spectrum of properties and engineering uses. In the cases cited, two opposing factors appear to be in operation: (1) those that favor strong intermolecular bonding—strong intermolecular bonding forces and structural regularity, and (2) those that favor weak intermolecular bonding—weak intermolecular bonding forces and structural irregularity. The nature of the intermolecular bonding forces is listed as a fundamental factor, and structural regularity is one of the considerations of molecular architecture, another fundamental factor. The arrangement of the molecules en masse, as we have seen, depends on these two factors which, one might say, involve the *molecular state*. On the other hand, the conformation plus the morphology and order characteristic of a collection of molecules form the basis of the third fundamental factor, and this involves consideration of the *physical state* of the system.

209 Conformation

Here we are concerned with the form a molecule assumes through rotation about primary valence bonds. The conformations available to a polymer molecule may be manifold or severely restricted, depending on various steric factors as well as on whether the system as a whole is ordered (crystalline) or unordered (amorphous). There are two levels at which we can discuss molecular conformations: (1) the local level, where we are concerned with the relative position of neighboring groups in the polymer chain, and (2) long range, where we are concerned with describing the size and shape of the entire molecule.

One of these local levels is the planar *zigzag conformation* with which the student is already familiar. This conformation occurs to the largest extent in the crystalline form of many simple molecules like polyethylene, poly(vinyl alcohol), and the polyamides, where alignment and packing are not complicated by the presence of large, bulky side groups. One of the most important forms that arises in the isotactic and syndotactic α-olefin polymers is the *helical conformation*. In many instances, the side group is too bulky to be accommodated in a zigzag conformation as previously pictured, for steric crowding prevents it. (The planar zigzag form was used previously only as a visual aid in defining the various tactic forms.) This steric crowding is relieved by rotation of the molecules in the main chain, all in the same direction to form either a right- or left-handed helix. Some typical helical structures are depicted in Figure 2-6. This form appears almost exclusively in the crystalline form of stereoregular polymers with bulky side groups—for example, isotactic polypropylene, polyisobutylene, and polystyrene.

Further, it is important to note how differences in conformational shapes

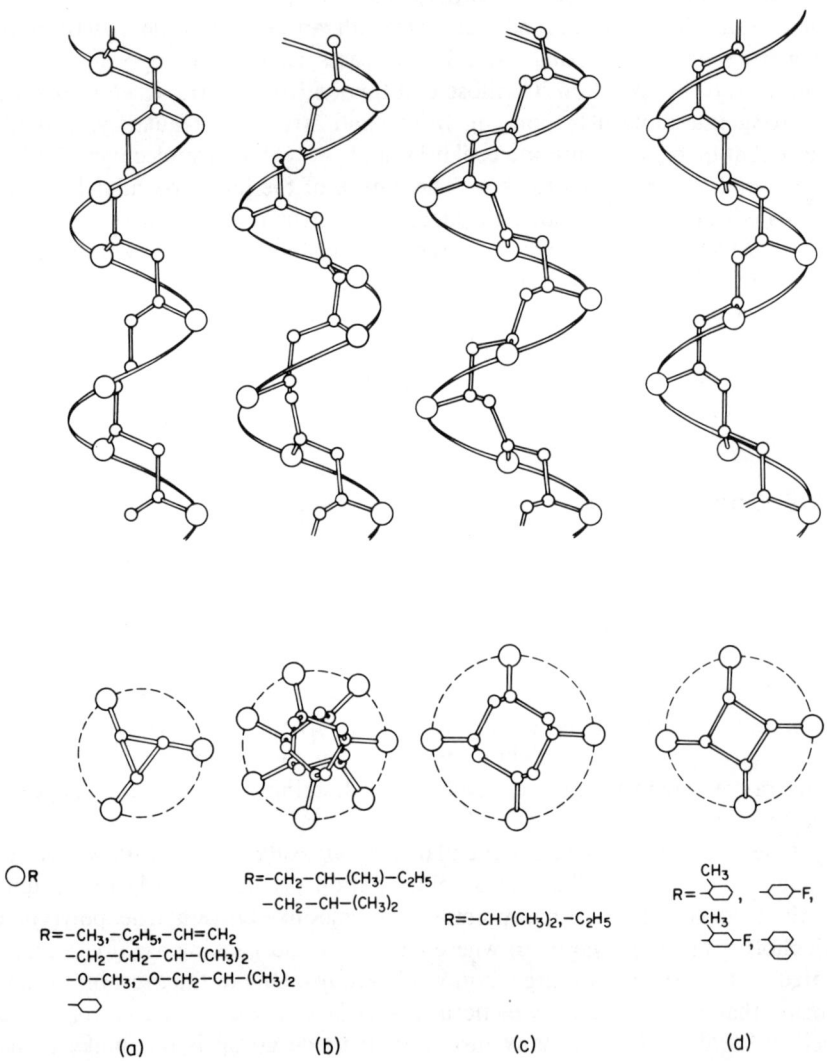

Figure 2-6. Helical conformations of isotactic vinyl polymers. [N. G. Gaylord, and H. Mark, *Linear and Stereoregular Addition Polymers*, Interscience, New York, 1959.]

may affect properties and how in such instances we would have to consider the effect of superposing fundamental factors. An instance of how one can make inaccurate predictions is illustrated by considering the melting points of polyethylene, polypropylene, and polyisobutene. First ignoring differences in conformation and considering only structural factors, as discussed in the preceding section, we note that polypropylene has a higher melting point than polyethylene (165 vs. 135 °C). One can reason that polypropylene, has a higher melting point because it has a methyl group close to the chain backbone to stiffen it. We can reason further than if we put another methyl group in close to the chain, opposite to the original as in polyisobutylene, we should further raise T_m. Actually T_m is reduced in that $T_m = 45$ °C for polyisobutylene. We have however ignored the fact that the conformations for all these species are different. In particular, the polyisobutylene helix requires eight units in five turns to complete itself which results in a loose, open structure compared to the tight, compact helix of polypropylene which requires 3 units in 1 turn. Such an example illustrates the need for caution in predicting properties from a knowledge of structure.

Furthermore, the materials in Table 2–6 were selected with care to be certain that the chains had similar conformations. Thus, we have also shown in Table 2–6 the number of monomer units per number of turns required for the various helices to complete themselves.

The final local form is the *random conformation*, where rotation about primary valence bonds is relatively free but nevertheless restricted by bond length, bond angles, and by steric crowding of side groups. This is the form most polymers assume in such amorphous systems as solutions, melts, or some solids. This randomness is induced by thermal fluctuations.

In systems where the local conformation is random, the conformation of the molecules as a whole must also be random or in the form of a random coil, as shown in Figure 1–2. In crystalline systems, the long–range mole-

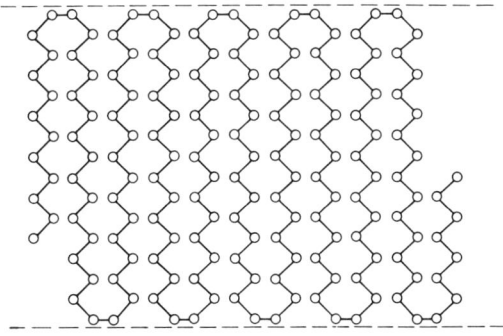

Figure 2–7. Chain folding of a single polymer molecule within a fold plane.

cular conformation is quite regular. In addition to the zigzag or helical order imposed at the local level, the polymer chains are further ordered in that they fold back and forth on themselves in a regular planar arrangement. This conformational habit is shown in Figure 2–7 in an oversimplified manner for a single chain with a zigzag conformation. The plane in which the entire molecule lies is known as the *fold plane*, and in the context of this morphology we speak of *chain folding*. The plane of the zigzag may or may not lie in the fold plane.

210 Morphology and Order

This subject deals with the structure and forms exhibited by aggregates of molecules in the bulk.

The notion that a single molecule assumes random conformations leads to the premise that polymer chains in an aggregated, unordered (amorphous) state also assume random conformations. The rationale is as follows: Since a molecule is completely surrounded by like segments, there will be no net interactive forces; and, therefore, the molecule will assume random conformations. This reasoning can only lead to the premise that polymer chains are entangled under crowded conditions. In the melt, for instance, polymer chains are somewhat graphically pictured as cooked spaghetti in a bowl or as worms in a can.

Polymer single crystals are flat, platelike, well-ordered structures on the order of 100 Å thick, in which the fold planes are found packed in an orderly fashion. Bulk crystalline polymer is composed of stacks of these lamallaelike crystallites. As already noted, the importance of the crystalline form resides in the close molecular packing, which permits strong intermolecular bonding and, as a result, good mechanical-strength properties. Mechanical strength can be further enhanced by orientation of the crystallites as induced by stretching and rolling. This technique is used in fiber formation, and the phenomenon of alignment is encountered in the stretching of natural rubber.

PART II

Polymer Synthesis

The processes by which a polymer is synthesized or modified are the most important steps in producing a useful polymer artifact. It is here that all the capabilities or potential capabilities of the polymer system must be imparted. Substantially different materials can result from polymerizing the same monomer by different methods. In other instances, the nature of the monomer may be such that it can be polymerized by only one process. In general, polymerization reactions, as well as reactions of polymers, are among the most complicated and least understood of all chemical reactions.

Reaction control presents more problems in the synthesis of polymers than it does in the synthesis of many low-molecular-weight materials. Let us cite a few examples.

1. Polymer reaction environments may exhibit a rapid rise in viscosity as the reaction progresses because of increasing molecular weight and/or solids content. In such an instance, the transport properties of the reaction mixture are certain to change, thereby complicating analysis of the kinetics of the reaction system.
2. Certain reactions take place at relatively low temperatures (on the order of -100 °C) in a heterogeneous medium at rapid rates.
3. In the preparation of a resin from a set of polyfunctional reactants, conditions within the reaction vessel must be established such that solidification will not occur in the reaction vessel. Not only would it be

difficult to remove such a polymer from the reactor, but it would also be virtually useless for further application.
4. Ultrahigh purity of the reaction ingredients is usually required for high molecular weights or for reaction to occur at all.

Molecular architecture, especially molecular weight and its distribution, becomes an even more important consideration in polymer synthesis. In low-molecular-weight compound synthesis, the molecular weights of both the reactants and products are generally monodispersed and easily defined. In polymer reactions, the molecular weight of both the product molecules and the reactant molecules (which in themselves may be product molecules) are very often changing in molecular weight as well as in molecular-weight distribution as the reaction progresses. In polymer reactor design, the engineer must consider the problem of selecting the polymerization process and optimizing the reaction conditions in terms of the desired molecular architecture of the product. Many of the molecular processes that occur during synthesis are controllable; more often than not, they are random processes that one lives with rather than controls.

CLASSIFICATION OF POLYMERS AND POLYMERIZATION MECHANISMS

Carothers (1929) suggested classifying polymers and polymerization reactions according to the stoichiometry of the reaction as either condensation or addition. Condensation polymers were formed in a process (condensation polymerization) that resulted in the splitting out of a small molecule; addition polymers were formed in a process (addition polymerization) where a small molecule did not split out. On the other hand, P. J. Flory placed the emphasis on the reaction mechanism by which the polymer was formed. Condensation polymers are usually formed by the stepwise intermolecular condensation of reactive (functional) groups; addition polymers usually result from chain reactions promoted through an active center. The preceding considerations lead to the modern terminology for these types, namely, step-reaction and chain-reaction polymerization. Inconsistencies in both systems of nomenclature arise occasionally, but these are easily recognized and cause little confusion.

The distinguishing mechanistic features of the chain-reaction and step-reaction polymerizations are tabulated below. They are presented here somewhat prematurely, but the full implications of these features will become apparent in chapters three and four. Subsequent reference should be made to them.

A particularly illuminating way to differentiate between chain-and step–reaction polymerization is to construct a "matrix" of monomer units and

Part II Synthesis of Polymers   45

DISTINGUISHING FEATURES OF CHAIN AND STEP-REACTION
POLYMERIZATION MECHANISMS[a]

Chain Polymerization	Step Polymerization
Only those molecular species containing active centers may add monomer units.	Any two potentially reactive molecular species present can react and combine.
The monomer concentration decreases steadily during the progress of the reaction.	Rapid monomer depletion occurs very soon after the reaction begins.
High-molecular-weight polymer is formed at once.	Molecular weight increases slowly as the run progresses. A high degree of conversion is required for high molecular weight.
The reaction contains at any time during the run only monomer, inactive high polymer, and about 10^{-8} parts active centers (i.e., growing chains).	At any stage of the reaction, all sizes of molecular species are present in a calculable distribution.

[a] After Billmeyer, *Textbook of Polymer Science*.

then show how the units become connected with the passage of time. Shown below are two 5 × 5 such "monomer-unit-matrices." The arrowheads indicate the way in which reaction occurred. Adjacent opposing arrowheads indicate step reaction, while a series of single arrowheads all pointing in the same direction along the reaction "path" indicates chain reaction.

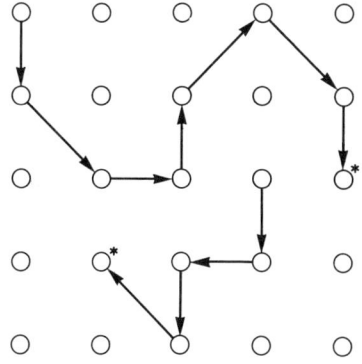 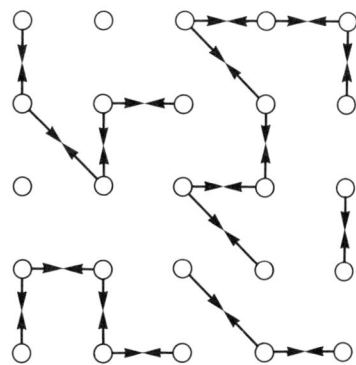

Asterisk denotes an active center.

3
Step-Reaction Polymerization

301 Chemistry of Step-Reaction Polymerization

Much of the early work of elucidating the chemistry of polymer reactions was done by Carothers. He pointed out the fact that the chemistry of condensation polymerization did not vary significantly from that of low-molecular-weight condensations.

Several common functional groups frequently arise in organic step reactions; representative of those participating in linear-condensation polymerization are the hydroxyl (—OH), carboxyl (—COOH), amine (—NH$_2$), ethylene oxide (—CH—CH$_2$), and isocyanate (—NCO). Some representative
$\diagdown\!\!\diagup$
O

functional group reactions are shown in Table 3-1. The interunit linkages are shown as the product of each reaction. The name for the polymer type is derived from the interunit linkage. Note that the formation of polyurethanes and polyureas involves hydrogen-transfer reactions without the emission of a low-molecular-weight by-product; they qualify as condensation polymers because of the reaction mechanism. Variations in the formation of polyesters and polyamides are introduced if the corresponding acid chlorides are substituted for the carboxyl groups.

If A and B represent two interacting functional groups on nonreacting groups R, R', or R'', the various reaction systems introduced in Chapter 1 can be represented as follows:

System Type	Generalized Representation
Bifunctional	ARB
Bi-bifunctional	$RA_2 + R'B_2$
Polyfunctional	$RA_2 + R'B_2 + R''B_b,\ b \geq 3$

These simple representations are useful in theoretical analyses in that reference to specific chemical systems is avoided and the results are generally applicable. Combinations other than those shown are, of course, possible.

One of the interesting facets of condensation polymerization is the enormous variety of polymers that can be generated, starting only with the few functional group reactions illustrated. This variety is obtained by manipulating the following:

1. The combinations of reacting function groups—a wide selection of A and B combinations is available.
2. The system type—use of bifunctional, bi-bifunctional, or polyfunctional systems leads to linear, branched and crosslinked networks, or copolymer systems.
3. The stoichiometric proportions—leads to control of $\overline{DP}_n$ as well as degree of branching and crosslinking.
4. The structure and compositions of the nonreactive portion—the R groups may be aromatic or aliphatic.

It is not always possible or desirable to use direct reaction, and in these

TABLE 3-1

SOME REPRESENTATIVE POLYMER FUNCTIONAL GROUPS REACTIONS

Reactants Functional Groups	Product Interunit Linkage	Polymer Type
—OH + —COOH	$-\overset{O}{\underset{\|}{C}}-O-$	Polyester
—COOH + —NH$_2$	$-\overset{O}{\underset{\|}{C}}-NH-$	Polyamide
—COOH + —COOH	$-\overset{O}{\underset{\|}{C}}-O-\overset{O}{\underset{\|}{C}}-$	Polyanhydride
—NCO + —OH	$-O-\overset{O}{\underset{\|}{C}}-NH-$	Polyurethane
—NCO + —COOH	$-O-\overset{O}{\underset{\|}{C}}-NH-$	Polyurethane
—NCO + —NH$_2$	$-NH-\overset{O}{\underset{\|}{C}}-NH-$	Polyurea
—OH + —OH	—O—	Polyether

—CH——CH$_2$
  \\   /
    O
	—O—	Polyether

TABLE 3-2
SOME REPRESENTATIVE POLYMER INTERCHANGE REACTIONS

Reactants Functional Groups	Product Interunit linkage	Polymer Type
—OH + —COOR	—C(=O)—O—	Polyester
—NH$_2$ + —C(=O)—NHR	—NH—C(=O)—NH—	Polyurea
—NH$_2$ + —COOR	—C(=O)—NH—	Polyamide
—OH + —OR	—O—	Polyether or Polyacetal

instances indirect or interchange reaction may be used. Some of the more common reactions are listed in Table 3–2.

Exchange reactions may be used in instances where it is difficult to purify or manipulate the reactants for direct reaction. Such reactions are of considerable commercial importance. An example of where this method is used is in the synthesis of poly(1,4-ethylene terephthalate) from 1,4-terephthalic acid (TA) and ethylene glycol (EG). Although EG can be easily purified to the required degree by distillation, TA is a rather intractable waxy solid that is difficult to purify. For the most part, TA is reacted with a low-boiling monohydric alcohol, such as methyl alcohol, to form dimethyl terephthalate (DMT), which is easily purified. The purified forms of DMT and EG are then reacted in an ester interchange reaction where the by-product, methyl alcohol, is removed by volatilization.

Formaldehyde is a commonly used reactant in several polyfunctional polymerization systems. The most commercially important types are shown below

System

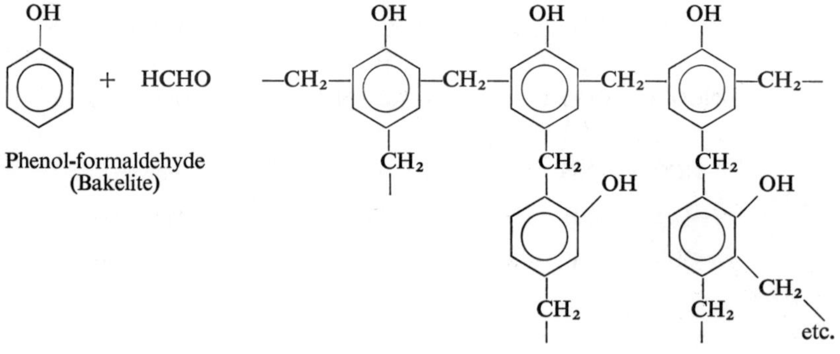

Phenol-formaldehyde (Bakelite)

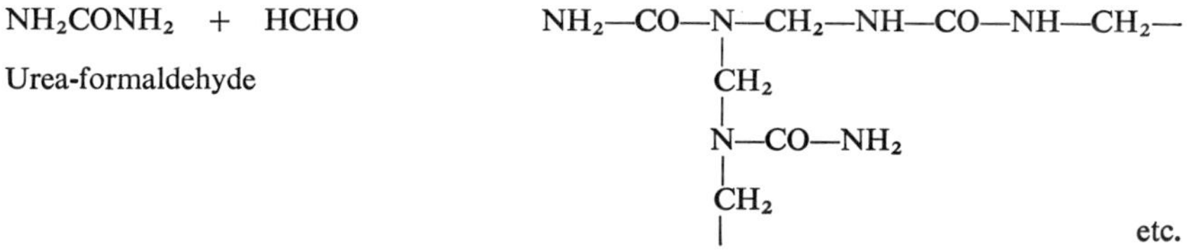

NH$_2$CONH$_2$ + HCHO

Urea-formaldehyde

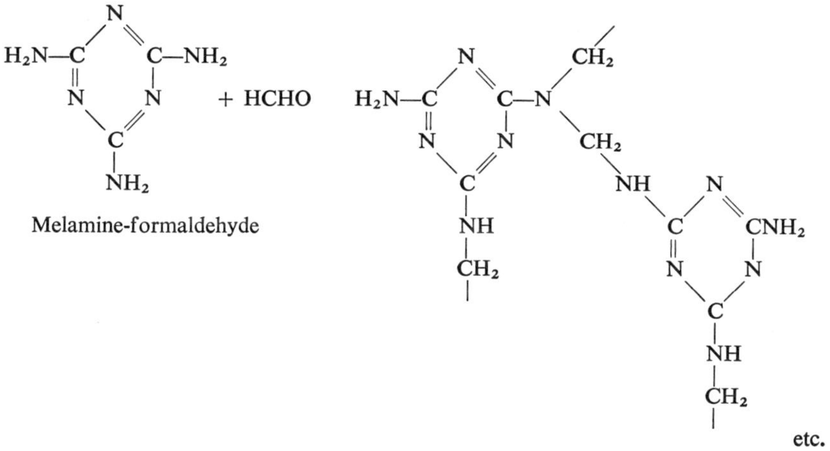

NH$_2$—CO—N—CH$_2$—NH—CO—NH—CH$_2$—
           |
          CH$_2$
           |
          N—CO—NH$_2$
           |
          CH$_2$
           | etc.

H$_2$N—C≡N—C—NH$_2$ (melamine ring with NH$_2$) + HCHO

Melamine-formaldehyde

[Melamine-formaldehyde polymer structures shown] etc.

302 Ring Versus Chain Formation

A minor but interesting consideration arises with bifunctional linear monomers, such as the hydroxy or amino acids of the following structure:

$$\text{HO—(CH}_2)_x\text{—COOH}$$

$$\text{H}_2\text{N—(CH}_2)_x\text{—COOH}$$

where x may be 3, 4, or 5. With such materials, ring formation predominates over linear polymerization. In general, if strain-free ring structures can form, they will. With $x < 3$, the rings would not be strain free, whereas with $x > 5$, the probability of ring closure is negligible because of the distance of separation of the reactive groups.

The foregoing limitations are of little consequence because the rings that do form can be reacted in a process called *ring-scission polymerization*. The most important example is the formation of polycaprolactam, Nylon 6, from caprolactam.

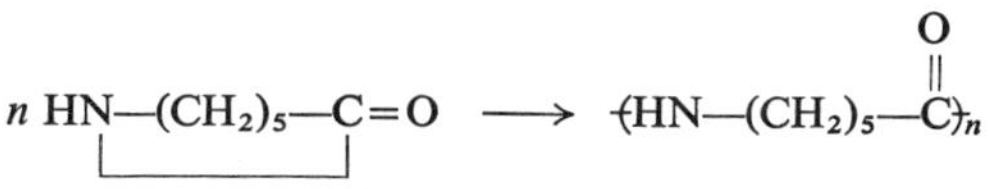

$$n\,\text{HN—(CH}_2)_5\text{—C=O} \longrightarrow \{\text{HN—(CH}_2)_5\text{—C}\}_n \quad (\text{C=O})$$

The polymerization of cyclic ethers is also important. Mechanistically, these are hybrid reactions, for they have characteristics of both condensation and addition polymerizations.

THEORETICAL ANALYSIS OF POLYCONDENSATION REACTIONS

As in all chemical-processing operations, it is advantageous to have accurate expressions for the various kinetic parameters to aid in process design and product control. In condensation polymerization, three main approaches have been taken: kinetic, stochiometric, and statistical. In principle, the three approaches can be applied to all types of polycondensation reactions, but, in practice, serious drawbacks arise in each. The application of elementary kinetic analysis has met with little success, and the reactions are far more complex than is usually appreciated. For the most part, complete analysis of the kinetic behavior of step-reaction polymerization remains one of the unsolved problems of modern polymer science. The stoichiometric approach is universally applicable, but it is limited to obtaining a single parameter, $\bar{X}_n$. The application of elementary statistical theory to these systems has led to some of the most intriguing and elegant results in the literature.

Here we shall cite some examples to illustrate the three approaches and to further acquaint the reader with the fundamentals of polycondensation reactions as well as the nature of the reaction products.

303 Mechanisms of Polycondensation Reactions

The reaction conditions employed in condensation polymerization generally fall into one of two categories: high temperature (> 200 °C) and low temperature (< 100 °C). The *high-temperature reactions* are by far the most commercially important and best understood, and they will be the subject of primary interest here. Considerable future promise may be associated with the development of the low-temperature reactions. Not a great deal is known about them, but they will be treated briefly in a later section.

Ordinarily, the high-temperature reactions are reversible, and equilibrium exists between the polymeric species and the by-product molecules. In order to drive the reaction to completion, it is necessary to remove the by-product molecules. This problem becomes compounded as the reaction progresses, for the viscosity rises as molecular weight increases. The problem can be alleviated to some degree by heat, stirring, and reduced pressures; in addition, thin-film reactors may be used. The rates are low where removal of by-product is hampered, as in a large kettle reactor, and reaction times are on the order of 20 hours or so. In thin-film reactors, high-surface areas are generated and the distance of diffusion for the by-product is drastically reduced to affect much faster reactions. However, generation of sufficient surface area and control of film thickness for uniform reaction in large scale production units imposes serious problems. The rates may be increased by an increase in temperature, a change in solvent polarity, or an increase in catalyst concentration.

It has long been thought that high-temperature polycondensations proceed via mechanisms exactly parallel to those for their low-molecular-weight, monofunctional counterparts. Flory's* pioneering studies indicated that the rate of reaction, energies of activation, heats of reaction, kinetics and effective catalysts were much the same. It was surprising that there should be such a close similarity. It was reasoned that these giant molecules with their retarded mobility should have considerably lower reaction rates. Originally it was argued that the collision rates for these large molecules should be low as predicted by their low, kinetic-theory velocity and that these rates should be further suppressed by the high viscosity of the reaction medium. Shielding of the reactive group within the coiling chain was also suggested as a steric factor affecting the rate. However, reaction studies on homologous series show that the rate constant, measured under comparable conditions, approaches an asymptotic limit as the chain length increases, generally after three or four —CH_2— groups in esterification. These results lead to Flory's *equal reactivity principle: The intrinsic reactivity of all functional groups is constant, independent of molecular size*. It now remains to explain this phenomenon of equal reactivity of functional groups, which prevails in spite of large molecular size and high, reaction-medium viscosity. There are three factors to consider here: the collision rate and shielding of the functional group, plus the bulk diffusion of the molecule as a whole.

The mobility of the terminal functional group is much greater than would be indicated by the macroscopic viscosity. The terminal group is relatively free to move about, for it is attached to the molecule only in one place. Consequently, while the chain is constantly altering its conformation due to thermal motion, the chain end may diffuse over a considerable region even though the gross mobility of the molecule as a whole is low. The actual frequency of collision of reactive groups may be as great as in their low-molecular-weight counterparts. Thus Flory concluded that the actual collision frequency was independent of the molecular mobility or the macroscopic viscosity.

A pair of neighboring functional groups on different molecules may collide many times before they diffuse apart or before they react. The lower

---

* P. J. Flory, *Principles of Polymer Chemistry*, (Ithaca, New York: Cornell University Press, 1953).

the diffusion rate for both the chain as a whole and the functional group, the greater will be the number of collisions and hence the greater the chance of a fruitful collision. As it turns out, only one collision in 10^{13} leads to chemical reaction. On the other hand, once the functional groups separate, a long time will be required before a new functional group is encountered. Flory concludes that over a long-enough time period, the decreased mobility will alter the time distribution of collisions experienced by a functional group but not the average number of collisions.

Flory disposed of the shielding problem by proposing that in concentrated solutions the molecules are extensively intertwined and show no preference for segments of its own chain over those of others. The functional group is thus shielded equally by all segments and they only act as so much diluent.

304 Kinetics of Polycondensation Reactions

Unfortunately, kinetic data appearing in the open literature are meager. Much information undoubtedly exists in industrial files, but it remains unpublished for reasons of proprietary interest. The early data obtained by Flory in his studies of polyesterification will, however, suffice for our discussion. It is also believed that the conclusions based on these data are quite general.

Esterification reactions are acid catalyzed. In the absence of an added catalyst, it is postulated that one of the reacting acid groups acts as such. Accordingly, a third-order rate equation should hold:

$$\frac{-d[\text{COOH}]}{dt} = k[\text{COOH}]^2[\text{OH}] \tag{3-1}$$

where the concentrations are expressed in terms of moles of functional groups. Note the inclusion of k as a rate constant independent of molecular size. If the reaction is equimolar, this equation becomes upon integration

$$2kt = \frac{1}{C^2} - \frac{1}{C(0)^2} \tag{3-2}$$

where C = concentration of unreacted acid (or hydroxyl) groups at time t
$C(0)$ = initial concentration of acid groups

If we write $C = C(0)(1-p)$, where p is the fractional conversion, then Eq. (3-2) becomes

$$2C(0)^2 kt = \frac{1}{(1-p)^2} - 1 \tag{3-3}$$

A plot of $1/(1-p)^2$ versus t should be linear if Eq. (3-3) correctly represents the kinetic behavior of the system.

Acid-catalyzed systems are believed to follow second-order reaction kinetics. For equimolar reactants,

$$-\frac{dC}{dt} = k'C^2 \tag{3-4}$$

where the rate constant k' includes the catalyst concentration. This equation integrates to

$$\frac{1}{C} - \frac{1}{C(0)} = kt \tag{3-5}$$

When $C = C(0)(1-p)$ is substituted, Eq. (3-5) becomes

$$C(0)k't = \frac{1}{1-p} - 1 \tag{3-6}$$

This result indicates that a plot of $1/(1-p)$ versus t should be linear if second-order kinetics prevail.

Figures 3-1 and 3-2 show typical esterification data for uncatalyzed and catalyzed systems, respectively, plotted as indicated by Eqs. (3-3) and (3-6). At first glance, the data seem fairly well represented, at least on the time scale. However, if one solves for p, the extent of reaction, at the limits of the linear portion of the data, one sees that the data correspond to the indicated relationships only from about 80 to 93 percent esterification for the uncatalyzed systems and from 93 to 100 percent for the catalyzed systems. Evidently these are not simple second- or third-order reaction kinetics. The apparent agreement between theory and experiment in this and similar situations has resulted in a misconception which has prevailed for some time in the literature. The correspondence here between theory and data can only be regarded as fortuitous. Advanced treatments will have to consider the reverse reaction of hydrolysis as well as possible diffusion-controlling mechanisms.*

---

* The following paper is especially pertinent in this regard: G. A. Campbell, E. F. Elton, and E. G. Bobolek, *J. Appl. Polym. Sci.*, **14** (4), 1025, (1970).

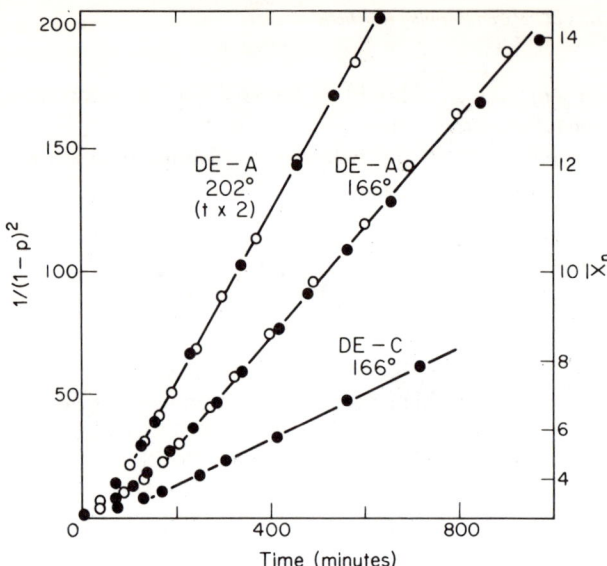

Figure 3–1. Reactions of diethylene glycol with adipic acid (DE-A) and of diethylene glycol with caporic acid (DE-C). Time values at 202°C have been multiplied by two. [P. J. Flory, J. Am. Chem. Soc., 62, 2261 (1940). Reprinted with permission of the copyright owner, The American Chemical Society.]

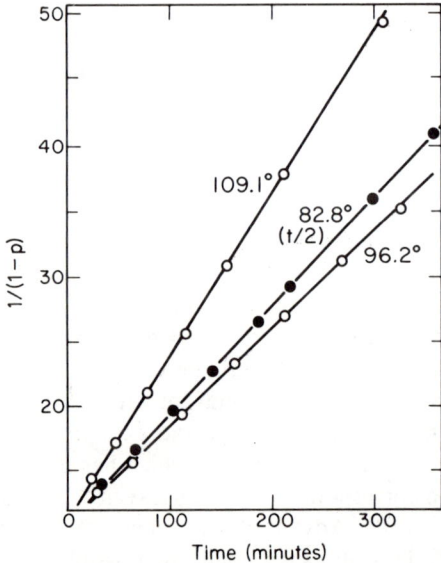

Figure 3–2. Reaction of decamethylene glycol with adipic acid at the temperature indicated, catalyzed by 0.10 equivalent percent of p–toluenesulfonic acid. [P. J. Flory, J. Am. Chem. Soc., 61, 3334 (1940). Reprinted with of the copyright owner, The American Chemical Society.]

305 Stoichiometry of Linear Systems

The conversion or extent of reaction is expressed as the fraction of one of the species of functional groups that has reacted, and it is usually represented by the symbol p. If we consider reacted A groups, the extent of reaction of A is represented by the symbol p_A, which is given by

$$p_A = \frac{F_A(0) - F_A(t)}{F_A(0)} \tag{3-7}$$

where $F_A(0)$ = the number of A functions present initially and $F_A(t)$ = the number of unreacted A functions present at any time t.

In a bifunctional system ARB or in a stoichiometrically balanced bi-bifunctional system $RA_2 + R'B_2$, since the reaction is 1:1 in A and B, $p = p_A = p_B$.

In systems that are not stoichiometrically balanced, $p_A \neq p_B$, but p_A will be related to p_B as

$$\frac{F_A(0)}{F_B(0)} = \frac{p_B}{p_A} = r \tag{3-8}$$

In practice, p is found through a combination of chemical end-group-analysis and stoichiometric balance of the reactants.

In an absolutely pure, stoichiometrically balanced reaction system, it is theoretically conceivable for the reaction to proceed indefinitely and to produce a polymer of infinite molecular weight. If there is stoichiometric imbalance, the molecular weight will not be infinite, and the polymerization reaction will be *blocked*. As a limiting example, consider a bifunctional system where the stoichiometric proportions are 2:1 as

$$2RA_2 + R'B_2 \longrightarrow ARA-BR'B-ARA$$

The product is not a high polymer at all, for the maximum average degree of polymerization is only three. Ordinarily blocking imposes unacceptable limitations in high-polymer synthesis. However, it can be employed to advantage, if controlled, to regulate molecular weight or to prevent the reaction from proceeding to the gel point in polyfunctional systems.

In order to take a general approach to representing the effect of stoichiometry on $\bar{X}_n$, consider the following reaction system:

$$RA_2 + (xs)R'B_2$$

where the B functions will be in excess and $r < 1$.

The total number of monomer units present initially can be expressed as

$$N(0) = \frac{F_A(0) + F_B(0)}{2} = \frac{F_A(0)(1 + 1/r)}{2} \qquad (3\text{-}9)$$

For $p_A < 1$ there are three possible types of molecules that can form according to the terminal functional group:

$$A \sim\!\sim\!\sim\!\sim B$$
$$B \sim\!\sim\!\sim\!\sim B$$
$$A \sim\!\sim\!\sim\!\sim A$$

Since B is in excess, it is highly unlikely that any of the last type will form. Thus the number of chain ends at any point in the reaction is, to a very good approximation, two times the number of unreacted A functions, plus the initial excess of B functions. The total number of polymer molecules, $P(t)$, at any extent of reaction of A is given by

$$P(t) = \frac{2F_A(0)(1 - p_A) + [F_B(0) - F_A(0)]}{2} \qquad (3\text{-}10)$$

where the first term represents the number of chain ends associated with molecules beginning in A and ending in B (stoichiometric equivalence) and the second term represents the number of chain ends associated with molecules beginning and ending in B (stoichiometric excess). Upon rearrangement, Eq. (3–10) gives

$$P(t) = \frac{F_A(0)[2(1 - p_A) + (1 - r)/r]}{2} \qquad (3\text{-}11)$$

Now recall that $\bar{X}_n = [N(0) - N(t)]/P(t)$, and observe that in step-reaction polymerization $N(t) \ll N(0)$ very soon after the reaction begins; therefore we obtain $\bar{X}_n$ as

$$\bar{X}_n = \frac{N(0)}{P(t)} \qquad (3\text{-}12)$$

for step-reaction polymerizations. Combining Eqs. (3-9) and (3-11) with (3-12), we obtain the desired result, which was first derived by Flory.

$$\bar{X}_n = \frac{1 + r}{1 + r - 2rp_A} \qquad (3\text{-}13)$$

A number of important lessons are to be learned from Eq. (3-13). First consider a system that is stoichiometrically balanced ($r = 1$). Then Eq. (3-13) reduces to

$$\bar{X}_n = \frac{1}{1-p} \qquad (3\text{-}14)$$

This is a result originally obtained by Carothers* in a slightly different fashion. It is one of the fundamental analytic representations of step-reaction polymerization. Furthermore, consider the tabulation below:

p	0	0.5	0.8	0.9	0.95	0.99	0.999	1.0
$\bar{X}_n$	1	2	5	10	20	100	1000	∞

Note that in order to achieve reasonably high molecular weights in linear-condensation polymers, the extent of reaction must exceed 99 percent. In terms of the normally anticipated industrial and laboratory yields, this is a rather stringent requirement. In its day, this was an extremely important result in that chemists were not observing good mechanical properties in their synthetic products compared to those of natural polymers. At best, resins with oil-like consistencies were obtained and subsequently discarded. Carothers showed that in order to obtain useful synthetic products, it was necessary to drive the reaction to completion.

As another limiting case, consider B functions 1 percent in excess such that $r = 100/101$. From Eq. (3-13) we find that $\bar{X}_n$ is limited to a value of 201 for $p_A = 1$. This result shows the importance of stoichiometric balance to obtain high molecular weights. As indicated previously, blocking may be used to advantage in the deliberate preparation of low-molecular-weight polymers for subsequent treatment in another process. Such materials are referred to as *prepolymers*. Also, blocking may be used to prevent gelation in polyfunctional systems, although in this case Eq. (3-13) would not be applicable.

306 Size Distribution in Linear-Condensation Systems

Flory was among the first to apply elementary statistical concepts to the analysis of polymer systems. Based on the assumption of chemical equilibrium and stoichiometric balance, he was able to develop an analytic expression to describe the size distribution in linear-polymerization systems.

* *Collected Papers of Wallace H. Carothers on Polymerization*, op. cit.

Before beginning with the derivation *per se*, the student should recall that a probability is exactly analogous to a number fraction. We make use of this point immediately in stating that the probability that any given functional group in a reaction system has reacted is simply equal to the extent of reaction of that functional group. For A groups, this would be p_A; and since equimolar quantities are present, this probability would also be equal to p_B or simply p. Suppose that we have an ARB (bifunctional) monomer to condense to polymer, and we start with one mole of reactants—that is, one mole of A groups and one mole of B. If we come to a point where the extent of reaction, p, is, say, 0.75, we know that the hypothetical chances of our encountering a reacted A or B group in the reaction vessel is 0.75 or 3 out of 4. The hypothetical chances of encountering an unreacted A or B group is simply one out of four or $(1-0.75)$. Establishment of the condition of equilibrium is necessary to remove the possibility of nonequal probability for the functional groups.

Consider the following molecule and the probability that it has formed:

$$\text{ARB} \overline{}_1 \text{ARB} \overline{}_2 \text{ARB} \overline{}_3 \ldots \overline{} \text{ARB} \overline{}_{x-1} \text{ARB}_x$$

The probability that the 1st B group has reacted is p. The probability that the 2nd B group has reacted is p, and so forth up to the $(x-1)^{\text{th}}$ B group. However, the probability that the xth B group has not reacted must be accounted for, and this probability is $(1-p)$. The probability that groups 1 and 2 have reacted as $p \times p = p^2$. The probability that groups 1 through 3 have reacted is $p \times p \times p = p^3$ and so forth. The probability that groups 1 through $(x-1)$ have reacted $= p^{x-1}$. Finally, the probability, $P(x)$, that the molecule in question is exactly x units long is given by

$$P(x) = p^{x-1}(1-p) \tag{3-15}$$

which is simply the probability of all the independent events considered occurring in a consecutive fashion. By definition, this $P(x)$ is the number (or mole) fraction, n_x, of those particular molecules in the reaction system at the extent of reaction p. Hence

$$n_x = p^{x-1}(1-p) \tag{3-16}$$

This equation gives the number (or mole) fraction of each size species present in the polymer system. If $P(t)$ = the total moles of molecules at p greater than zero and $N_x(t)$ = the moles of x-mers, the following holds:

$$N_x(t) = P(t)(1-p)p^{x-1} \tag{3-17}$$

We can carry this derivation one step further to get the weight-fraction-distribution function w_x. With the weight fraction of any x-mer given by $xN_x(t)/N(0)$ and with $P(t) = N(0)(1-p)$, we obtain

$$w_x = x(1-p)^2 p^{x-1} \tag{3-18}$$

The distributions given by Eqs. (3-16) and (3-18) are known as the *most probable distributions*. They are shown in Figures 3-3 and 3-4 for several values of p. Note that on a mole fraction basis, the smaller species are always the most abundant. On a weight fraction basis, however, w_x passes through a maximum, which one can show occurs at $\bar{X}_n$ as $p \to 1$. They have been verified experimentally and are important aids in helping us to visualize the development of the size distributions alluded to in Chapter 1.

Other important results arise in the application of these distributions to the computation of $\bar{X}_n$ and $\bar{X}_w$. With $\bar{X}_n = \sum_{x=1}^{\infty} n_x x$ and $\bar{X}_w = \sum_{x=1}^{\infty} w_x x$, one can (via the binomial expansion theroem) show for linear-condensation systems

$$\bar{X}_n = \frac{1}{1-p} \tag{3-19}$$

and

$$\bar{X}_w = \frac{1+p}{1-p} \tag{3-20}$$

The number average is a result obtainable from stoichiometric analysis, but distribution functions are always required to compute higher-order averages.

Another important result is obtained by comparing $\bar{X}_w$ and $\bar{X}_n$ as

$$\frac{\bar{X}_w}{\bar{X}_n} = 1 + p \tag{3-21}$$

This ratio gives a simple quantitative idea of the spread of the distribution curve, for as $p \to 1$, $\bar{X}_w/\bar{X}_n \to 2$ (provided that there are no side reactions).

In principle, the most probable distributions can be derived for more complicated systems by following the procedure outlined above. However, the difficulty of this task increases rapidly with increasing complexity of the reaction systems.

307 Crosslinking in Step-Reaction Systems

Phenomenologically, the *gel point* of a polyfunctional system is classically defined as that point in the reaction where the viscosity becomes so large that

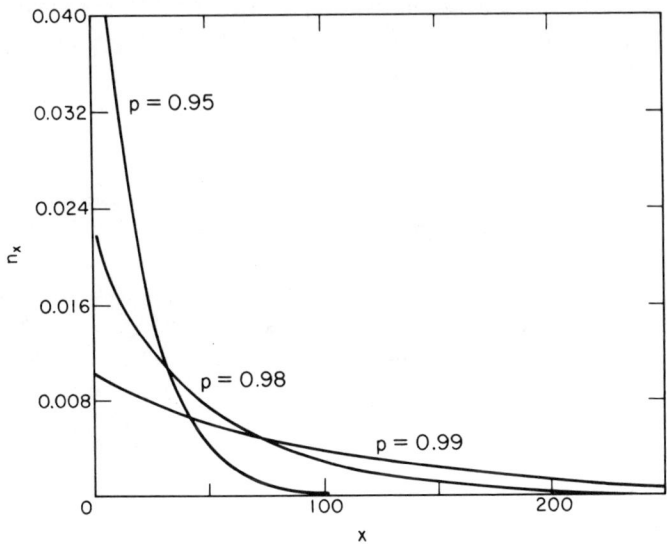

Figure 3-3. Mole fraction distribution of chain molecules in a linear condensation polymer for several extents of reaction p. [P. J. Flory, J. Am. Chem. Soc., *58*, 1877 (1936). Reprinted with permission of the copyright owner, the American Chemical Society.]

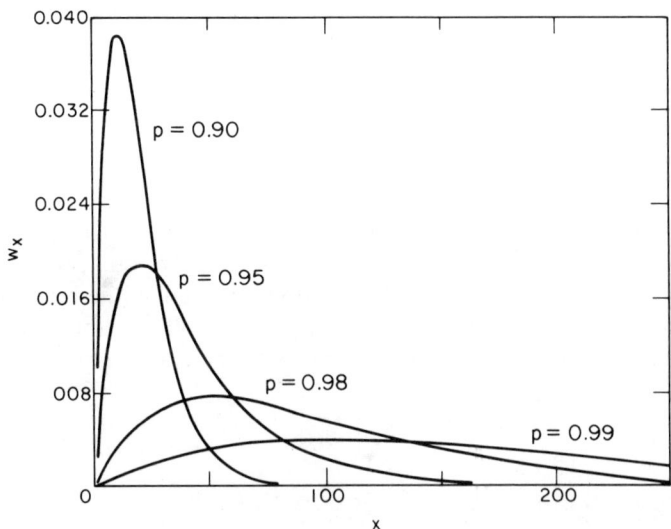

Figure 3-4. Weight fraction distribution of chain molecules in linear condensation polymers for several extents of reaction p. [P. J. Flory, op. cit. Figure 3-3.]

the reaction medium loses all fluidity and bubbles cease to rise through it. At this point, *gelation* is said to have occurred. It is obviously important to be able to predict the onset of gelation. Figure 3-5 shows how the viscosity and extent of reaction typically vary with time in a polyfunctional system. There is, at first, a gradual rise in viscosity and then a sudden enormous rise as the gel point is approached. Also note that the reaction proceeds very slowly as the gel point is approached. These two factors make it difficult to predict the gel point well in advance by following the reaction in process with conversion or viscosity data.

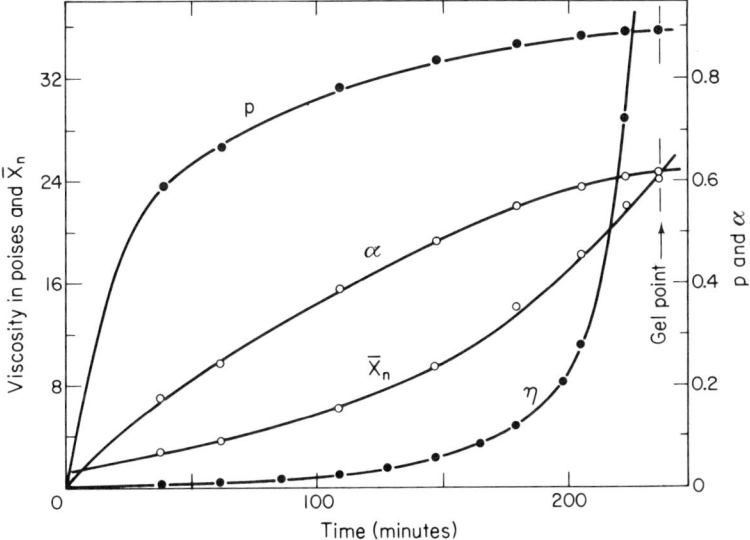

Figure 3-5. The course of typical three-dimensional polyesterification. [P. J. Flory, J. Am. Chem. Soc., *63*, 3083 (1941). Reprinted with permission of the copyright owner, the American Chemical Society.]

Gelation takes place long before all the reactants are bound together, so that reaction continues even after gelation. It is common knowledge that at the gel point not all of the material is insoluble, but the gelled portion represents only a small percentage of the total reaction mixture. To amplify: if one were to attempt to dissolve in a suitable solvent the gelled polymer mass just as it exists at its gel point, all but a small portion would dissolve. Only a few very large molecules are required to induce gelation. The portion that is soluble is referred to as the *sol*, whereas the part that is insoluble is referred to as the *gel*.

It will be recalled that in order to attain a three-dimensional network, one reactant must possess a functionality of 3 or more. A monomer unit possessing such a degree of functionality will be called a *branch unit*. That portion of a polymeric molecule lying between two branch units or between a branch unit and an unreacted bifunctional unit will be called a *chain section*. The chain sections may vary in length, but this factor is not important at this juncture. A point in a molecule from which three or more chain sections emanate will be known as a *branch point*.

A key assumption for valid statistical treatment is that the probability that any particular functional group has reacted is independent of and uninfluenced by the number and configuration of other groups in the molecule to which this functional group is attached. This assumption is not always true; the secondary hydroxyl on glycerol is, for instance, less reactive than either of the primary hydroxyl groups. Also, intramolecular reaction will be assumed to be totally absent, an assumption that prevents the results from being exact.

308 Criteria for Incipient Formation of Infinitive Networks

As a model of the infinite network for the development of Flory's gel point theory, consider a polyfunctional system containing trifunctional branch units: RA_3 reacting with bifunctional molecules $R'A_2$ and $R''B_2$—not necessarily in equimolar quantities. A schematic representation of a portion of the network which could result is shown in Figure 3–6. Assume that a chain section has been selected at random from the resulting gelled polymeric

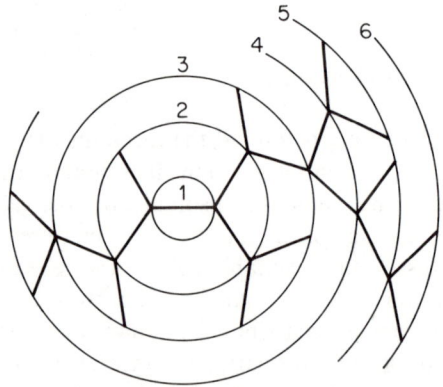

Figure 3–6. Schematic representation of a trifunctionally branched network polymer. [P. J. Flory, op. cit. Figure 3–5.]

structure and that it lies within the first envelope of Figure 3-6. The immediate problem is to determine the probability that this chain section is part of an infinite network and, therefore, that this chain section is part of a gelled polymer system. In Figure 3-6 this chain section in the first envelope happens to give rise to branch units, one at each end. The four new chains that result happen to lead to three new branch points (on envelope 2) and one terminal bifunctional group. The resulting six new chain sections happen to lead to two new branch units and four terminal bifunctional groups on envelope three, and so on. The fact that the chain sections are depicted as being equal in length will not affect our analysis. We are interested in the existence of such chain sections regardless of their lengths.

Assume that there exists a *branching probability* α that any functional group of a branch unit leads via bifunctional units to a branch unit rather than a bifunctional terminal group. Obviously α will depend on the relative concentrations of the branch and bifunctional units as well as the fraction of A and B groups that have reacted. The probability of the other alternative—that is, the probability that the functional group of the branch unit will lead to an unreacted bifunctional terminal group—is $(1-\alpha)$.

Based on the model described above, consider the ith envelope from the randomly selected chain section. Suppose that there are Y_i branch units on the ith envelope. If all the chain sections emanating from these branch units (or branch points) ended in branch units on the $i+1$th envelope, then there would be $2Y_i$ of them. However, there is only a certain probability, α, that chain sections beginning in branch units (actually all chain sections in the network begin in branch units) will end in branch units. Therefore the expected number of branch units on the $i+1$th envelope, Y_{i+1}, will be $2Y_i\alpha$. The criterion for continuous expansion of the network is that the number of chain sections emanating from the $i+1$th envelope, $2Y_{i+1}$, be greater than the number of chain sections emanating from the ith envelope, $2Y_i$—that is, $Y_{i+1} > Y_i$. This, in turn, requires that $\alpha > \frac{1}{2}$. When $\alpha < \frac{1}{2}$ an infinite network will not be generated. Obviously *the critical value of α*, namely α_c, is $\frac{1}{2}$. A rationale similar to the preceding can one be devised for polyfunctional systems containing higher functional branch units. Induction leads to the following result:

$$\alpha_c = \frac{1}{b-1} \qquad (3\text{-}22)$$

where b equals the functionality of the branch unit. In a situation where more than one species of branch units is encountered, α_c would be computed using an average b.

309 The Probability α

It now remains to compute the actual value of α for a specific blend of reactants. As a typical example of a polyfunctional system, continue to consider the previously noted set of reactants. We must now compute the probability that a given functional group of a branch unit leads via a chain section to another branch unit rather than a bifunctional terminal unit. First we define a new fractional parameter, ρ, which will also subsequently serve as a probability—that is

$$\rho = \frac{\text{number of A functions in branch units}}{F_A(0)}$$

Note that we are not concerned with functional equivalence in this development. Consider the following chain section of a gel network and the probability α' that the chain section at position 1 will lead via bifunctional units to the branch unit at position 6:

$$\begin{array}{c} A \\ \diagdown \\ A \diagup \end{array} R A \underset{1}{-} BR'' B \underset{2}{-} AR' A \underset{3}{-} BR'' R \underset{4}{-} AR' A \underset{5}{-} BR'' B \underset{6}{-} A R \begin{array}{c} \diagup A \\ \diagdown A \end{array}$$

The probability that A at position 1 will have reacted with a bifunctional B is p_A. The probability that B at position 2 will have reacted is p_B. However, we must also consider the probability that this B reacted with a bifunctional unit and not a trifunctional unit,* and this probability is simply $(1-\rho)$, the number fraction of A's on bifunctional units. The product of these probabilities gives the desired result, $p_B(1-\rho)$, which is the probability that B at position 2 reacted with a bifunctional A. The appropriate probabilities for positions 3, 4, and 5 are similarly obtained. For position 6, we need the probability that this B has reacted with a trifunctional A unit, which is $p_B \rho$. Thus the probability α' is given by

$$\alpha' = p_A [p_B(1-\rho)p_A]^2 p_B \rho \qquad (3\text{-}23)$$

In general, these sections can be represented as

$$\begin{array}{c} A \\ \diagdown \\ A \diagup \end{array} RA + BR''B - AR'A + BR''B - AR \begin{array}{c} \diagup A \\ \diagdown A \end{array}$$

---

* The probability that A at position 1 reacted with a bifunctional B is obviously equal to one.

If $\alpha(i)$ is the probability that the foregoing section, containing i (BR″B—AR′A) units, has formed, then

$$\alpha(i) = p_A p_B \rho [p_A p_B (1-\rho)]^i \tag{3-24}$$

We are, however, interested in a total probability, α, that is independent of the length of the chain section, as reflected in the value of i. Accordingly, $\alpha = \sum_{i=0}^{\infty} \alpha(i)$ or

$$\alpha = \sum_{i=0}^{\infty} [p_A p_B (1-\rho)]^i p_A p_B \rho \tag{3-25}$$

$$\alpha = \frac{p_A p_B \rho}{1 - p_A p_B (1-\rho)} \tag{3-26}$$

With $\quad p_B/p_A = r$

$$\alpha = \frac{r p_A^2 \rho}{1 - r p_A^2 (1-\rho)} \tag{3-27}$$

$$\alpha = \frac{p_B^2 \rho}{r - p_B^2 (1-\rho)} \tag{3-28}$$

Equations (3-27) and (3-28) are the working equations for computing the gel point for a particular stoichiometry or for the reverse situation, computing stoichiometry for a desired gel point. A number of special cases arise: when $F_A(0) = F_B(0)$, $r = 1$ (functional equivalence); when $\rho = 1$ (only RA_3 is present—no $R'A_2$), or when both $F_A(0) = F_B(0)$ and $\rho = 1$.

310 Gel Point Observations

The gel point has been classically designated as the time when the reaction mixture suddenly loses fluidity or when bubbles cease to rise through it. In the light of the preceding development, it is natural to postulate that this experimental gel point should correspond to the theoretical critical point, p_c, for infinite network formation.

Flory made measurements of gel points in reactions of diethylene glycol with succinic or adipic acid with varying proportions of the tribasic acid, tricarballylic acid. The extents of reaction were determined by titration and plotted as in Figure 3-5. The number-average degree of polymerization was calculated from Eq. (3–31); one can see that it is not very large and that it does not increase rapidly at the gel point. Flory's gel point observations are shown in Table 3-3, and the results of one experiment are plotted in Figure 3-5. Invariably the observed gel point was higher than the theoretical gel point. The discrepancy was attributed to the failure of the theory to account for a minor amount of intramolecular condensation.

All experimental studies, where the loss of fluidity has been taken as marking the gel point, have shown that α_c observed was higher than α_c calculated. This fact was taken to confirm the hypothesis that gelation of polyfunctional systems was attributable to the formation of infinitely large molecules.

TABLE 3-3

GEL POINTS FOR POLYMERS CONTAINING TRICARBALLYLIC ACID[a]

Additional Ingredients Diethylene Glycol and	$r = \dfrac{[COOH]}{[OH]}$	ρ	p at Gel Point		α Observed at Gel Point
			Observed	Calculated	
Adipic acid	1.000	0.293	0.911	0.879	0.59
Succinic acid	1.000	0.194	0.939	0.916	0.59
Succinic acid	1.002	0.404	0.894	0.843	0.62
Adipic acid	0.800	0.375	0.9907	0.955	0.58

[a] P. J. Flory, *J. Am. Chem. Soc*, **63**, 3083, (1941). Reproduced with permission of the copyright owner, The American Chemical Society.

In a recent study, Bobalek et al.* explored the effects of chemical composition, extent of reaction, and possible side reactions on the gel point behavior of a model resin system. For their model system, they used glycerol, phthalic anhydride, and lauric acid with a reference composition of 1:1:0.4, respectively. Their results indicate that a redefinition of the experimentally observed gel point is in order.

The proposed concept is shown in Figure 3-7. As the extent of reaction, and the molecular weight increase, the probable gel point, as predicted by Flory's theory, is reached. Here the formation of macroscopic three-dimensional network molecules occurs. These molecules exhibit limited solubility because of their large size and complex structure; as a result, they precipitate to form microgel particles that remain suspended as a disperse phase in the continuous phase formed by the lower molecular-weight component. Thus the discrete macroscopic networks that begin to form collapse because of limited solubility before they can interract with each other to form a single infinite network. The mechanism of gel-particle stabilization

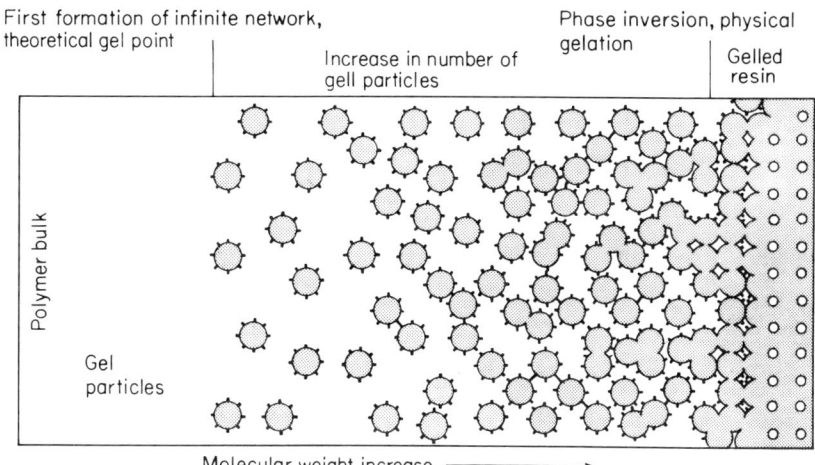

Figure 3-7. Proposed model for gelation process. [E. G. Bobalek, E. R. Moore, S. S. Levy and C. C. Lee, *J. Appl. Polym. Sci.*, **8**, 625, (1964).]

is not clear; but because of this stabilization, they are isolated and unable to react with one another. Also, reaction within these particles is retarded because of this phase heterogeneity.

As the reaction progresses beyond the theoretical gel point, the concentration of gel particles increases until a point is reached where the concentration becomes sufficiently high for the particles to interreact. Phase inversion occurs and, the flocculated microgel becomes the continuous phase, causing the steep rise in viscosity and immobilization of the fluid.

This new gelation concept can be summarized as follows: At the theoretical gel point, a number of macroscopic three-dimensional networks form and undergo phase separation. The microgel particles so formed remain suspended and increase in number as reaction continues. At the classically observed gel point, the concentration of microgel particles reaches a critical value and causes phase inversion as well as the subsequent steep rise in viscosity.

311 $\bar{X}_n$ in Polyfunctional Systems

The number-average degree of polymerization *before* gelation for a polyfunctional system $RA_b + R'A_2 + R''B_2$ can be computed stoichiometrically as follows: The total number of monomer units present initially is given by

$$N(0) = \frac{F_A(0)(1-\rho)}{2} + \frac{F_A(0)\rho}{b} + \frac{F_B(0)}{2} \qquad (3\text{-}29)$$

The number of molecules $P(t)$ present at the extent of the reaction, p_A, equals the number of monomer units present initially, $N(0)$, less the number of linkages formed, $p_A F_A(0)$. Accordingly, with

$$\bar{X}_n = \frac{N(0)}{N(0) - p_A F_A(0)} \tag{3-30}$$

the number-average degree of polymerization below the gel point becomes

$$\bar{X}_n = \frac{b(1-\rho+1/r)+2\rho}{b(1-\rho+1/r-2p_A)+2\rho} \tag{3-31}$$

In principle, the molecular-weight-distribution functions for such systems can be derived for the polymer up to the gel point, but this is somewhat more difficult than for the linear case. Said functions would depend on the functionality and the stoichiometry.

312 Size Distributions in Polyfunctional Systems*

The simplest case of a polyfunction system is represented by the self-condensation of RA_b, giving structures as indicated in Figure 3-8 for $b = 3$. For simplicity, we will restrict our discussion of size distributions in polyfunctional systems to such structures. Extension to systems containing bifunctional groups is given in advanced treatments, but it is observed that the general characteristics of the size distribution are retained. Discussion will fall naturally into pregel and postgel periods.

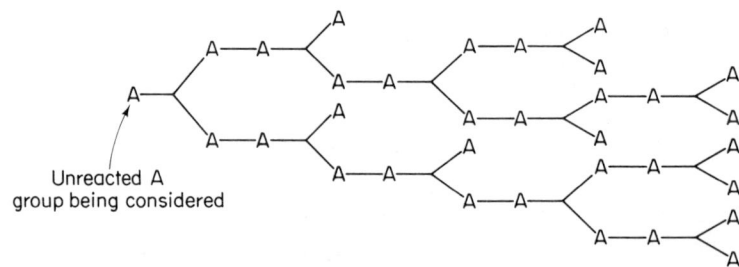

Figure 3–8. Self-condensation polyfunctional system.

* Sections 312 and 313 consist of a condensation and generalization of the development by P. J. Flory in *Principles of Polymer Chemistry*, pp. 365–378. op. cit.

First note that the branching probability α is simply equal to p, the extent of reaction, and that α_c is equal to p_c, the extent of reaction at the gel point. In the ensuing discussion, it will prove convenient to visualize structures with $b = 3$, although generality will be maintained by retaining the symbol b in our analytic developments. We must also forbid the possibility of intramolecular reaction, which is probably not a serious consideration in the pregel period.

Since the number of molecules present at any extent of reaction is given by the monomer units present initially, minus the number of linkages formed,

$$P(t) = N(0) - \frac{N(0)b\alpha}{2}$$

$$= N(0)\left(1 - \frac{\alpha b}{2}\right) \tag{3-32}$$

Also, since $\bar{X}_n = N(0)/P(t)$,

$$\bar{X}_n = \frac{1}{1 - \alpha b/2} \tag{3-33}$$

For an x-mer of the sort shown in Figure 3-8, the number of A—A bonds that have formed is $(x-1)$, while the number of A groups that have not reacted is $[(b-2)x+1]$. From the network for $\alpha < \alpha_c$, select an unreacted A and consider the remaining monomer units distinguishable. The probability $P'(x)$ that a particular x-mer has formed according to a specific pattern is

$$P'(x) = \alpha^{x-1}(1-\alpha)^{(b-2)x+1} \tag{3-34}$$

Let Ω_x be the number of ways such a molecule could have formed. Thus the probability $P(x)$ that any unreacted A group is part of an x-mer is

$$P(x) = \Omega_x \alpha^{x-1}(1-\alpha)^{bx-2x+1} \tag{3-35}$$

To compute Ω_x, consider the construction of an x-mer in which we first make each monomer unit distinguishable by numbering one of its A groups. We will then select a numbered A group as a point of reference and continue taking numbered A groups to react with the unnumbered groups which are already part of the network we are constructing. When the x-mer is finally constructed, there will remain one unreacted numbered A group, and $(x-1)$ numbered A groups will have reacted with $(x-1)$ unnumbered A groups. To begin with, there will be $(b-1)x$ unnumbered A groups, out of which $(x-1)$ will react. The number of ways these $(x-1)$ unnumbered indistinguishable A's can be selected for reaction is

$$\frac{\{[bx-x]\}\{[bx-x]-1\}\{[bx-x]-2\}\ldots\{[bx-x]-[(x-1)-1]\}}{(x-1)!}$$

Multiplying and dividing by $\{[bx-x]-[(x)-1]\}!$, we obtain

$$\frac{(bx-x)!}{(bx-2x+1)!(x-1)!}$$

which is equivalent to the number of ways of choosing $(x-1)$ indistinguishable items from $(bx-x)$ items. Now, the $(x-1)$ numbered A's can react with the $(x-1)$ unnumbered A's in $(x-1)!$ ways, and $x!$ distinguishable structures can be so formed (since the x numbered A's are distinguishable). Therefore Ω_x is given by

$$\Omega_x = \frac{(bx-x)!}{(bx-2x+1)!(x-1)!}\frac{(x-1)!}{x!}$$

and, finally,

$$\Omega_x = \frac{(bx-x)!}{(bx-2x+1)!x!} \tag{3-36}$$

In this case, P(x) is not the mole fraction of x-mers, but

$$P(x) = \frac{\text{number of unreacted A's on } x\text{-mers}}{\text{total unreacted A's}}$$

This can be described stoichiometrically as

$$P(x) = \frac{[(b-2)x+1]N_x}{(1-\alpha)[N(0)b]} \tag{3-37}$$

By equating Eqs. (3–35), (3–36), and (3–37), we obtain

$$N_x = N(0)\frac{(1-\alpha)^2}{\alpha}\Omega'_x\beta^x \tag{3-38}$$

where

$$\Omega'_x = \frac{b(fx-x)!}{(bx-2x+1)!x!} \tag{3-39}$$

and

$$\beta = \alpha(1-\alpha)^{b-2} \tag{3-40}$$

With $n_x = N_x/P(t)$ and $w_x = xN_x/N(0)$, we obtain

$$n_x = \frac{(1-\alpha)^2}{\alpha(1-\alpha b/2)} \Omega'_x \beta^x \qquad (3\text{-}41)$$

$$w_x = \frac{(1-\alpha)^2}{\alpha} x \Omega'_x \beta^x \qquad (3\text{-}42)$$

Since we have w_x, $\bar{X}_w$ can be computed through the use of standard mathematical procedures. It is found that

$$\bar{X}_w = \frac{1+\alpha}{1-\alpha(b-1)} \qquad (3\text{-}43)$$

Comparing $\bar{X}_w$ and $\bar{X}_n$ from Eq. (3–33), we observe that

$$\frac{\bar{X}_w}{\bar{X}_n} = \frac{(1+\alpha)(1-\alpha b/2)}{1-\alpha(b-1)} \qquad (3\text{-}44)$$

The weight-fraction distribution, given by Eq. (3–42), is shown plotted in Figure 3–9 for $b = 3$. Curves are shown only for $\alpha \leq \alpha_c$, or below the gel point. Since intramolecular reaction has been precluded, the preceding equations will begin to deviate as the gel point is approached—that is, as the tendency

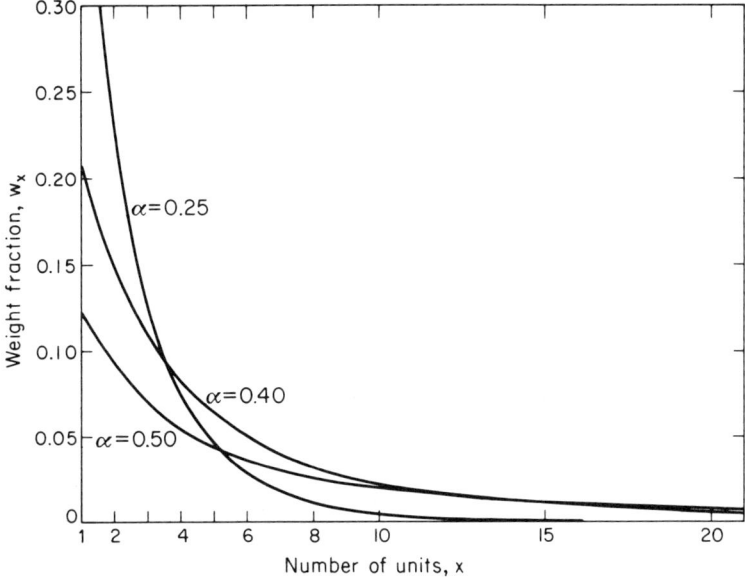

Figure 3–9. Weight fraction distribution for a branched polymer prepared from a simple trifunctional monomer at the α's indicated. [P. J. Flory, Chem. Revs., *39*, 137 (1946).]

toward intramolecular reaction increases. If applied to the *entire* reaction mass after gelation, the predicted results will certainly be in error since the gel portion will, by its very nature, contain many intramolecular connections. The equations should apply, however, to the sol portion during *all* stages of reaction. At the gel point, $\bar{X}_n$ is finite (= 4 for $b = 3$), while $\bar{X}_w$ is infinite. The ratio $\bar{X}_w/\bar{X}_n$ will also be infinite, indicating an extreme size heterogeneity at the gel point.

313 Postgel Relations in Polyfunctional Systems

Development here centers on the apparent anomaly that summation of either Eqs. (3–41) or (3–42) does not lead to unity for $\alpha > \alpha_c$. This point arises because the sums depend on β, which is double valued in α. That is, as α ranges from 0 to 1, β ranges from 0 to 0, passing through a maximum when $\alpha = \alpha_c$. This difficulty is easily dispensed with by appending the results with a statement that α should be taken as the lowest root of β up to $\alpha = \alpha_c$. For $\alpha \leq \alpha_c$, Σw_x and Σn_x do equal unity.

Nothing in the derivations, however, implies that Eqs. (3–41) or (3–42) should not be applicable for finite (soluble) species above the gel point. This is indeed the case. Considering the summation of w_x for $\alpha > \alpha_c$, it can be shown that

$$\sum_{\text{all finite species}} w_x = w_s = \frac{(1-\alpha)^2 \alpha'}{(1-\alpha')^2 \alpha} \tag{3-45}$$

where w_s is the weight fraction of soluble constituents and α' is the lowest root of β.

Since the sum of the weight fraction of gel w_g and sol w_s must be unity, we obtain an expression for w_g

$$w_g = 1 - \frac{(1-\alpha)^2 \alpha'}{(1-\alpha')^2 \alpha} \tag{3-46}$$

The weight fraction of gel is plotted in Figure 3–10 with the weight fractions of the soluble, finite species as a function of α for $b = 3$. Notice that w_g increases rapidly as reaction proceeds past the gel point. This abrupt onset of gelation is characteristic of polyfunctional systems.

Let us now review briefly the role of α, α', and p as the reaction progresses. With complete generality, α represents the branching probability determined by p, the extent of reaction, and α' is the lowest root of Eq. (3–40) for the same value of β. If $\alpha \leq \alpha_c$, $\alpha = \alpha'$; and according to Eqs. (3–45) and (3–46), $w_s = 1$ while $w_g = 0$. For $\alpha > \alpha_c$, $\alpha > \alpha'$, and $w_s < 1$.

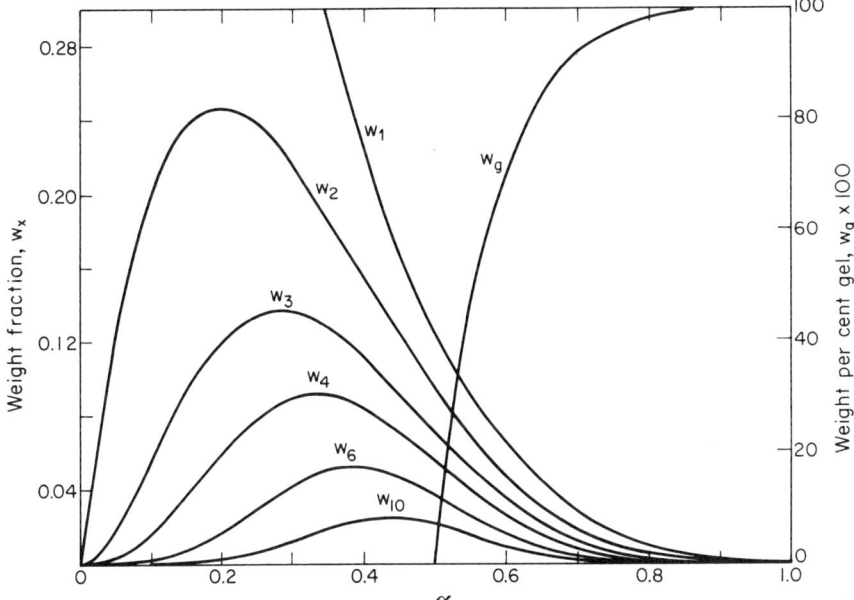

Figure 3-10. Weight functions of various finite species and of gel in a simple trifunctional condensation of p. Curves have been calculated from Eqs. 3-42 and 3-46. [P. J. Flory, op. cit. Figure 3-9.]

Another pertinent result arises if we compare Eqs. (3-42) and (3-45) as the ratio w_x/w_s, which we define as w'_x. In this context, w'_x represents the weight fraction of sol that exists as x-mer regardless of the extent of reaction. For $\alpha < \alpha_c$, w'_x equals w_x. Finally,

$$w'_x = \frac{(1-\alpha')^2}{\alpha'} x\Omega'_x \beta^x \qquad (3\text{-}47)$$

This result is identical to Eq. (3-42) for w_x except that α has been replaced by α', the lower root of β. In these systems, therefore, the size distribution of the sol fraction for $\alpha > \alpha_c$ is the same as prevailed for the entire mixture when $p = \alpha'$. Thus the curves of Figure 3-10 for $\alpha = 0.25$ and 0.40 apply to the sol fraction at $\alpha = 0.75$ and 0.60, respectively.

The behavior of polyfunctional systems over the entire range of polymerization may be summarized based on the development in the preceding sections. As reaction progresses, the size distribution increases in dispersity and large molecules form at the expense of the smaller, although the smaller are always more abundant than the larger species. Maximum heteroegeneity occurs at the gel point. At the gel point, infinite networks form and gelation occurs abruptly. The proportion of gel increases rapidly with condensation beyond α_c. The size distribution of the soluble portion and the averages characterizing it undergo retroversion over the same course it followed up to the gel point.

314 Low-Temperature Polymerization

Low-temperature methods employ reactions that are capable of proceeding at rapid rates at room temperature to give quantitative yields. The best known and most widely used of these reactions utilize an organic acid halide and a compound containing active hydrogens. Several examples are shown in Table 3-4. Many hydrogen transfer reactions (such as isocyanates with

TABLE 3-4
EXAMPLES OF LOW-TEMPERATURE CONDENSATION REACTIONS

Reaction	Product
$-NH_2 + -\overset{O}{\underset{\|}{C}}-Cl \longrightarrow -\overset{O}{\underset{\|}{C}}-NH-$	Polyamide
$-NH_2 + \overset{O}{\underset{\|}{C}}-Cl \longrightarrow -NH-\overset{O}{\underset{\|}{C}}-NH-$	Polyurea
$-NH_2 + -O-\overset{O}{\underset{\|}{C}}-Cl \longrightarrow -NH-\overset{O}{\underset{\|}{C}}-O-$	Polyurethane
$-OH + -\overset{O}{\underset{\|}{C}}-Cl \longrightarrow -\overset{O}{\underset{\|}{C}}-O-$	Polyester
$-OH + Cl-\overset{O}{\underset{\|}{C}}-Cl \longrightarrow -O-\overset{O}{\underset{\|}{C}}-O-$	Polycarbonate

diamines or diols, epoxides with diamines, and cyclic anhydrides with amines) also give quantitative yields at room temperature with a high-reaction rate.

These reactions, in contrast to the high-temperature processes, are irreversible, and one of the reactants is an acid acceptor, such as NaOH. The chemistry of a typical system is

$$-\overset{O}{\underset{\|}{C}}-Cl + -NH_2 \xrightarrow{NaOH} -\overset{O}{\underset{\|}{C}}-NH- + NaCl + H_2O$$

Two methods of low-temperature polycondensation are used—*interfacial polycondensation* and *solution polycondensation*. In the interfacial method, the two intermediates are dissolved in a pair of immiscible liquids, one of which is usually water. The aqueous phase contains the diamine or diol and the alkali. The organic solvent, usually carbon tetrachloride, dichloromethane, xylene, or hexane, contains the diacid halide. According to its name, reaction takes place at or near the interface. The system may be stirred or unstirred. In the unstirred systems, the polymer can be pulled off at the interface as a continuous film or filament as shown in Figure 3–11.

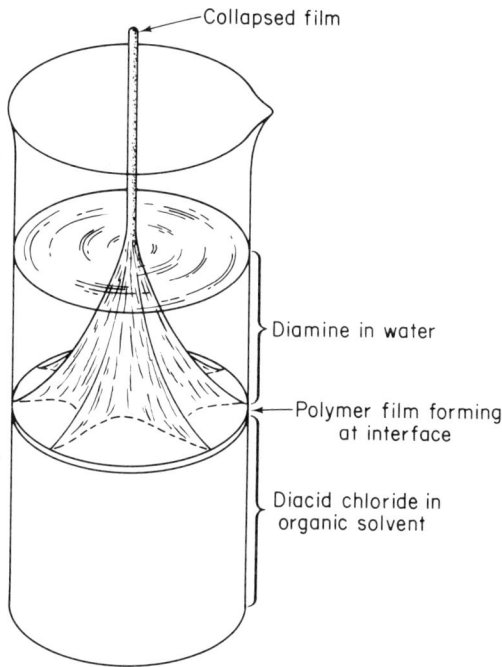

Figure 3–11. Formation of a polyamide by interfacial polymerization. [F. W. Billmeyer, op. cit.]

Solution polycondensation is carried out in a single inert solvent, such as one of the many common organic solvents without reactive functional groups. In this case the acid acceptor is a tertiary amine. Reaction may start with all reactants fully dissolved, and the polymer may remain in solution or precipitate as the reaction progresses.

Table 3-5 compares the high- and low-temperature processes. The primary disadvantages of the low-temperature method are as follows: (1) the formation of large quantities of by-product salt, which must be removed from the polymer, (2) the need to handle, recover, and purify for reuse large quantities of solvent, and (3) the high cost of the intermediates, particularly the diacidhalide, can more than offset other savings.

Most of the well-known condensation polymers can be prepared by the low-temperature process, and over 1200 have been prepared in this fashion.

It is particularly applicable in situations where the reactants or products are thermally unstable. Unstirred interfacial reactions are used to produce polyamide fibers useful for textiles. In addition, the production of polycarbonates from bisphenol A is of commercial importance.

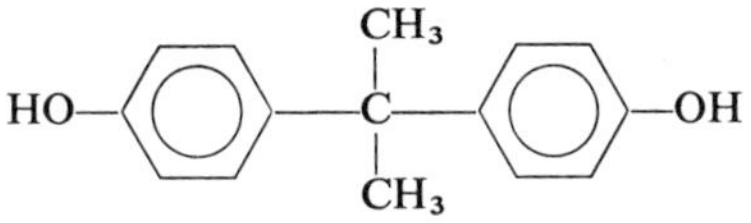

$$HO-\underset{}{\bigcirc}-\underset{\underset{CH_3}{|}}{\overset{\overset{CH_3}{|}}{C}}-\underset{}{\bigcirc}-OH$$

(Bisphenol A)

$$+ \; Cl-\underset{}{\overset{O}{\underset{\|}{C}}}-Cl \; \xrightarrow[\text{acceptor}]{\text{acid}} \; -O-\underset{}{\bigcirc}-\underset{\underset{CH_3}{|}}{\overset{\overset{CH_3}{|}}{C}}-\underset{}{\bigcirc}-O-\underset{}{\overset{O}{\underset{\|}{C}}}-$$

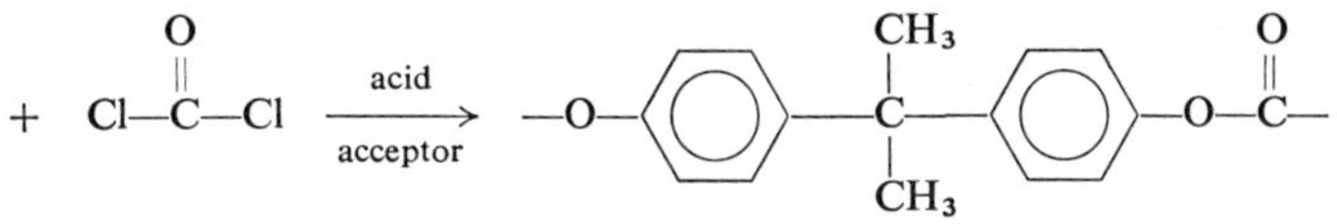

Polycarbonate

TABLE 3-5

COMPARISON OF REQUIREMENTS FOR LOW- AND HIGH-TEMPERATURE POLYCONDENSATION METHODS[a]

	Low Temperature	High Temperature
Intermediates		
Purity	Moderate to high	High
Stoichiometric equivalence	Often tolerant of wide deviations	Necessary
Stability to heat	Unnecessary	Necessary
Structure	Wide range but limited by re-activity requirement	Limited to thermal stability and a lower-reactivity requirement
Polymerization Conditions		
Time	Several minutes	1–25 hours
Temperature	0–40°C	>200°C
Pressure	Atmospheric	High and low
Equipment	Simple, open	Special, sealed
Products		
Yield	Low to high	High
Structure	Extremely wide	Limited by stability to heat and feasibility
By-products	Salt	Water and volatile organic compounds

[a] P. W. Morgan, *Condensation Polymers* (New York: Interscience, 1965).

A particularly interesting application is found in the shrink-proof treatment of wool. The fabric is first wet with a dilute solution of hexamethylenediamine, the excess liquid is wrung out, and the fabric is passed through a dilute solution of sebacyl chloride in a water-immiscible organic solvent to produce a poly(hexamethylene sebacamide) coating on the wool via an interfacial mechanism.

315 Copolymerization

The generation of numerous varieties of copolymers is a relatively easy matter in step-reaction polymerization compared to addition copolymerization. Because of the equal reactivity principle, random copolymers can be synthesized simply by mixing bifunctional or bi-bifunctional monomer units and allowing the reaction to proceed in the usual way. On the other hand (also because of the equal reactivity principle), alternating copolymers cannot be made. Only certain vinyl monomer pairs in addition copolymerization can form such materials, although one could view polymers produced from bi-bifunctional monomer unit pairs as their polycondensation analogs. Block copolymers can be made by blending two different linear systems, which have been reacted to an intermediate stage in separate reactors, and then continuing the reaction after mixing. If the reaction mixture were maintained at reaction conditions for a sufficiently long time and equilibrium conditions were thereby established, one would obtain a random copolymer. Under the usual conditions, however, the reaction would proceed to completion faster than equilibrium could be established. The synthesis of graft copolymers requires the presence of active functional groups distributed along the backbone chain but which do not participate in the initial polymerization. The hydrogen atom of the amide nitrogen of polyamides is easily removed, and polyamide chains, for example, can be reacted with liquid ethylene oxide,

$$\begin{array}{c} CH_2{-}CH_2 \\ \diagdown \diagup \\ O \end{array}$$

to produce polyoxyethylene side chains. The necessary reactive group might be introduced into the backbone by copolymerization of the primary monomer with a minor amount of a comonomer carrying the necessary group; or the comonomer may be introduced with the pendant group already attached. Grafts of polyoxyethylene can be introduced in this manner in poly(ethylene terephthalate) according to the following scheme (top p. 78). This material can then be copolymerized with a mixture of ethylene glycol and dimethyl terephthalate.

$$CH_3OOC-\underset{\underset{ONa}{|}}{C_6H_3}-COOCH_3 + Cl\text{\textendash}(CH_2CH_2-O)_x CH_3$$

$$CH_3OOC-\underset{\underset{O-(CH_2-CH_2-O)_x CH_3}{|}}{C_6H_3}-COOCH_3$$

The number of such copolymers that one can generate is limited by the intrinsic chemical reactivity of the desired constituents as well as by one's imagination.

PRODUCTION OF CONDENSATION POLYMERS

Condensation polymers are typically produced in high-temperature processes in the absence of solvent—a process referred to as *bulk* or *mass polymerization*. The major polyfunctional systems based on formaldehyde are begun in an aqueous solution—a process referred to as *solution polymerization*. As the reaction however, progresses, the water is removed, so that the reaction becomes a bulk polymerization.

316 Bulk Polymerization of Condensation Polymers

Since the reactions are not highly exothermic and since the reactants are of low activity, high temperatures (typically 260–275 °C) are required. As reaction progresses, the molecular weight of the reactants increase, and the viscosity changes from a waterlike consistency (about 1 cp) to well over 1000 poise as measured at the elevated reaction temperatures employed. This high viscosity of the reaction medium during the later stages of reaction causes considerable difficulty in the removal of the volatile by-products. The kinetics may thus change from a chemically-controlled regime to a diffusion-controlled regime as reaction progresses. To aid in the elimination of water, one may employ vigorous agitation (high viscosities will frustrate this approach), high vacuum, nitrogen purging, and thin-film reactors. If sufficiently thin films are generated, polymerization may be completed within a few seconds; but creating high surface areas in commercial-scale reactors and controlling film thickness for uniform reaction time lead to major problems. These polymerizations are often classified as catalyzed, "second-order" reactions. Kinetic analysis is faced, however, with two problems. The reaction changes from a chemically controlled to a diffusion-controlled mechanism, and the extent to which this process occurs depends on the reactor design. Furthermore, at elevated temperatures, side reactions occur and interfere with accurate kinetic analysis. Reaction rate and time must be accurately controlled to ensure generation of the proper molecular-weight material.

For the most part, batch reactors are used, but there is an increasing trend toward the use of continuous reactors. Batch reactors are typically stainless steel vessels equipped with a heavy-duty anchor-type agitator. During the low-viscosity stage of reaction, the pressure may be near atmospheric; but as the viscosity increases, a high vacuum is applied. An inert gas blanket may be used to purge the system of oxygen, which reacts with the polymer to cause decomposition and discoloration. At the desired extent of reaction, the molten polymer is pressure extruded through a die in the bottom of the vessel. The polymer may then be quench-cooled to stop reaction and then diced into $\frac{1}{8}$ by $\frac{1}{8}$-in. pellets, or it may be fed in the molten state to other processing equipment. Complete removal of residual polymer from the reaction vessel, so that none enters the next batch, is a serious problem.

317 Polymerization of Polyfunctional Systems

The major thermosetting resins are formed by the reaction of formaldehyde with phenol, urea, or melamine. Reaction typically passes through three stages. The A stage is conducted in aqueous solution where the reactants are of such low degree of polymerization (< 10) that they are soluble. The B stage commences when the phase separation occurs because of low-reactant solubility. A vacuum is applied, water is removed, and reaction progresses into the melt phase. The degree of polymerization will increase to the range of 100 to 200 to produce branched polymer. If reaction were continued, excessive crosslinking and gelation would occur, thereby marking the onset of the C stage. In practice, the B stage resin—still a thermoplastic material—is removed from the reactor by pressure extrusion. This product may be quench-cooled for compounding into a molding powder, or it may be passed onto other processing equipment in the molten state.

The second phase in the manufacture of formaldehyde thermosetting resins is the formulation of a *molding powder*. This material consists of the powdered B-stage resin intimately mixed (at room temperature) with a variety of inorganic or organic fillers, pigments, colorants, accelerators, plasticizers, stabilizers, and mold lubricants. The molding powder is then placed in a mold, and high pressure and temperature are applied to reactivate and mold the B-stage resin into its immutable C-stage shape.

4
Chain Reaction Polymerization

Chain polymerization occurs with monomer units containing double bonds—namely, the vinyl, divinyl, and 1,3–diene monomers. An example of each of these monomers is

Vinyl	Styrene	$CH_2{=}CH\phi$
Divinyl	Divinyl benzene	$CH_2{=}CH-\phi-CH{=}CH_2$
1,3–diene	Butadiene	$CH_2{=}CH-CH{=}CH_2$

where the symbol ϕ is used to represent the phenyl group. Mechanistically, ring-scission polymerization may also be included, but it is often treated as a hybrid of step and chain polymerization. In this chapter we treat only homopolymerizations; however, much of the discussion is also germane to the material in Chapter 5 on addition copolymerization.

Addition polymerization involves the rapid successive addition of unsaturated monomer units, one at a time, to the end of a growing chain through an active center. The activity of the growing chain is transferred to the monomer units as they add to the chain so that the active center is always located at the end of a growing chain, The creation of an active center rapidly leads to the generation of a high-molecular-weight polymer molecule. Monomer molecules may add at the rate of 10^3 to 10^4 per second. The active centers, which are present in very low concentrations, may be free radicals, carbanions, carbonium ions, or coordination complexes between the growing chain, catalyst surface, and the adding monomer. Addition polymerization can be subdivided into the categories of *free radical, anionic, cationic,* or *coordination polymerization* respectively. Although there are many important similarities among these types, there are almost as many important differences, which become especially apparent when the reaction mechanisms are studied in detail. A completely general treatment of chain polymerization is therefore not possible. Nevertheless, we begin with such an attempt while taking care to point out some of the important differences.

As the discussion develops, the student should refer to the introduction of Part II and relate the material presented there with the material presented here.

401 Kinetic Processes of Chain Polymerization

In generalizing, we characterize all chain polymerizations in terms of four chemical kinetic processes:* (1) initiation, (2) propagation, (3) chain transfer, and (4) termination. *Initiation* involves the generation of the active species. The manner in which these species are generated differs markedly from system to system, and for this reason it constitutes one of the most important distinguishing features of an addition polymerization. We denote a primary active species by A_0^* and a vinyl monomer by $CH_2{=}CHX$, where X may be a halogen, alkyl, ester, phenyl, or some other group. The initiation of the actual polymerization process may be represented as

$$A_0^* + CH_2{=}CHX \longrightarrow A_0{-}CH_2{-}C^*\begin{smallmatrix}H\\ \\X\end{smallmatrix}$$

where the asterisk denotes an active center. Species that produce the primary active sites are referred to as *initiators* if they are consumed in the reaction, and *catalysts* if they are regenerated.

In *propagation* the monomer units are successively joined together through an active center to form polymer, sometimes referred to as "active polymer" to distinguish it from the final inactive polymer product. Superficially, all propagations occur in the same manner. For a growing vinyl polymer chain, this process may be schematically represented as

$$\sim\sim\sim CH_2{-}\overset{H}{\underset{X}{C^*}}{+}CH_2{=}CHX \longrightarrow \sim\sim\sim CH_2{-}CHX{-}CH_2{-}\overset{H}{\underset{X}{C^*}}$$

---

*It is necessary to distinguish the kinetics of chemical processes from those of physical processes, such as diffusion or adsorption, which may be important under certain operating conditions.

Notice how the head of the active chain adds to the tail of the monomer to produce a linear, head-to-tail linkage, and how the activity is transferred via the opening of the double bond to the head of the adding monomer. The repetition of this process many times over leads to linear high polymer. A concise representation of this process for a single chain is

$$A_1^* + M \longrightarrow A_2^*$$
$$A_2^* + M \longrightarrow A_3^*$$
$$\vdots \quad \vdots \quad \vdots$$
$$A_{n-1}^* + M \longrightarrow A_n^*$$

where A_1^* is the product of the initiation step and A_n^* represents an active species with n monomer units.

Because of the combined effects of polarity, steric hindrance, and resonance stabilization, head-to-tail addition is highly favored in vinyl polymerization. In certain ionic and coordination polymerizations, stronger directing influences are exerted through the existence of a strong intermediate between incoming monomers and the active center. The polymers generated in these processes are stereoregular.

Propagation of the dienes may proceed in either of two ways: either as a 1,4–addition (cis or trans)

$$\sim\sim\sim \overset{H}{\underset{H}{C^*}} + CH_2 = C\overset{X}{-} CH = CH_2 \longrightarrow \sim\sim\sim CH_2 - CH_2 - \overset{X}{C} = CH - \overset{H}{\underset{H}{C^*}}$$

where unsaturation is incorporated into the chain, or as a 1,2–addition

$$\sim\sim\sim \overset{H}{\underset{H}{C^*}} + CH_2 = \overset{X}{C} - CH = CH_2 \longrightarrow \sim\sim\sim CH_2 - CH_2 - \overset{X}{\underset{\underset{CH_2}{\overset{\|}{CH}}}{C^*}}$$

where unsaturation is incorporated as a pendant vinyl group. The prevailing mode of addition depends on the specific process.

In *chain transfer*, the activity is transferred from an initially active polymer chain to a previously inactive species. This process may be depicted as

$$A_n^* + TH \rightarrow P_n + T^*$$

where TH is the transfer species, P_n is the inactive polymer product, and T^* is the newly created active species. In the reaction shown, the transfer of an atom of hydrogen triggers the actual shift in activity. The presence of active hydrogen is most important in chain polymerization, but transfer can occur with other active atoms as well. The species TH, to which the reactivity is transferred, may be monomer, solvent, or a reagent deliberately added to promote chain transfer, in which case TH would be referred to as a *chain transfer agent*, Such hydrogen-transfer reactions may also occur to the backbone of other polymer chains thus providing sites for branch growth. The major effect of chain transfer is to modify molecular weight and/or molecular-weight distributions.

In *termination*, the kinetic chain-growth process is stopped via the annihilation of the active species. This process is to be distinguished from chain transfer, where the kinetic chain growth is transferred but not stopped. The manner in which termination occurs also differs markedly from system to system and is, therefore, another important feature of the individual chain polymerizations.

We should now consider how the fact that all four of these reactions (initiation, propagation, transfer and termination) may occur simultaneously will effect the molecular weight distribution of the polymer being generated. Initiator is usually generated, or regenerated in the case of a catalyst, continuously during the process; hence, initiation will also occur continuously throughout the life of the polymerization. Growth of the polymer chains, as measured by the rate of addition of monomer, at each active site may be assumed to occur at equal rates, regardless of chain length. However, transfer and termination will occur in a random fashion with units of various lengths. Assuming for the sake of simplicity that initiation occurs at a uniform rate, we see that the random nature of the last two processes inevitably leads to the generation of random length polymer chains, or, in other words, a molecular weight distribution.

Another characteristic of all chain-growth polymerizations is their susceptibility to rate retardation or inhibition by traces of other substances. If a substance merely slows a reaction down, it is a *retarder*. If a substance stops a reaction for a period of time and then allows it to proceed normally, it is an *inhibitor*. Some serve as both, first inhibiting the reaction, then allowing it to proceed at a slower than normal rate. These characteristics are illustrated in Figure 4-1. These substances may be present as impurities, or they may be added deliberately to prevent polymerization during shipping and storage. Such common species as water, oxygen, or selected solvents may serve as inhibitors. Under such circumstances these reagents must be purged from the reaction mixture. Hydroquinone or *t*-butyl catechol are inhibitors commonly added to the monomer by the manufacturer to prevent inadvertent polymerization.

This terminates our general introductory discussion, and we now move on to some specific chemistry, citing a number of examples for each mechanism. Covering the complete range of possibilities is far beyond our intent,

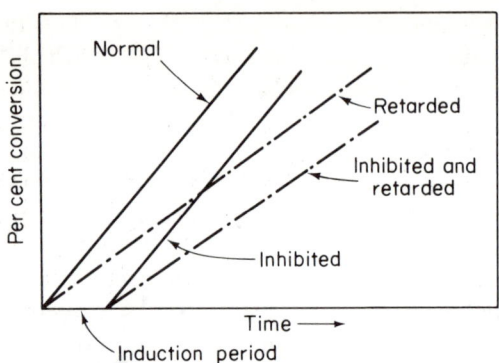

Figure 4-1. Effect of inhibitors and retarders on conversion.

for the number of variations that can be worked on each mechanism is considerable. Our purpose here will be to discuss a few classical examples to serve as a basis for further study.

402 Effect of Substituents

The active center or the mechanism by which a monomer polymerizes best depends on the nature, number, and spatial arrangement of the substitutents to the double bond. Thus we must consider the effect the substituents have on the polarity of the double bond, on the resonance stabilization of the active species that is formed, and on steric hindrance. Ethylene, the unsubstitued case, is relatively unresponsive toward most attempts to cause polymerization. Generally a substituted monomer is much more responsive toward at least one mode of activation than is ethylene.

Where monomer reactivity for polymerization via a particular mechanism is enhanced by the presence of a certain substituent, the presence of a second identical substituent, or one manifesting the same effect, will further enhance this reactivity, provided it is placed unsymmetrically to the double bond and provided steric hindrance does not exert a dominating effect. For many vinyl polymerizations, this generalization can be represented as

$$CH_2=CX_2 \text{ or } CH_2=CXY > CH_2=CHX > CH_2=CH_2 > CHX=CHX$$

where X and Y may be halogen, alkyl, ester, phenyl, or other groups. Table 4-1 summarizes, for several important olefin monomers, the mechanisms by which they may be polymerized. Most may be polymerized by more than one mechanism.

TABLE 4-1

APPLICABILITY OF VARIOUS TYPES OF INITIATION MECHANISMS
TO THE POLYMERIZATION OF OLEFIN MONOMERS[a]

Olefin Monomer	Monomer Structure	Free Radical	Anionic	Cationic	Coordination
Ethylene	$CH_2=CH_2$	+	−	+	+
Propene	$CH_2=CHMe$	−	−	−	+
Butene-1	$CH_2=CHEt$	−	−	−	+
Isobutene	$CH_2=CMe_2$	−	−	+	−
Butadiene-1, 3	$CH_2=CH-CH=CH_2$	+	+	−	+
Isoprene	$CH_2=C(Me)-CH=CH_2$	+	+	−	+
Styrene	$CH_2=CHPh$	+	+	+	+
Vinyl chloride	$CH_2=CHCl$	+	−	−	+
Vinylidene Chloride	$CH_2=CCl_2$	+	+	−	−
Vinyl fluoride	$CH_2=CHF$	+	−	−	−
Tetrafluoroethylene	$CF_2=CF_2$	+	−	−	+
Vinyl ethers	$CH_2=CHOR$	−	−	+	+
Vinyl esters	$CH_2CHOCOR$	+	−	−	−
Acrylic esters	$CH_2CHCOOR$	+	+	−	+
Methacrylic esters	$CH_2=C(Me)COOR$	+	+	−	+
Acrylonitrile	$CH_2=CHCN$	+	+	−	+

[a]Symbols: + monomer can be be polymerized to high-molecular-weight polymer by this form of initiation; − no polymerization reaction occurs or only low-molecular-weight polymers or oligomers are obtained with this type of initiator. Robert W. Lenz, *Organic Chemistry of High Polymers* (New York: Interscience, 1967, p. 244).

403 Free Radical Polymerization

Of all the addition polymerization processes, the chemistry of free radical polymerization is the best understood. For the most part, the nature of the kinetic processes have been fully elucidated. Radical polymerization can be conducted in several reaction environments. A wide variety of polymers are produced, especially copolymers, with correspondingly innumerable applications. Hence this process is of considerable practical significance.

The first step is the generation of the primary free radicals. The primary radicals are normally produced by the thermal or photochemical decomposition of organic peroxides or hydroperoxides, or azo or diazo compounds. These compounds should be referred to as *initiators*, and not as catalysts, since they are consumed in the reaction. Widely encountered means of generating radicals are from benzoyl peroxide (Bz_2O_2)

$$\text{C}_6\text{H}_5\text{-C(=O)-O-O-C(=O)-C}_6\text{H}_5 \longrightarrow 2\ \text{C}_6\text{H}_5\text{-C(=O)-O}\cdot \longrightarrow 2\ \text{C}_6\text{H}_5\cdot + 2\text{CO}_2$$

from azobisisobutyronitrile (AIBN)

$$(\text{CH}_3)_2\text{-C(CN)-N=N-C(CN)-(CH}_3)_2 \longrightarrow 2\text{CH}_3\text{-C(CN)}\cdot + \text{N}_2$$

from *t*-butylhydroperoxide

$$(\text{CH}_3)_3\text{-COOH} \longrightarrow (\text{CH}_3)_3\text{-CO}\cdot + \cdot\text{OH}$$

and from potassium persulfate

$$\text{K}_2\text{S}_2\text{O}_8 \longrightarrow 2\text{K}^+ + 2\text{SO}_4^-\cdot \quad \text{(a radical ion)}$$

Such decomposition reactions may be represented as

$$\text{I} \longrightarrow 2\text{A}_0\cdot$$

where I is the initiator species and $\text{A}_0\cdot$ is the primary free radical. The next step in the initiation process involves reaction of the primary radical with monomer, as previously depicted. However, the dissociation of the initiator is a slow reaction requiring high-activation energy, and kinetically, for radical initiation, one expresses the total process in terms of this dissociation. The concentration of the initiator is usually low, on the order of 0.1 to 0.5 percent by weight based on monomer content, and moderately elevated temperatures, are employed, usually 50 to 80 °C.

Free radicals can also be generated in redox decomposition reactions. In this regard, the ferrous ion is a frequently encountered reagent; several examples of its use are given below.

$$\text{H}_2\text{O}_2 + \text{Fe}^{++} \longrightarrow \text{Fe}^{+++} + \text{OH}^- + \cdot\text{OH}$$

$$\text{S}_2\text{O}_8^{--} + \text{Fe}^{++} \longrightarrow \text{Fe}^{+++} + \text{SO}_4^{--} + \text{SO}_4^-\cdot$$

$$(\text{CH}_3)_3\text{-COOH} + \text{Fe}^{++} \longrightarrow (\text{CH}_3)_3\text{-C-O}\cdot + \text{OH}^- + \text{Fe}^{+++}$$

In contrast to the thermal generation of radicals, redox generation proceeds at rapid rates even at room temperature.

Propagation proceeds as previously depicted. Chain transfer can be quite extensive in some systems, leading either to reduced molecular weight or to chain branching. Transfer with several varieties of reagents takes place.

Termination occurs by mutual annihilation of two radicals. Two routes are possible. In *combination*, two species combine to form one large, deactivated polymer chain

$$A_0\text{+}CH_2\text{—}CHX\text{)}_{\overline{n}}\ CH_2\text{—}\underset{X}{\overset{H}{\underset{|}{C}}}\cdot\ +\ A_0\text{+}CH_2\text{—}CHX\text{)}_{\overline{m}}\ CH_2\text{—}\underset{X}{\overset{H}{\underset{|}{C}}}\cdot\ \longrightarrow$$

$$A_0\text{+}CH_2\text{—}CHX\text{)}_{\overline{n}}\ CH_2CHX\text{—}CHX\text{—}CH_2\text{+}CHX\text{—}CH_2\text{)}_{\overline{m}}\ A_0$$

This is by far the most common mode of termination. Note that termination is head-to-head and that the initiator residue remains to form the end groups. In *disproportionation*, termination occurs by hydrogen transfer and two deactivated molecules are formed, one with terminal saturation, the other with terminal unsaturation. These product molecules appear as

$$A_0\text{+}CH_2\text{—}CHX\text{)}_{\overline{n}}\ CH=CHX\ +\ A_0\text{+}CH_2\text{—}CHX\text{)}_{\overline{m}}\ CH_2\text{—}CH_2X$$

IONIC POLYMERIZATION

Of the ionic polymerizations, both cationic and anionic, less is known than for radical polymerization. The situation is complicated by the fact that the intimate mechanisms of specific reaction systems can be markedly different even within a class of ionic reactions. In this context, and unlike radical polymerization, the solvent plays a significant role. Because of the ionic nature of the reaction, the dielectric constant of the solvent is usually of overwhelming importance, but also important is its solvating power for the ion pairs. Additional factors are the counterion type, temperature, and monomer structure as manifested in resonance, steric, and polar effects.

404 Cationic Polymerization

Monomers having electropositive substituents are most readily polymerized by cationic mechanisms. In a monomer with electropositive substituents, the molecule becomes polarized in such a way that the double bond has an excess electronic charge. It is thus highly susceptible to attack by an electron acceptor, which functions as the activator for cationic polymerization. Also, electropositive substituents will provide a high degree of resonance stabilization to the active species, once formed. By reference to Table 4–1, we see that isobutene can be polymerized only by cationic initiation, whereas monomers, such as the vinyl and vinylidene halides or acrylonitrile, with their electronegative substituents, will not yield at all to cationic initiation.

Catalysts for cationic polymerization fall into three classes: (1) classical protonic acids, such as HCl, H_2SO_4, $HClO_4$, Cl_3COOH, (2) Lewis acids, such as BF_3, $AlCl_3$, $TiCl_4$, $SnBr_4$, $SbCl_3$, $BiCl_3$, and (3) other cationic generators, such as t-$BuClO_4$, I_2, and ϕ_3CCl. The first two types are the most important. The last types are initiators rather than true catalysts, for they are consumed, and will not be considered further here. Pure Lewis acids are usually not active catalysts, and other reagents, containing the necessary proton, are required to activate them. Examples of such catalyst-co-catalyst system are

$$AlCl_3 + HCl \longrightarrow H^{\oplus}AlCl_4^{\ominus}$$

$$BF_3 + HOH \longrightarrow H^{\oplus}BF_3OH^{\ominus}$$

Cationic catalyst systems of the first two types can be represented as $H^{\oplus}A_0^{\ominus}$. Using this representation to illustrate, initiation proceeds as

$$H^{\oplus}A_0^{\ominus} + \begin{matrix} H \\ \diagdown \\ H \end{matrix} C=C \begin{matrix} X^{\delta\oplus} \\ \diagup \\ \diagdown \\ X \\ \delta\oplus \end{matrix}^{\delta\oplus} \longrightarrow H_3-C-\overset{X}{\underset{X}{C}}{}^{\oplus} \cdots A_0^{\ominus}$$

Reaction is conducted in a medium with a low-dielectric constant such that the cation and its counter ion (gegenion) are closely associated as an ion pair. Cases are encountered at the extremes where the pair may be closely bound or completely ionized.

Propagation proceeds as outlined previously. Mechanistically, it is pictured as a concerted push-pull attack of the carbonium ion-counter ion pair on the monomeric double bond.

$$\sim\!\!\sim\!\!\sim CH_2-\overset{X}{\underset{X}{C^{\oplus}}} \cdots \cdots A_0^{\ominus}$$
$$H_2C_{\delta\ominus}\!\!\cdots\!\!C\diagup^{X^{\delta\oplus}}_{\diagdown X_{\delta\oplus}}$$

Chain transfer is prevalent in cationic polymerization, and effective transfer to monomer, counter ion, polymer, or solvent occurs readily. Transfer to the counter ion regenerates the original initiating ion pair, and it is often cited as a terminating mechanism. This latter reaction occurs by transfer of a proton from the polymer to the ion pair, leaving the inactive chain with terminal unsaturation.

$$\sim\!\!\sim\!\!\sim CH_2-\overset{X}{\underset{X}{C^{\oplus}}}\cdots A_0^{\ominus} \longrightarrow \sim\!\!\sim\!\!\sim CH=CX_2 + H^{\oplus}A_0^{\ominus}$$

Some cationic polymerizations of olefinic monomers seem to lack a simple kinetic termination mechanism; and where it does occur, the actual mechanism is, for the most part, undefined.

As already noted, the dielectric constant of the reaction medium has a profound influence on the course of the polymerization. In solvents with relatively high (but still low) dielectric constants, the ion pair is relatively separated such that initiation and propagation will be aided, while termination or transfer will be suppressed. The overall effect of such a solvent is to increase the rate of polymerization and molecular weight simultaneously.

The possibility of simultaneously achieving very high rates and molecular weights is an outstanding characteristic of cationic polymerization. For instance, isobutylene polymerizes with $HAlCl_4$ or HBF_3OH to molecular weights on the order of several million within a few seconds. Another of the unusual features is that, in contrast to radical polymerization, the rates and molecular weights both increase with decreasing reaction temperature and that these temperatures tend to be rather low. In the preceding case, temperatures of about $-100\ °C$ are maintained. These features will be explained later on a kinetic basis, but a chemical argument would proceed as follows: The ion-molecule interaction of the initiation, and the propagation steps requires virtually no activation energy. Termination involves complex ion-pair rearrangement in a medium with a low-dielectric constant, and some activation energy is required. Thus lowering the temperature suppresses the reactions with high-activation energy and leaves relatively unaffected the propagation and initiation reactions.

Because of the high energy of the ionic species, a number of side reactions can compete with propagation. This problem is mitigated somewhat by the use of the lowest possible temperatures. Also because of energetic considerations transfer is suppressed by a reduction in temperature. The reaction systems are often heterogeneous in that inorganic catalysts are used to polymerize organic monomers in a solvent in which the polymer is insoluble.

405 Anionic Polymerization

In instances where a monomeric substituent is electronegative, the double bond will suffer a depletion of electrons and will be susceptible to attack by an electron donor. The intermediate species will carry the extra electron characteristic of anionic polymerization. The stability of the intermediate will be enhanced by resonance. Here the electron transfer process is merely opposite in direction to that of cationic polymerization.

Initiation may be accomplished either by direct attack of a base on the monomer to form carbanion or by transfer of an electron from an active donor molecule to the monomeric double bond to form an anion radical. In base initiation, the strength of the base required diminishes with increasing electronegativity of the monomeric substituents. The electronegativity of some selected substituents follows as —CN > —COOR > —ϕ $\cong$ —CH=CH$_2$ ≫ —CH$_3$. Examples of some reactive bases are the organolithium compounds, such as n-BuLi, ϕCH$_2$Li, the organosodium compounds, such as NaNH$_2$, ϕCH$_2$Na, CH$_2$=CH—CH$_2$Na, the Grignard reagents, such as EtMgBr, ϕCH$_2$MgBr, and, finally, KNH$_2$. Such bases can, be represented by A$^\oplus$B:$^\ominus$, and initiation can be represented as

$$A^\oplus B:^\ominus + \underset{H}{\overset{H}{}}C=C\underset{X\ \delta^\ominus}{\overset{X\ \delta^\ominus}{}} \longrightarrow B-CH_2-\underset{X}{\overset{X}{C}}:^\ominus \cdots A^\oplus$$

Initiation by electron transfer is based on the ability of the alkali metals to supply electrons to the double bonds. Transfer of an electron yields a positively charged, alkali-metal counter ion, and an anion radical. Pairs of anion radicals quickly undergo combination to form a dimeric dianion. Two examples of this mode of initiation are (1) direct attack of the alkali metal M· on the monomer as

$$M\cdot + CH_2=CX_2 \longrightarrow \cdot CH_2-\underset{X}{\overset{X}{C}}:^\ominus \cdots M^\oplus$$

which after radical combination becomes

$$M^\oplus \cdots {}^\ominus:\underset{X}{\overset{X}{C}}-CH_2-CH_2-\underset{X}{\overset{X}{C}}:^\ominus \cdots M^\oplus$$

and (2) attack of the metal through an intermediate compound such as naphthalene

$$M\cdot + \text{[naphthalene]} \longrightarrow \text{[naphthalene radical anion]} \cdots M^\oplus$$

$$\text{[naphthalene radical anion]} \cdots M^\oplus + CH_2=\overset{*}{C}X_2 \longrightarrow \cdot CH_2-\underset{X}{\overset{X}{C}}:^\ominus \cdots M^\oplus$$

$$+ \text{[naphthalene]}$$

$$2 \cdot CH_2-\underset{X}{\overset{X}{\underset{|}{\overset{|}{C}}}}:^{\ominus} \cdots M^{\oplus} \longrightarrow M^{\oplus} \cdots ^{\ominus}:\underset{X}{\overset{X}{\underset{|}{\overset{|}{C}}}}-CH_2-CH_2-\underset{X}{\overset{X}{\underset{|}{\overset{|}{C}}}}:^{\ominus} \cdots M^{\oplus}$$

As before, propagation occurs by a concerted push-pull mechanism with the dielectric medium of the solvent assuming a dominant role in determining the course of the reaction. Chain transfer is virtually nonexistent in systems initiated with organometallic compounds and in systems utilizing inert solvents. There is no facile chemical termination step in anionic polymerization; and in the presence of highly pure reactants and inert solvents, the active end groups have indefinite lifetimes. Such systems are referred to as *living polymers*,* and if one were to recharge a monomer-exhausted system with fresh monomer, polymerization would resume. Termination can certainly be brought about by introducing suitably reactive materials, such as water, alcohol, or ammonia.

This lack of a termination or transfer step introduces fascinating possibilities in building two model polymer systems: monodispersed molecular-weight polymer and block copolymers. In preparing the monodispersed polymer, sodium and biphenyl are first suitably combined in an inert solvent like tetrahydrofuran to form the very reactive intermediate species

If monomer is added with thorough mixing such that the monomer is rapidly distributed throughout the mixture, consumption of the sodium biphenylide to form anion radicals will be nearly instantaneous. Pairs of anion radicals, in turn, will rapidly combine to make the initiation process complete and irreversible. Growth will proceed normally, in the absence of termination or transfer (provided the necessary precautions have been taken to ensure the proper level of reactant purity). The important points here are that (1) creation of the growth centers is irreversible and virtually instantaneous, (2) growth proceeds at each active site at an equal rate, and (3) since there is no randomly occurring termination, all polymer chains will be of equal length, or, in other words, a uniform molecular-weight polymer will be produced.† The opportunities for forming such materials in organic polymer

---
* Certain ring-scission polymerizations (for example, ethylene oxide polymerizations) also constitute living polymer systems.

† It is theoretically impossible to produce a perfectly monodisperse molecular weight material, although the distribution may be extremely narrow. See Section 412.

synthesis are rare, and this technique assumes paramount importance for the preparation of standard materials for analytical work.

Block copolymers can be formed simply by first following the foregoing procedure and then adding a second monomer when the supply of the first is exhausted. This discussion has glossed over many of the specific limitations of these techniques, but several types of each polymer have been prepared for use as laboratory standards and model-system studies. Monomers that have been used in various ways are styrene, α-methyl styrene, methyl methacrylate and isoprene. (See Section 511.)

406 Coordination Polymerization

One of the most important and most recent advances in polymerization technology has been the development of processes leading to the generation of stereoregular polymer structures. The key to all these processes are the catalyst systems. Virtually countless varieties are known, and almost all are crystalline solids so that the polymerization processes are heterogenous.

Karl Ziegler,* in his studies with polyethylene, was the first to realize the importance of the relation between the product and the composition and structure of the catalyst. Giulio Natta† extended the use of the catalyst combinations of Ziegler and recognized that the more crystalline catalysts produced highly crystalline poly(α-olefins). As previously noted, for their contribution, accomplished mainly in the 1950s, Ziegler and Natta were awarded the Nobel Prize in chemistry in 1963. The general class of catalyst systems that they discovered and developed is commonly referred to as *Ziegler-Natta catalysts*. A preferred designation for the overall process would be *coordination polymerization*, for, as we shall see, the nature of the mechanism is reflected in this term.

The most widely used Ziegler-Natta (or coordination) catalyst systems consist of the reaction products formed from an organometallic compound and a transition metal compound. The classical example is the reaction product of triethyl aluminum and titanium tetrachloride. The specific nature of the product or products is not known, but the catalyst site is thought to consist of the crystalline form of either an alkylated transition metal halide or a complex formed between the transition metal and the organometallic compound. Each leads to two proposals for the propagation mechanism: the former, a monometallic mechanism; and the latter, a bi-

---

*K. Ziegler, E. Holykamp, H. Briel, and H. Martin. *Angew. Chem.* **67**, 426, 541 (1955).
†G. Natta, P. Pino, P. Corradini. F. Danusso, E. Mantica, G. Mazzanti, and G. Moraghio, *J. Am. Chem. Soc.*, **77**, 1708 (1955); G. Natta, *Makromal. Chem.* **16**, 213 (1955) and *J. Polymer Sci.*, **16**, 143 (1955); **35**, 94 (1960).

metallic mechanism. There is insufficient experimental evidence to differentiate between the two. Each of the mechanisms is outlined below. For the monometallic mechanism, the catalyst is formed as

$$TiCl_4 + Al(C_2H_5)_3 \longrightarrow Cl_3Ti\overset{Cl\cdots}{\underset{CH_2}{\diagup}}\overset{C_2H_5}{\underset{C_2H_5}{\diagdown}}Al\overset{C_2H_5}{\underset{C_2H_5}{\diagdown}} \longrightarrow$$
$$\overset{|}{CH_3}$$

$$Cl_3TiC_2H_5 + Al(C_2H_5)_2Cl$$

and the propagation proceeds as

$$\underset{Cl}{\overset{Cl}{\diagdown}}\underset{|}{\overset{C_2H_5}{|}}\underset{Cl}{Ti} + CH_2=CH-R \longrightarrow \underset{Cl}{\overset{Cl}{\diagdown}}\underset{|}{\overset{C_2H_5}{|}}\underset{Cl}{Ti}\cdots\overset{H\ R}{\underset{CH_2}{\overset{\diagdown\diagup}{C}}} \longrightarrow$$

$$\underset{Cl}{\overset{Cl}{\diagdown}}\underset{|}{\overset{C_2H_5\cdots}{|}}\underset{Cl}{Ti}\overset{H\ R}{\underset{CH_2}{\overset{\diagdown\diagup}{C}}} \longrightarrow \underset{Cl}{\overset{Cl}{\diagdown}}\underset{|}{Ti}\underset{Cl}{-CH_2-CHR-C_2H_5}$$

In the bimetallic mechanism, it is thought that the following complex serves as the catalyst

$$\underset{Cl}{\overset{Cl}{\diagdown}}Ti\overset{Cl\cdots}{\underset{CH_2}{\diagup}}\overset{C_2H_5}{\underset{C_2H_5}{\diagdown}}Al\overset{C_2H_5}{\underset{C_2H_5}{\diagdown}}$$
$$\overset{|}{CH_3}$$

and that propagation proceeds as

$$\underset{Cl}{\overset{Cl}{\diagdown}}Ti\overset{Cl\cdots}{\underset{CH_2}{\diagup}}\overset{C_2H_5}{\underset{C_2H_5}{\diagdown}}Al\overset{C_2H_5}{\underset{C_2H_5}{\diagdown}} + CH_2=CHR \longrightarrow$$
$$\overset{|}{CH_3}$$

$$\underset{CH_2=CHR}{\overset{Cl}{\underset{|}{Cl-Ti}}}\overset{Cl\cdots}{\underset{C_2H_5}{\diagup}}Al\overset{C_2H_5}{\underset{C_2H_5}{\diagdown}} \longrightarrow \underset{\overset{|}{CH_2}\ \overset{|}{CH_2}}{\overset{Cl}{\underset{|}{Cl-Ti}}}\overset{Cl\cdots}{\underset{\ }{\diagup}}Al\overset{C_2H_5}{\underset{C_2H_5}{\diagdown}} \longrightarrow$$
$$\overset{|}{CHR}\ \overset{|}{CH_3}$$

$$\underset{Cl}{\overset{Cl}{\diagdown}}Ti\overset{Cl\cdots}{\underset{CH_2}{\diagup}}\overset{C_2H_5}{\underset{C_2H_5}{\diagdown}}Al\overset{C_2H_5}{\underset{C_2H_5}{\diagdown}}$$
$$\overset{|}{CHR}$$
$$\overset{|}{C_2H_5}$$

In each case, the ethyl group associated with the aluminum becomes the chain end. The further addition of monomer will proceed in an analogous manner with the ethyl group replaced by the growing chain. The intimate details of the reaction mechanisms are, of course, not revealed in such simplistic representations.

The essential feature of any mechanism is that directing forces must be present to cause each monomer to always add in a fashion that is identical to its predecessors. This replication is accomplished through the formation of a coordination complex, which involves oriented adsorption of the monomer to form an anionic intermediate species with bonding between the vacant d-orbitals of the transition metal and the π-electrons of the monomeric double bond. It is also believed that stereoregular polymers result only when the catalyst is part of a solid crystal surface. Differences in crystal perfection would result in varying stereoregulation and reactivity at different sites. This variation in site activity explains the broad molecular-weight distributions sometimes found in stereoregular polymers. Termination is brought about by the addition of a compound containing a hydrogen atom, which is reactive toward organometallic compounds—for example, water or alcohol.

In order to prepare a highly stereoregular polymer, one must exercise careful control in the preparation of the catalyst. The catalyst is slurried in a dry, oxygen-free, inert hydrocarbon solvent, such as heptane, along with the monomer to be polymerized. Polymerization is normally conducted at temperatures up to 90 °C and monomer pressures up to 5 atm. Both the polymer that forms and the catalyst are insoluble in the suspending medium. The polymer coats the catalyst as a viscous gel-like mass such that the reaction becomes monomer diffusion controlled. Investigations of the catalyst mechanism, as well as control of the reaction, are made difficult by such catalyst occlusion.

407 Branching Mechanisms

We are concerned here with reactions the polymer chain may undergo during a polymerization that result in branched growth or even cross-linked structures. Branching occurs in many polymerization processes. However, the degree of branching that does occur varies markedly from system to system, from the presence of a rare branch to its common occurrence. The dependence of the degree of branching on various reaction parameters will be discussed after some general branching mechanisms have been outlined.

One of the most common mechanisms of branching is initiated by the chain transfer of an α-hydrogen from the backbone of a polymer chain, either active or inactive, to an actively growing species. The mechanism is illustrated below with a growing chain attacking the α-hydrogen of a randomly selected repeat unit.

Sec. 407 Branching Mechanisms

$$\begin{array}{c}\text{~~~CH}_2-\underset{X}{\overset{H}{\underset{|}{C}}}\text{~~~} \quad + \quad \overset{*}{\cdot}\underset{X}{\overset{H}{\underset{|}{C}}}-\text{CH}_2\text{~~~} \quad \longrightarrow \\ \\ \text{~~~CH}_2-\underset{X}{\overset{*}{\underset{|}{C}}}\text{~~~} \quad + \quad \underset{X}{\overset{}{\underset{|}{CH_2}}}-\text{CH}_2\text{~~~}\end{array}$$

Presumably, the newly created active site is as reactive as the old, and the addition of monomer proceeds normally to generate a branched species. Termination of the branched growth also occurs in the normal manner. The most important characteristic of this type of branching is that the number-average molecular weight is not affected, whereas the weight average can increase severely if branching is extensive. Rationalization of this situation is left as a student exercise. Branching, by this mechanism, is less likely to occur in systems that do not have α-hydrogens, as in α-methyl styrene or methyl methacrylate. Where the reactivity of the active species is exceptionally high, as with radical polymerized vinyl acetate, the most accessible hydrogen atoms of the pendant methyl group are removed.

Branching can also occur if residual unsaturation exists in the chain, as is the case with some termination mechanisms or with the polymerization of dienes. This mechanism is outlined for butadiene as

$$\text{~~~CH}_2-\text{CH}=\text{CH}-\text{CH}_2\text{~~~} \quad + \quad ^*\text{CH}_2-\text{CH}=\text{CH}-\text{CH}_2\text{~~~}$$

$$\text{~~~CH}_2-\underset{\underset{*}{|}}{\text{CH}}-\overset{\overset{\text{CH}_2-\text{CH}=\text{CH}-\text{CH}_2\text{~~~}}{|}}{\text{CH}}-\text{CH}_2\text{~~~}$$

Growth occurs at the new active center, but here we have a case with two branches at virtually the same point. If this type of transfer occurs to a high degree, it is obvious that crosslinking would also occur.

The amount of branching that arises in any particular reaction environment or by any particular mode of polymerization will depend on such kinetic parameters as (1) reactivity of the active species, (2) the presence of α-hydrogens or other reactive transferable elements. (3) concentration of the various species present, and (4) temperature. Branching is rather common in radical polymerizations, less so in cationic polymerization, rare in coordination polymerization, and virtually absent in anionic polymerizations.

To indicate how important branching can be, with regard to the polymer product, we compare some properties of branched and linear polyethylene in Table 4-2. The details of the processes in which each is synthesized are discussed later in the chapter; but, briefly, branched polyethylene is produced in a free radical process at relatively high pressures and temperatures, several hundred atmospheres and 175 to 200 °C, whereas linear polyethylene is pro-

TABLE 4-2

COMPARISON OF BRANCHED AND LINEAR POLYETHYLENE

	Linear	Branched
% Crystallinity	90	50–60
Melting temperature	135°C	115°C
Density	0.95–0.97	0.91–0.94
Modulus of elasticity	100,000 psi	20,000 psi
$\bar{M}_w/\bar{M}_n$	2	20 to 50

duced via a Ziegler-Natta process at only moderately elevated pressures and temperatures, 15 to 20 atmospheres and 135 to 150 °C. Striking differences are noted in the modulus of elasticity and the molecular weight distributions as evidenced in the $\bar{M}_w/\bar{M}_n$ ratios.

4.08 Steady-State Kinetic Analysis

Once the chemical processes for any system are known, we can proceed to undertake a kinetic analysis. Most elementary treatments for even homogeneous environments suffer from the fact that it is necessary to assume a steady-state concentration of active species. Such treatments are therefore valid under limited conditions. Still, considerable insight into the nature of polymerization reactions can be gained through such analyses, so they are worth pursuing. As in step-reaction polymerization, we are primarily interested in deriving relations for the rates of polymerization, the average molecular weights and/or the molecular-weight distributions. Rate expressions are easily derived. Somewhat less success is met in molecular-weight-distribution analysis of chain polymerizations compared to linear step-reaction polymerizations.

We will illustrate the general approach for homogeneous rate analysis by considering free radical polymerization, since this is the best known. First, we summarize the chemistry of the reactions just considered with a notation that is convenient to use in representing the various kinetic processes. Similar expressions can be written for systems not considered here.

1. *Initiation*

    Chemical: $I \xrightarrow{k_d} 2A_0$ (rate controlling); $A_0 + M \xrightarrow{k_i} A_1$

    Catalytic: $HA + M \xrightarrow{k_i} A_1$

2. *Propagation*

    $A_1 + M \xrightarrow{k_p} A_2$

$$A_2 + M \xrightarrow{k_p} A_3$$

$$\cdot \quad \cdot \quad \cdot$$
$$\cdot \quad \cdot \quad \cdot$$
$$\cdot \quad \cdot \quad \cdot$$

$$A_{n-1} + M \xrightarrow{k_p} A_n$$

3. *Transfer*

$$A_n + TH \xrightarrow{k_{tr}} P_n + T^*$$

4. *Termination*

Combination $\quad A_n + A_m \xrightarrow{k_t} P_{(n+m)}$

Disproportionation $\quad A_n + A_m \xrightarrow{k_t} P_n + P_{nu}$

Unimolecular $\quad A_n \xrightarrow{k_t} P_{nu}$

where I = a chemical initiator
 A_0 = a primary reactive species (no added monomer)
 HA = a catalyst
 TH = a transfer agent
 M = a monomer molecule
 A_n = an active species containing n monomer units
 P_n = an inactive polymer species containing n monomer units
 P_{nu} = P_n with terminal unsaturation
 A = any active species, regardless of chain length

The rate constants are identified as
 k_d = Decomposition
 k_i = Initiation
 k_p = Propagation
 k_{tr} = Transfer
 k_t = Termination

Brackets around a chemical species, [ ], will indicate the concentration of that species in some convenient units, usually moles/liter. The letter v followed by the subscripted letters $d, i, p, tr,$ or t, will indicate the appropriate rate processes.

409 Kinetics of Homogeneous Free Radical Polymerization

The basic assumption here is that the concentration of active species is constant: $\Sigma_n[A_n]$ = constant. In order for this to be true, the rate of initiation v_i must equal the rate of termination, $v_i = v_t$. The first step is to write the rate expressions for each of the kinetic processes based on the general expressions presented in the preceding section.

1. *Initiation.* v_i = the rate of generation of primary active species. For free radical initiation, the rate of decomposition of initiator is rate controlling and first order with respect to the initiator concentration. Hence

$$-\frac{d[\text{I}]}{dt} = k_d[\text{I}] \qquad (4\text{-}1)$$

$$[\text{I}] = [\text{I}_0]\exp(-k_d t) \qquad (4\text{-}2)$$

$$v_i = 2\epsilon k_d[\text{I}]^* \qquad (4\text{-}3)$$

where ϵ is the initiator efficiency, which usually ranges from 0.6 to 1.0. The factor of 2 enters in Eq. (4–3) because two active species result from the decomposition of each initiator molecule. In many instances, the rate constant is so low ($\approx 10^{-4}$ to 10^{-5} min^{-1}) that very little initiator is consumed in polymerizing to 100 percent yield so that [I] is essentially constant, and v_i can also be considered constant. Initiator concentrations for radical polymerization are typically in the neighborhood of 0.1 to 0.5 weight percent based on the monomer content.

2. *Propagation.* v_p = the rate of generation of polymer; it is equivalent to the rate of loss of monomer.

$$v_p = -\frac{d[\text{M}]}{dt}$$

$$v_p = k_p[\text{M}]\Sigma[\text{A}_n] = k_p[\text{M}][\text{A}] \qquad (4\text{-}6)$$

We have also assumed that all species are equally reactive, regardless of chain length, and evaluated $\Sigma[\text{A}_n]$ as [A], the total concentration of active species.

3. *Transfer.* v_{tr} = the rate of transfer of activity

$$v_{tr} = k_{tr}[\text{TH}]\Sigma[\text{A}_n] = k_{tr}[\text{TH}][\text{A}] \qquad (4\text{-}7)$$

4. *Termination.* v_t = the rate of loss of active species. For bimolecular reactions,

$$v_t = 2k_t\Sigma[\text{A}_n]\Sigma[\text{A}_m] = 2k_t[\text{A}]^2 \qquad (4\text{-}8)$$

To obtain the desired expression for v_p, we solve for [A] by equating v_i and v_t and insert this into the equation for v_p. We must also assume that either trans-

---

* This expression should not be equated to the rate of change in the concentration of primary active species, $d[\text{A}_0]/dt$. We do not, at this point consider their fate, only the fact that they are being generated at a specified rate.

fer is absent, or that if it does occur, the transferred species is as reactive as the original species. Thus

$$v_p = k_p \left(\frac{\epsilon k_d [\text{I}]}{k_t} \right)^{\frac{1}{2}} [\text{M}] \qquad (4\text{-}9)$$

Equation (4–9) holds in the early stages of reaction for a large number of free radical polymerizations, but it should not be considered valid beyond 10 to 15 percent conversion, without experimental verification. Analogous equations can be derived for the cationic and anionic polymerizations

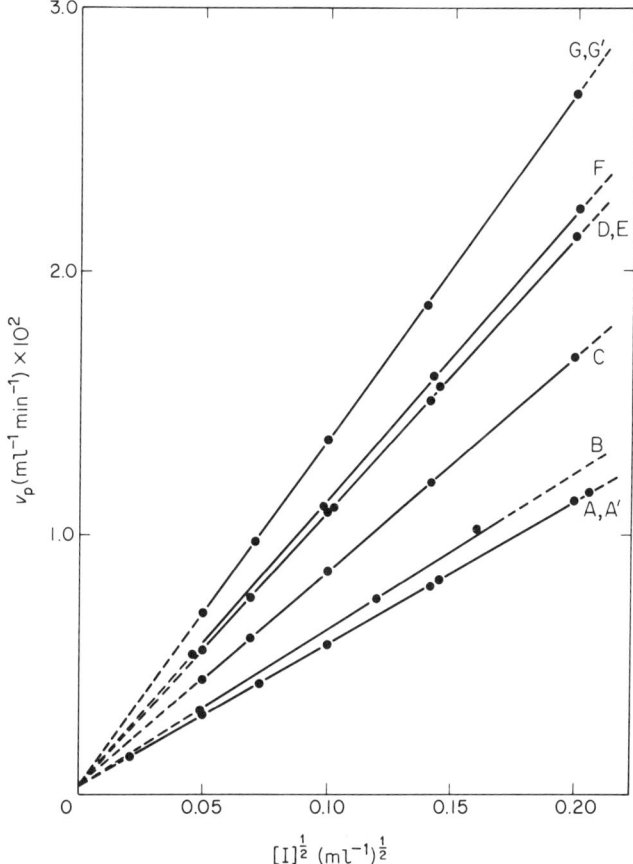

Figure 4–2. Plot of initial v_p vs. $[\text{I}]^{\frac{1}{2}}$ for the bulk polymerization of styrene at 70°C. A, bis(p-chlorobenzoyl)peroxide; A', Luperco BDB; B, benzoyl peroxide; C, acetyl peroxide in dimethyl phthalate; D, lauroyl peroxide; E, myristoyl peroxide; F, caprylyl peroxide; G, bis(2,4-dichlorobenzoyl)peroxide; G', Luperco CDB. [A. I. Lowell and J. R. Price, J. Polymer Sci., 43, 1 (1960).]

discussed, but they may be valid in even a more limited number of instances because the reactions proceed so rapidly that it is unlikely that steady state can be established.

The proportionality predicted by Eq. (4–9) between the rate of polymerization and the square root of the initiator concentration has been abundantly confirmed for low extents of reaction for numerous monomer-initiator pairs. Figure 4–2 is a log-log plot of the initial rates of polymerization against initiator concentration for styrene with various peroxide initiators. If the initiator efficiency for these reactions is constant, the ratio of $v_p/[M][I]^{\frac{1}{2}}$ should be constant for initial rates at various stages of dilution with an inert solvent. This ratio usually decreases but only slowly even over a tenfold dilution range. If this fact is coupled with a close adherence to kinetics that

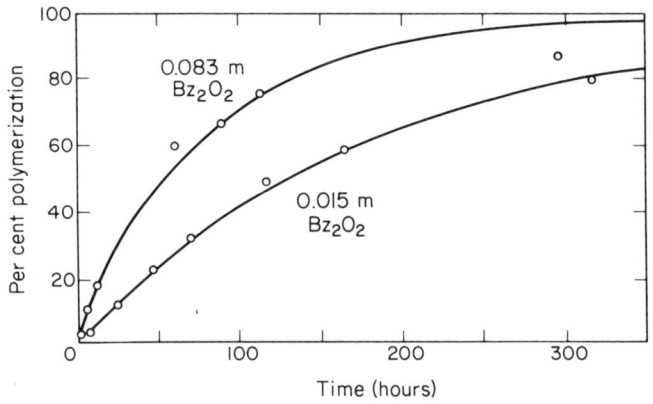

Figure 4–3. Polymerization of 40 percent styrene in toluene at 50°C in the presence of the amounts of benzoyl peroxide indicated. [G. V. Schulz and E. Husemann, Z. physik. Chem., *B39*, 246 (1938).]

are first order in monomer, an efficiency of unity is indicated for polymerization in undiluted monomer. Such an adherence to first-order monomer kinetics, however, is rare, especially for extents of reaction in the range of practical interest.

Figure 4–3 shows data for styrene polymerized with benzoyl peroxide, 40 percent monomer in toluene. In this instance, the assumptions of a constant initiator concentration and 100 percent radical efficiency are well justified over a broad extent of reaction. We note, in general, that v_p should decrease as monomer is consumed. In the polymerization of certain monomers, either undiluted or in concentrated solution, a point in the reaction is reached where there is a marked increase in both reaction rate and molecular weight. This effect may be termed the *autoacceleration*, the *Trommsdorff*, or the *gel effect*. It is shown in Figure 4–4 for methyl methacrylate at various concentrations of monomer in benzene. It is particularly pronounced with methyl methacrylate, methyl acrylate, and acrylic acid. It occurs independently of initiator and is caused by a physical factor that we have heretofore ignored: the rise in viscosity brought about by the continuous generation of high-molecular-weight polymer. This rise in viscosity causes the reaction to become diffusion, rather than chemically, controlled, and valid rate expressions must reflect this situation.

Physically, autoacceleration involves a complex series of events that develop as the reaction enters the diffusion-controlled regime. As the viscosity of the reaction medium increases, monomer diffusion decreases much less

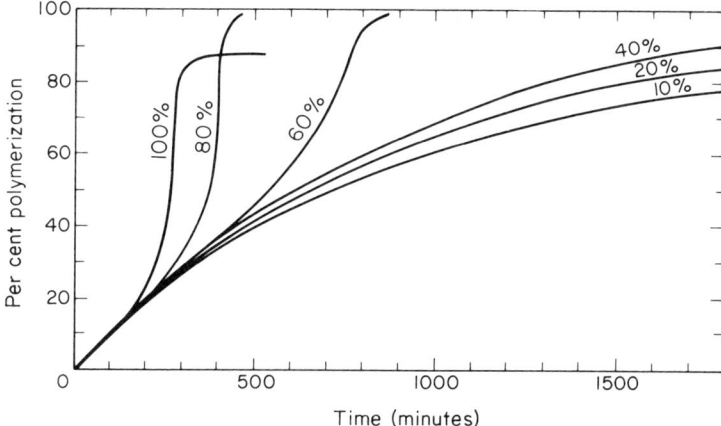

Figure 4-4. The course of polymerization of methyl methacrylate at 50°C in the presence of benzoyl peroxide at various concentrations of monomer in benzene. [G. V. Shulz and G. Harborth, Macromol. Chem., *1*, 106 (1947). Reprinted with permission; Hüthig and Werpf, publishers.]

rapidly than does diffusion of the growing polymer chains. Thus propagation at a particular site diminishes slightly, if at all, but its chances of termination by another active species decrease markedly. At the same time, radicals are being generated at a near normal rate. Since the overall rate of termination is decreasing, the result is an increase in the number of active sites—an unsteady state condition—with an overall increase in the rate of polymerization. Since many chain polymerizations are highly exothermic, this polymerization rate increase is most capable of setting off another round of rate increases through a rapid rise in temperature. This is obviously a potentially dangerous situation, and precautions must be taken, especially on the industrial scale, to limit the risk of a runaway reaction and subsequent explosion.

Reference to Figure 4-4 shows that the rates decrease at conversions on the order of 90 percent. At this point, the polymeric mass takes on a glasslike consistency below 90 °C, and the propagation rate becomes monomer diffusion controlled.

Other manifestations of a gel effect, or of a diffusion-controlled mechanism, are encountered in heterogeneous reactions as well. The phenomenon is thus generally important. Reference to these will be made at the appropriate places in the sections on the production of addition polymers.

4.10 Instantaneous Number-Average Degree of Polymerization

The problem of deriving analytic expressions for molecular weight and molecular-weight distributions in chain polymerization is much more cumbersome than for linear condensation polymerization. This situation is true even for the simplest case of linear polymerization under steady state conditions. In practice, the problem is compounded by the frequently encountered phenomenon of chain branching. In this section we develop an expression for the number-average degree of polymerization that is being generated within a short time interval.

Recall that $\bar{X}_n$ is given by

$$\bar{X}_n(t) = \frac{M(0) - M(t)}{P(t)}$$

where $M(0)$ = moles of monomer present initially
$M(t)$ = moles of monomer present at any time t
$P(t)$ = moles of polymer present at any time t

Differentiating with respect to time, we get

$$P(t)\frac{d\bar{X}_n(t)}{dt} + \bar{X}_n(t)\frac{dP(t)}{dt} = \frac{-dM(t)}{dt}$$

During a short time interval, the *instantaneous number-average degree of polymerization*, $\bar{X}_{ni}$, will be constant, and we will obtain

$$\bar{X}_{ni} = \frac{-dM(t)/dt}{dP(t)/dt} \tag{4-10}$$

In the absence of chain transfer and for termination by combination, $dP/dt = \frac{1}{2}v_t$, whereas for termination by disproportionation or unimolecular termination, $dP/dt = v_t$. For termination by combination,

$$\bar{X}_{ni} = \frac{2v_p}{v_t} = \frac{2v_p}{v_i} \tag{4-11}$$

whereas for termination by disproportionation or unimolecular termination.

$$\bar{X}_{ni} = \frac{v_p}{v_t} = \frac{v_p}{v_i} \qquad (4\text{-}12)$$

Eq. (4–11) and (4–12) are completely general for all chain polymerizations. These equations are, however, usually introduced through the concept of a *kinetic chain length*, $v = v_p/v_i$, which represents the number of monomer units reacting with a given active center from its initiation to its termination.

Equations (4–11) and (4–12) can be utilized to indicate directions of change, even if not the magnitude, for unsteady conditions. In the following section, we will investigate the thermal dependence of molecular weight based on these equations. Returning to the autoacceleration effect, it is easily recognized that a significant decrease in k_t with no change in k_p is in line with the observed rapid rise in $\bar{X}_n$.

Some homogeneous systems with radical polymerization conducted in undiluted monomer obey Eqs. (4–11) and (4–12), but verification is rare. Since $\bar{X}_{ni}$ is lower than the predicted value in virtually all deviations, one must conclude that chain transfer is occurring. Depending on the reactants available, transfer may be with monomer, initiator, or solvent. Transfer may be occurring with polymer, but as previously indicated, this has no bearing on the kinetic chain length. We thus rewrite the expression for the kinetic chain length as

$$v = \frac{v_p}{\text{(rates of all reactions leading to dead polymer)}^*} \qquad (4\text{-}13)$$

and assume that all transfer species are as reactive as the original. Taking into account reactions that may occur, Eq. (4–13) becomes

$$v = \frac{v_p}{v_t + k_{tr,M}[M][A] + k_{tr,SH}[SH][A] + k_{tr,I}[I][A]} \qquad (4\text{-}14)$$

where SH represents a solvent molecule that effects transfer through an atom of H, $k_{tr,M}$ is the rate constant for transfer to monomer, $k_{tr,SH}$ is the rate constant for transfer to solvent, and $k_{tr,I}$ is the rate constant for transfer to initiator.

The propensity for transfer in a system is generally tabulated in terms of a *transfer constant*, defined as

$$C_{TH} = \frac{k_{tr,TH}}{k_p} \qquad (4\text{-}15)$$

---

* Excluding transfer to polymer.

Eq. (4–14) becomes

$$\frac{v_p/k_p}{v_t/k_p + C_M[M][A] + C_{SH}[SH][A] + C_I[I][A]} \qquad (4\text{-}16)$$

To use Eq. (4–16) it is necessary to insert the proper expressions for v_t and [A] for the system being studied as well as the proper values for the transfer constants. For example, in radical polymerization, termination by combination, we obtain

$$\frac{1}{\bar{X}_{ni}} = C_M + \frac{C_{SH}[SH]}{[M]} + \frac{k_t}{k_p^2}\frac{v_p}{[M]^2}$$

$$+ C_I \frac{k_t}{k_p^2 \epsilon k_d}\frac{v_p^2}{[M]^3} \qquad (4\text{-}17)$$

Fortunately, the situation need not be as complicated as Eq. (4–17) indicates, for transfer under many conditions will be dominated by one species. For instance, in radical solution polymerization, transfer to solvent is often most prevalent, or, in another vein, transfer to Bz_2O_2 is apparently more important than transfer to AIBN. In some instances, a material may be deliberately added to effect transfer to control molecular weight and viscosity; such a material is referred to as a *chain transfer agent* or a *regulator*. For example, aliphatic mercaptans, such as dodecyl mercaptan, are suitable agents for a number of radical polymerized systems, the most notable being styrene-butadiene copolymerization. In this case, the regulator serves to promote the copolymerization (by some unknown mechanism) as well as control molecular weight and, most importantly, the degree of crosslinking.

411 Thermal Dependence of Kinetic Expressions

The thermal behaviors of v_p and $\bar{X}_{ni}$ can be ascertained by expressing the rate constants in the respective equations in their Arrhenius form

$$k_b = A \exp \frac{-E_b}{RT}$$

and then by differentiating the natural logarithms of these new expressions with respect to temperature. For radical polymerization, the following expressions are obtained for constant monomer and initiator concentration:

$$\frac{d \ln v_p}{dT} = \frac{\left(E_p - \frac{E_t}{2}\right) + \frac{E_d}{2}}{RT^2} \qquad (4\text{-}18)$$

$$\frac{d \ln \bar{X}_{ni}}{dT} = \frac{\left(E_p - \frac{E_t}{2}\right) - \frac{E_d}{2}}{RT^2} \qquad (4\text{-}19)$$

For many radical polymerizations $(E_p - E_t/2)$ is about 5 to 6 kcal/mole, whereas E_d is about 30 kcal/mole for such initiators as benzoyl peroxide and azobisisobutyronitrile. Thus $dv_p/dT > 0$ and $d\bar{X}_{ni}/dT < 0$; or, to put it in words, these results predict that the rate of polymerization increases with increasing temperature while the molecular weight decreases.

In the ionic polymerizations, v_p is proportional to $(k_p k_i/k_t)$, and $\bar{X}_{ni}$ is proportional to either (k_p/k_t) or (k_p/k_{tr}), depending on whether termination or transfer predominates. The temperature dependencies then appear as

$$\frac{d \ln v_p}{dT} = \frac{(E_p + E_i) - E_t}{RT^2} \qquad (4\text{-}20)$$

$$\frac{d \ln \bar{X}_{ni}}{dT} = \frac{(E_p - E_t)}{RT^2} \quad \text{or} \quad \frac{(E_p - E_{tr})}{RT^2} \qquad (4\text{-}21)$$

Since initiation involves the approach of a neutral molecule to a reactive catalyst and propagation involves the approach of a neutral molecule to an ion in a medium of low-dielectric constant, no activation energy is necessary for either process. On the other hand, termination or transfer involves rearrangement and ion activation. Thus $dv_p/dT < 0$ and $d\bar{X}_{ni}/dT < 0$, which shows that both the rate of polymerization and molecular weight increase with decreasing temperature.

412 Molecular-Weight Distributions

The simplest case for developing an analytic expression for a molecular-weight distribution in vinyl polymerization arises when termination of growth is unimolecular in polymer, either by actual termination or by transfer—transfer to polymer excluded. If p is the probability that a growing chain will propagate rather than terminate, then p is given by

$$p = \frac{v_p}{v_p + v_t + v_{tr}} \qquad (4\text{-}22)$$

The probability of forming an x-mer after $x-1$ propagations and one termination or transfer is then $p^{x-1}(1-p)$, which is identical in form to the result for linear condensation polymerization. The probability p, however, is not an equilibrium value, nor is it necessarily constant throughout the polymerization. The foregoing expression then represents the distribution of the polymer being formed over a short interval of time, and under most conditions it cannot be applied to the overall product distribution. For high polymer to form, $v_p \gg v_t + v_{tr}$ so that $p \to 1$ and $\bar{X}_w/\bar{X}_n \to 2$.

C. Tanford* has developed a distribution function for radical vinyl polymerization, termination by combination, which suffers from limitations similar to the above. The steady-state rate expressions for each species are

$$\frac{d[A_1]}{dt} = 0 = v_i - k_p[A_1][M] - k_t[A_1]\sum_{n=0}^{\infty}[A_n] \quad (4\text{-}23)$$

$$\frac{d[A_2]}{dt} = 0 = k_p[A_1][M] - k_p[A_2][M] - k_t[A_2]\sum_{n=0}^{\infty}[A_n]$$

$$\vdots$$

$$\frac{d[A_n]}{dt} = 0 = k_p[A_{n-1}][M] - k_p[A_n][M] - k_t[A_n]\sum_{m=0}^{\infty}[A_m]$$

Steady state in this case requires that the concentration of each species remains constant. This is a much more stringent requirement than that previously evoked. By summing over Eq. (4–23), we obtain a result that is already known.

$$v_i = k_t \sum_{n=0}^{\infty}[A_n]\sum_{m=0}^{\infty}[A_m] = 2k_t(\sum_{n=0}^{\infty}[A_n])^2 \quad (4\text{-}24)$$

We use this equation and substitute for $[A_m]$ in the last expression of Eq. (4-23) to obtain

$$[A_n] = \alpha[A_{n-1}] \quad (4\text{-}25)$$

where

$$\alpha = \frac{k_p[M]}{k_p[M] + (k_t v_i)^{\frac{1}{2}}}$$

By expanding Eq. (4-25), we can show that

$$[A_n] = \alpha^n[A_1] \quad (4\text{-}26)$$

*C. Tanford, Physical Chemistry of Macromolecules (New York: John Wiley & Sons, 1961), p. 596.

The rate of forming inactive x-mers, P_x, is given by

$$\frac{dP_x}{dt} = -\frac{1}{2}k_t \sum_{n=0}^{x}[A_n][A_{x-n}] \quad (4\text{-}27)$$

On substituting Eq. (4–26), we obtain

$$\frac{dPx}{dt} = -\frac{1}{2}k_t[A_1]^2 \sum_{n=0}^{x} \alpha^n \alpha^{x-n} = -\frac{1}{2}k_t[A_1]^2(x+1)\alpha^x \quad (4\text{-}28)$$

For steady state conditions, the mole fraction of x-mer being formed at any instant will be proportional to $(x+1)\alpha^x$—that is,

$$n_x = B(x+1)\alpha^x \quad (4\text{-}29)$$

Since $\sum_{x=0}^{\infty} n_x = 1$, B can be evaluated as $(1-\alpha)^2$.
Thus

$$n_x = (1-\alpha)^2(x+1)\alpha^x \quad (4\text{-}30)$$

With $w_x = xn_x/\sum_{x=0}^{\infty} xn_x$, we obtain

$$w_x = \tfrac{1}{2}(1-\alpha)^3 x(x+1)\alpha^x \quad (4\text{-}31)$$

It follows that

$$\bar{X}_n = \frac{2}{1-\alpha} \quad (4\text{-}32)$$

$$\bar{X}_w = \frac{2+\alpha}{1-\alpha} \quad (4\text{-}33)$$

and

$$\frac{\bar{X}_w}{\bar{X}_n} = \frac{2+\alpha}{2} \quad (4\text{-}34)$$

If $\bar{X}_n$ is to be large, we see from Eq. (4–32) that $\alpha \to 1$ and that $\bar{X}_w/\bar{X}_n \to 1.5$. For radical polymerized methyl methacrylate, this ratio is close to 1.5; but for styrene, the ratio is about 2.0. Eq. (4–31) is shown plotted in Figure 4–5 for $\bar{X}_n = 500$ with the appropriate curve for a linear condensation polymer (or a vinyl polymer with unimolecular termination) of the same molecular weight. The distribution for free radical polymerization is somewhat sharper.

The preceding results will usually be correct only for initial conditions of polymerization. We will now derive the distribution function for a living polymer system—one that does not suffer from the foregoing limitations.

Consider a living polymer system where each initiator species I gives rise to one primary active species A_0. Upon the introduction of monomer, the primary active centers are immediately converted to A_1 species and growth proceeds without termination. The governing kinetic equations are

$$\frac{d[A_1]}{dt} = -k_p[A_1][M]$$

$$\frac{d[A_2]}{dt} = k_p[A_1][M] - k_p[A_2][M] \qquad (4\text{-}35)$$

$$\vdots$$

$$\frac{d[A_n]}{dt} = k_p[A_{n-1}][M] - k_p[A_n][M]$$

In order to eliminate the time dependence, we substitute the propagation equation, $-d[M]/dt = k_p[I][M]$ to obtain

$$\frac{d[A_1]}{A_1} = \frac{d[M]}{[I]} \qquad (4\text{-}36)$$

$$\frac{-d[A_n]}{[A_{n-1}] - [A_n]} = \frac{d[M]}{[I]}$$

It is convenient to cast these equations into another form by substituting $-d([M(0)] - [M(t)]) = -d(\Delta[M])$ for $d[M]$ to obtain

$$\frac{d[A_1]}{[A_1]} = -d\left(\frac{\Delta[M]}{[I]}\right) = -d\xi$$

$$\frac{d[A_n]}{[A_{n-1}] - [A_n]} = d\left(\frac{\Delta[M]}{[I]}\right) = d\xi \qquad (4\text{-}37)$$

where $\Delta[M]/[I] = \xi$. Note that $\xi = \bar{X}_n$. Progressive integration of Eqs. (4–37) for $n = 1, 2, 3$, etc. leads to the following:

$$[A_1] = [I]e^{-\xi}$$

$$[A_2] = [I]\xi e^{-\xi}$$

$$[A_3] = [I]\left(\frac{\xi^2}{2}\right)e^{-\xi} \qquad (4\text{-}38)$$

$$[A_n] = [I]e^{-\xi}\frac{\xi^{n-1}}{(n-1)^x}$$

The weight fraction of n-mers is given by

$$w_n = \frac{n[A_n]}{\xi[I]}$$

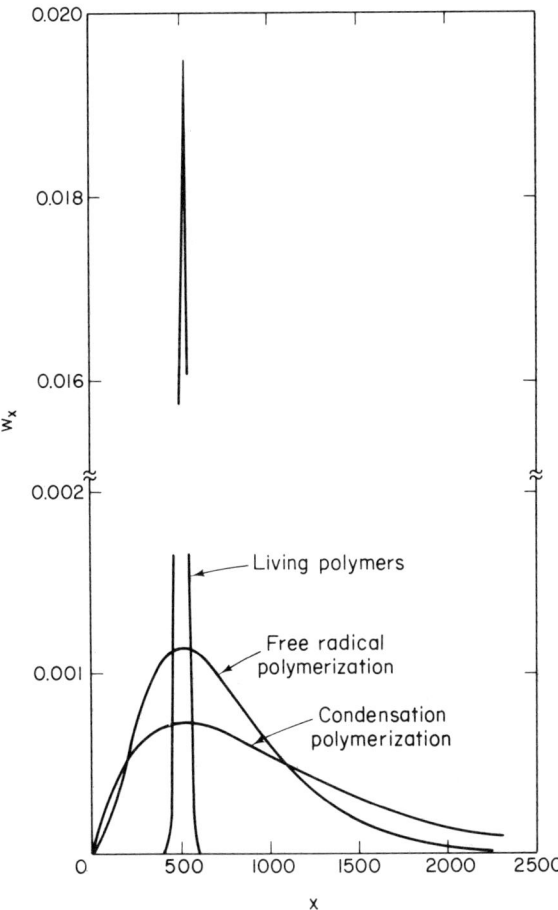

Figure 4-5. A comparison of the weight distribution functions which are predicted for three types of polymerization kinetics. [C. Tanford, op. cit.]

Substituting from Eq. (4–38), we obtain

$$w_n = \frac{\xi}{\xi+1} \cdot \frac{e^{-\xi} n \xi^{n-2}}{(n-1)!} \qquad (4\text{-}39)$$

Summation of $\bar{X}_w = \sum_{x=0}^{\infty} n w_n$ yields

$$\bar{X}_w = \frac{\xi^2 + 3\xi + 1}{\xi} \qquad (4\text{-}40)$$

The expression for $\bar{X}_n$, as previously stated, is

$$\bar{X}_n = \xi \qquad (4\text{-}41)$$

For $\bar{X}_w/\bar{X}_n$, one obtains

$$\frac{\bar{X}_w}{\bar{X}_n} = \frac{\xi^2 + 3\xi + 1}{\xi^2} \qquad (4\text{-}42)$$

For large values of ξ, $\bar{X}_w/\bar{X}_n \to 1$. Also shown in Figure 4–5 is Eq. (4–39) for $\xi = 500$ with n now identified as x. The sharpness of the distribution compared with those for free radical and step-reaction polymerizations is striking. Eqs. (4–38) and (4–39) are recognizable as *Poisson distributions*, and these results could also have been obtained through probabilistic considerations. This distribution represents the narrowest theoretical molecular-weight distribution that can possibly be achieved. These results and their relationship with the probability concepts of Poisson distributions were first obtained by Flory* for ethylene oxide polymerization.

PRODUCTION OF ADDITION POLYMERS

413 Homogeneous and Heterogeneous Polymerization

Broadly speaking, the production of addition polymers is conducted in either a homogeneous or heterogeneous environment. As the term implies, in *homogeneous polymerization*, all the reactants (for example, monomer, initiating species, and solvent) are mutually soluble and compatible with the polymer product. Reactions of this sort are chiefly characterized by high

* P. J. Flory, J. Am. Chem. Soc., **62**, 1561 (1940).

viscosities—at times as high as 10^5 to 10^6 poise—which makes mixing and heat transfer major considerations in reactor design. The heat-transfer problem is compounded by the generally low thermal conductivities and heat capacities characteristic of organic polymers. In addition, some polymer systems exhibit rather high heats of reaction—on the order of 20 to 25 kcal/mole.

In *heterogeneous polymerization*, the catalyst and polymer are not soluble in either the monomer or the solvent for the monomer. In this process, the polymer exists as a separate discrete phase, dispersed in droplet or globular form in an inert suspending medium. Each polymeric droplet will normally consist of many agglomerated polymer molecules swollen with imbibed monomer to a gel-like consistency. These particles may range in size from about 0.1 microns to several microns in diameter. The major locus of polymerization will be within these droplets, and the kinetic processes of initiation, propagation, transfer, and termination will operate more or less in the usual manner. The suspending medium is often water or an aliphatic hydrocarbon "solvent" such as heptane. Since the viscous polymer mass forms the discontinuous phase within the free-flowing phase of the suspending medium, the overall viscosity of the reaction environment is nearly that of the suspending medium. The severe problems of mixing and heat transfer encountered in homogenous polymerization are thereby mitigated. An important property of the suspended particles, in terms of the heat-transfer problem, is their relatively high surface-area-to-volume ratio. For this reason, the heat of reaction is readily transferred from the particles to the suspending medium. Transfer of heat from the suspending medium to the reactor walls or cooling coils is also readily achieved because of the fluid mobility of the suspension. A major problem is the stability of the suspended particles. Since they consist of a sticky polymeric mass, care must be exercised to guard against agglomeration or coagulation of the particles. Thus the level and type of agitation must be carefully controlled, and chemical agents may be added to enhance particle stability.

A useful way to view heterogeneous polymerization is as a suspension of individual chemical reactors in which polymerization is homogeneous or semihomogeneous (since there may be small concentration or thermal gradients within the particles). The process occurring within the particles will be quite like those occurring in a macroscopically homogenous environment. Therefore much of the information required to specify the state of a reaction environment and to design production scale reactors is the same as for homogenous conditions.

To summarize, high heats of reaction coupled with low-thermal conductivities, high viscosities, the Arrhenius-type temperature dependences make heat transfer, reactor size or geometry, and the state of dispersion highly important variables. Ideally, for design purposes, one should have complete data for diffusion, thermal conductivity, and heat capacity, as well as data for the viscosity as a function of temperature, concentration of monomer and polymer, and molecular weight.

414 Bulk or Mass Polymerization

This is generally a free-radical initiated process, conducted solely in the presence of monomer, initiator, and polymer product. Because of the acute heat-transfer problems, polymerization is restricted to materials with low reactivities and heats of reaction, such as styrene and methyl methacrylate. In batch processing, reaction is conducted in thin sections, such as sheets, slabs, or rods. In large samples with their low-surface-area-to-volume ratios, adequate temperature control is impossible to achieve, and local "hot spots" develop in the core of the polymerizing mass. In its mildest form, the development of "hot spots" will lead to charring of the material; but in its severest form, explosion can result. In continuous processes, the polymerizing mass is passed through specially designed, high-contact surface area, extruder-like reactors.

Removal of the unreacted monomer is a necessary and troublesome step in the manufacturing process. Most organic monomers are toxic and have highly objectionable odors. They also serve as plasticizers,* thereby altering the mechanical properties of the product. Monomer removal may be accomplished in devices known as vacuum extruders; that is, the polymeric product is melt-extruded as thin sections into a vacuum.

Because a highly pure product is obtained, it is an ideal process for the manufacture of optical grade poly(methyl methacrylate). Impact-resistant grade polystyrene—a graft copolymer with rubber—is manufactured in bulk in continuous processors (see Section 514).

415 Solution Polymerization

In this process, the monomer, polymer, and initiating species are all soluble in the selected solvent and the process is homogeneous. The viscosity of the reaction medium increases as the reaction progresses but not nearly as drastically as it does in bulk polymerization. Selection of the solvent is governed by the extent to which it may affect the rate of polymerization (in ionic reactions) or chain transfer. A convenient means of removing the heat of reaction, especially in radical initiated systems, is gained by running the

---

* See section 710.

reaction at the reflux temperature of the polymerizing mass. The vapors carry heat of reaction away as the heat of vaporization; the vapors pass into a condenser to be cooled, condensed, and returned to the reactor. Hence another basis for selection of a solvent is its boiling point. Some commonly used solvents are benzene, toluene, cumene, cyclohexane, heptane, chlorohydrocarbons, and water.

Use of this process is generally restricted to the production of specialty polymers or to situations where no other method can be employed. The major disadvantage associated with the process is the removal of solvent and unreacted monomer from the product. The costs and hazards involved in this step can be considerable.

416 Free-Radical Suspension Polymerization

This is a heterogeneous process in which water-insoluble monomer is first dispersed and suspended in water with the aid of chemical reagents and with the proper level and type of agitation. The particles or monomer globules are on the order of a micron or larger in diameter, so they are above the colloidal size range. An oil-soluble initiator like benzoyl peroxide is added, and polymerization is conducted at temperatures ranging from 50 to 90°C. Each of the suspended particles has all the characteristics of miniature bulk reactors, without a mechanism for internal mixing but with surface-area-to volume ratios adequate for heat removal. The main problem in this process is the maintenance of particle stability while the particles are being transformed from a mobile fluid through a viscous, sticky stage to a rigid solid. Monomer dispersion is initially achieved and preserved by mechanical mixing. Monomer dispersion is, furthermore, an important determinant of the final polymer particle size. Without proper mixing the reacting particles would readily agglomerate or settle out. Equally important is the proper use of chemical reagents, so-called *stabilizing agents*, to preserve stability. These reagents are employed in two ways. Materials like fatty acids or other surfactants, inorganic oxides, hydroxides, and water-insoluble salts locate at the interface to reduce surface tension and tackiness, while water-soluble polymers like poly(vinyl alcohol), methyl cellulose, or poly(acrylic acid) locate in the aqueous phase to raise the viscosity of the suspending medium.

After polymerization is complete, the particles are allowed to settle, the water is removed by filtration, and the particles are washed to remove the stabilizing agents.

This method is the most widely used for processing free radical polymers, both homopolymers and copolymers.

417 Suspension Polymerization

This process is similar to the free radical process, just described, but here the suspending medium is organic in character and suspending agents are not used. Moreover, the monomers may be gaseous—for example, ethylene or propylene.

The high-pressure production of polyethylene falls into a category somewhat between the two, but in spite of its free radical initiation and aqueous character it is more similar to polymer production in nonaqueous environments. Ethylene gas with a trace of O_2 is fed under a pressure of several hundred atmospheres to a stirred, silver-lined autoclave or to a continuous tubular autoclave containing a mixture of 10 parts water and one part benzene. The temperature is maintained within the range of 175 to 200 °C. The oxygen decomposes to produce free radicals and initiate polymerization. The polymer suspension that forms is continuously removed through a blow-down tank. The temperature and pressure are reduced, and the polymer is separated from the water-benzene mixture. The unreacted ethylene and the water-benzene mixture are recirculated to the reactor. The fine polyethylene powder is then washed, dried, and pelletized in an extruder. The polyethylene so produced is highly branched and is referred to as the low-density type, for its branched structure does not allow it to pack as closely as the linear in the crystalline state.

Linear polyethylene and stereoregular polymers are produced in a low-pressure process utilizing coordination-type catalysts. The catalysts are dispersed in the liquid phase in a stirred tank reactor with the monomeric gas being supplied at the relatively low pressure of 15 to 20 atm and the temperature being a rather moderate 135 to 150 °C. Heptane and cyclohexane are commonly used as the suspending mediums. A slurry containing 20 to 25 percent polymer is formed and treated in much the same way as in the high-pressure process.

418 Emulsion Polymerization

This process finds special importance in the production of synthetic rubbers of the styrene-butadiene copolymer type. It was developed during World War II to meet the needs of the times. The process finds application in instances where it is desired to have a water-dispersed product, such as those used to formulate latex paints, certain adhesives, or floor polishes. Emulsion polymerization is especially suited for such applications because it yields a product consisting of a stable dispersion of colloidal-size particles in contrast to the unstable dispersions of suspension polymerization. Separation of the polymer from the suspending medium and its subsequent purification are accomplished with difficulty. It is not necessary to purify the polymer for rubber-tire use. Emulsion polymerization products, or latexes as they are usually called, also find application in the formulation of polybends.

The conditions to be described here will always refer to those incurred in the very common emulsion polymerization reactor charge comprised of 180 parts of water, 100 parts of nearly insoluble vinyl monomer, 5 parts of fatty acid soap, and 0.5 parts of potassium persulfate—a water-soluble free radical initiator. Temperatures of 50 to 60 °C are ordinarily employed. The fundamental mechanism of emulsion polymerization, that we are about to describe, was first elucidated by W. D. Harkins and his associates in the 1940's*.

When a relatively water-insoluble vinyl monomer, such as styrene, is emulsified in water with the aid of anionic soap and adequate agitation, three phases result: (1) the aqueous, in which a small amount of monomer and soap is dissolved, (2) *emulsified monomer droplets* of microscopic (supercolloidal) size, and (3) submicroscopic (colloidal) micelles, which are saturated with monomer (see Figure 4-6).

It is essential that the solubility of the monomer in the aqueous phase be lower than about 0.004 mole/liter; otherwise the system will not be a typical emulsion polymerization. As will be seen, this slight solubility allows for transport of the monomer from the emulsified monomer reservoirs to the various reaction loci. Relative monomer insolubility will also ensure the fact that the acqueous phase is not a major locus of polymerization.

Two of the three phases shown in Figure 4-6 involve anionic soap, so some consideration must be given to the structure of the soap molecule in relation to these phases. The soap molecule has a long, oil-soluble portion attached at one end to a water-soluble group. Thus one end of the molecule is hydrophobic, whereas the other is hydrophilic. Soap anion molecules orient themselves at monomer-water, interfaces with their hydrophobic ends facing the monomer phase and their hydrophilic ends facing the aqueous phase. In this way a negative charge overcoats each emulsion droplet. Stability is imparted to the emulsified monomer droplets not only by the reduction of surface tension but also by the repulsive forces between the charged surfaces.

A small number of soap molecules group themselves into colloidal aggregates, called *micelles*, of 50 to 100 molecules. Each such group remains dispersed as a unit in the aqueous medium in equilibrium with soap anions that are soluble as molecular units. The actual structure of these micelles is still a highly controversial matter, but their molecules have been conjectured to arrange themselves in units of parallel monolayers with the hydrophobic ends facing each other and the hydrophillic ends facing the aqueous phase. The size of these micelles is limited by an equilibrium balance of the thermo-

---

*W. D. Harkins, *J. Am. Chem. Soc.*, **69**, 1428 (1947).

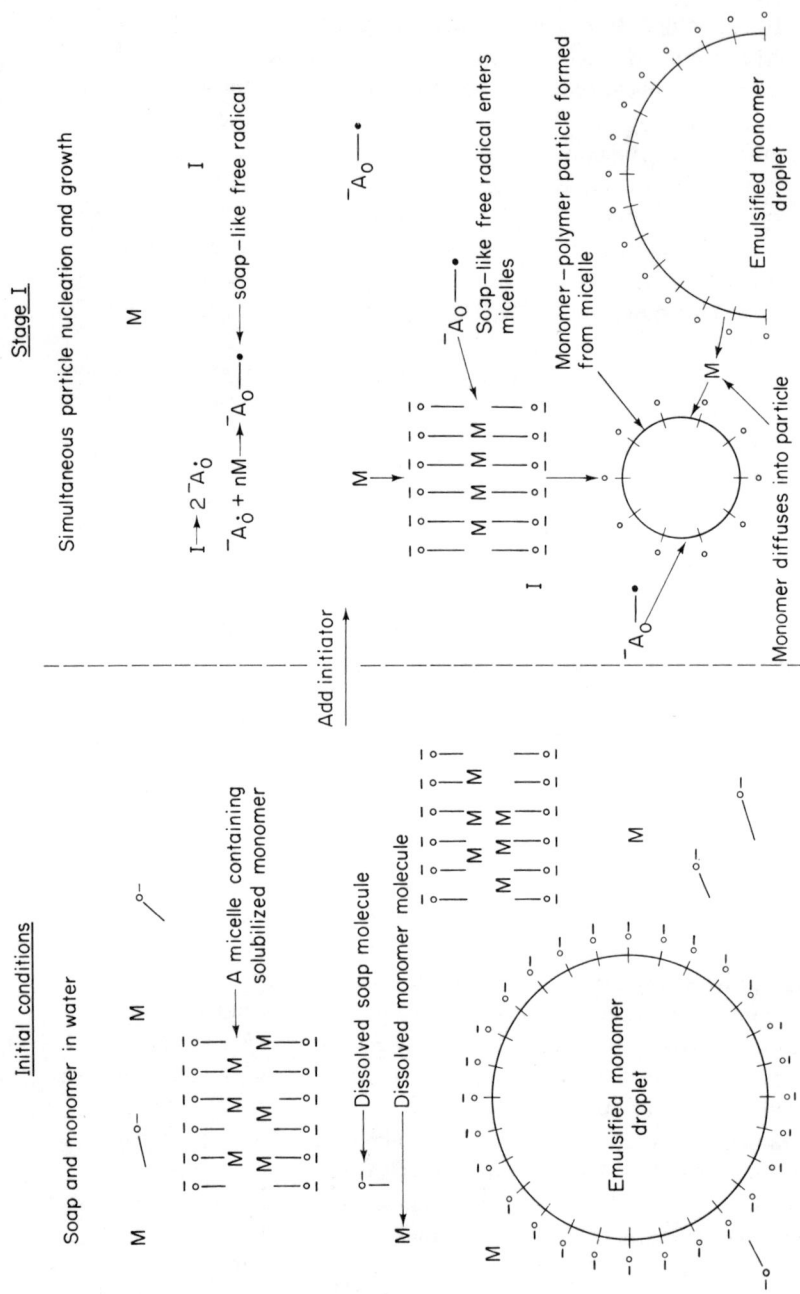

Figure 4–6. The phases present in emulsion polymerization.

dynamic forces accounting for the presence of the micelles. The coalescing factors—namely, the van der Waals attractive forces of the hydrocarbon chains for each other and the hydrophobic nature of the hydrocarbon chains forcing them to seek oleophilic surroundings—are just balanced by the ionic repulsive forces of the charged hydrophilic ends of the soap molecules. A critical soap concentration must be attained before micelles will form in sufficient numbers to be detected by physical and chemical tests. This critical concentration, which must be surpassed in ordinary emulsion polymerizations, is known as the *critical micelle concentration* (c.m.c.). The monomer that swells the micelles to saturation is said to be solubilized. The capacity of a particular soap to micellize and to solubilize monomer is a function of the molecular structure of the soap and of other conditions of the surrounding medium.

The emulsion polymerization reaction being described proceeds in three stages, and the corresponding rate curve is shown in Figure 4–7.

Stage I. Stage I begins (see Figure 4–6) when the free-radical-producing, water-soluble initiator, potassium persulfate, is added to the three-phase emulsion just described. Potassium persulfate decomposes thermally to form water-soluble, sulfate radical anions.

$$S_2O_8^= \xrightarrow[50-60°C]{heat} 2SO_4^- \cdot$$

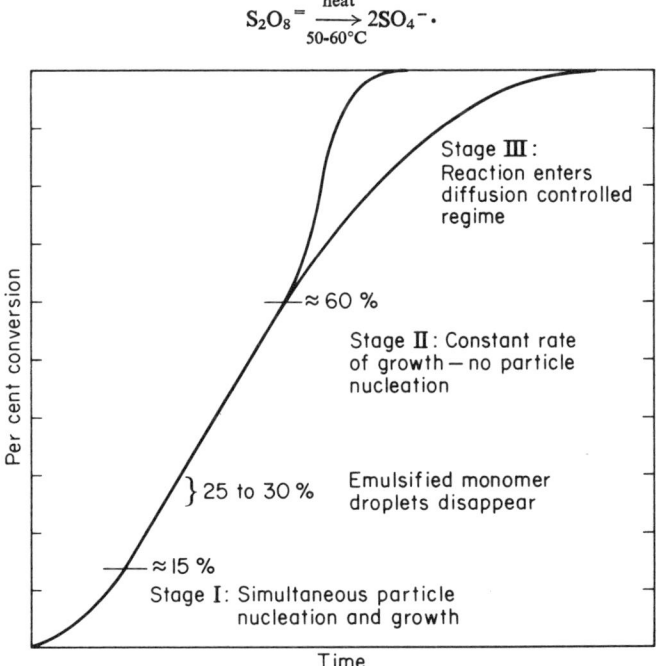

Figure 4–7. Schematic conversion-time curve, showing three stages of emulsion polymerization.

It is presumed that these free radicals react with monomer dissolved in the aqueous phase to form soap-type anionic free radicals:

$$SO_4^- \cdot + (n+1) M \xrightarrow[50-60°C]{} {}^-SO_4(CH_2-CX_2)_n CH_2-CX_2 \cdot$$

After a short time, while n is still small, these soap-type free radicals will become more hydrophobic and will seek oleophilic surroundings. The radicals then enter the interface of the oil phase as soap molecules, where they continue to polymerize. This mechanism of free radical entry into the oil phase is based on the contention that it would seem rather difficult for the negatively charged sulfate radical ions to penetrate the anionic interfacial barrier imposed by the oil phase by any other mechanism. One must also consider the fact that the primary sulfate radical anions are virtually insoluble in the oil phase.

Micelles capture many more of these soap-type free radicals than the emulsified monomer droplets and therefore serve as the primary loci for initial growth. There are two reasons: (1) The maximum dimension of the micelles is about 100 A, whereas the dimensions of the emulsified monomer droplets is on the order of 10,000 Å; so the micelles will have a much higher surface area to volume ratio than the monomer droplets and thus will be more efficient collectors. (2) There are many more micelles per milliliter of aqueous phase than monomer droplets (about 10^{18} versus 10^{11}). It is therefore convenient to neglect the number of free radicals that might be captured by the large droplets. The primary purpose of these monomer droplets will be to serve as reservoirs to keep the aqueous phase saturated with monomer.

As polymerization proceeds within a stung micelle, monomer within the micelles is used up, and it must be replenished from the aqueous phase. Before a second free radical can enter a stung micelle, the amount of monomer and polymer contained within the micelle causes it to swell to its dimensional limits of stability. A new phase, consisting of monomer-polymer particles, is thereby formed. These *monomer-polymer particles* adsorb the soap of their parent micelles to preserve their stability. They are still very much smaller— 300 to 700 Å in diameter—and more numerous than the emulsified monomer droplets; consequently, they compete with the remaining micelles for capture of free radicals. Monomer and free radicals are also supplied to these new active sites via diffusion from the aqueous phase as polymerization continues.

As these monomer-polymer particles continue to grow, they preferentially adsorb soap from existing micelles and emulsified monomer droplets, until at 13 to 20 percent conversion nearly all the soap will become adsorbed on the monomer-polymer particles, and the micelles will disappear. Since all new particles originate in the micelles, their disappearance markedly reduces the probability for nucleation of new monomer-polymer particles. At this point, Stage I is completed. Since this stage is characterized by simultaneous particle nucleation and particle growth, the rate of overall polymerization increases continuously as shown, and a particle-size distribution results.

Stage II. This stage is characterized by continued growth of the existing monomer-polymer particles and, in the absence of particle nucleation, a constant overall rate of polymerization as depicted in Figure 4–7. The particles are continuously supplied with free radicals from the aqueous phase and with monomer until the emulsified monomer droplets disappear at 25 to 30 percent conversion.

In the system under consideration, the rate of generation of free radicals is about 10^{13} per second per milliliter and the number of active sites is between 10^{14} to 10^{15} per milliliter. Therefore each site will capture, at the most, a free radical every 10 to 100 seconds, which means that the free radicals will generally enter the particles singly. The point is that a free-radical polymer chain within a monomer-polymer particle will react in an isolated state for a period of time (10–100 sec) until another free radical enters to terminate the reaction. The particle then remains inactive until another free radical enters to reinitiate polymerization. Polymerization within a monomer-polymer particle can thus be characterized by alternating periods of activity and inactivity.

Since a large number of these sites are available for polymerization and since polymerization within each site proceeds unhindered for a time, the emulsion polymerization system is featured by a rapid rate of reaction with a simultaneous generation of very high-molecular-weight polymer. Since styrene polymerizes at the rate of 10^3 molecules per second, degrees of polymerization on the order of 10^4 to 10^5 are expected and are achieved in practice.

Stage III. This stage begins when the overall rate of polymerization begins to deviate from linearity. As shown in Figure 4–7, this deviation may appear as a decrease in rate or an increase if the Trommsdorff effect is important. The monomer reservoirs will have long since disappeared and the ratio of monomer to polymer within the particle will have dwindled to the point where the reaction becomes diffusion controlled.

The principal chemical reactions, diffusion processes, and stages in emulsion polymerization are summarized in Table 4–3. Deviations from this mechanism of the emulsion polymerization reaction for this "classical" system are observed as a result of deviations from the formulation requirements outlined earlier. One such deviation is noted when monomers that are fairly water soluble, such as vinyl acetate, are used, thereby allowing a significant amount of polymerization to occur within the aqueous phase. Systems where the monomer is insoluble in its polymer, such as vinyl chloride, also exhibit deviations.

TABLE 4-3

PRINCIPAL CHEMICAL REACTIONS, DIFFUSION PROCESSES,
AND STAGES IN EMULSION POLYMERIZATION

A. Chemical Reactions

Aqueous phase

$$K_2S_2O_8 \rightarrow 2K^+ + 2SO_4^-\cdot$$

$$SO_4^-\cdot + nM \rightarrow {}^-SO_4 - M_n\cdot$$

Monomer-Polymer Particle

$$^-SO_4 - M_n\cdot + mM \rightarrow {}^-SO_4 - M_{m+n}\cdot$$

$$^-SO_4 - M_{m+n}\cdot + {}^-SO_4 - M_n\cdot \rightarrow P_{m+2n} \ (m \geqslant 2n)$$

B. Diffusion Processes

Monomer Transfer (Ceases 25–30% conversion)
  Emulsified Monomer Droplets→Aqueous Phase→
  Monomer-Polymer Particles (saturated)
Free Radical Transfer
  Aqueous Phase→Interfacial Zone→Interior of Monomer-Polymer Particle

C. Stages

I: 0 to about 15% conversion
   Simultaneous particle nucleation and growth with increasing rate.
II: 15 or 20% to about 55 or 60%.
   Constant rate growth of monomer-polymer particles.
III: Greater than 55 or 60% conversion.
   Non-linear growth rate with diffusion controlled regime; rate may decrease or increase depending on importance of gel effect.

419 Kinetics of Constant-Rate Styrene Emulsion Polymerization

Analysis here centers around the constant rate period observed during the stage II polymerization of styrene-like monomers.* If we consider the rate of polymerization per particle, R_{pp}, we can write with complete generality:

$$R_{pp} = k_p \bar{n}[M] \tag{4-43}$$

which is identical in form to Eq. (4–6), where $\bar{n}$ takes the place of [A]. We use $\bar{n}$ here because it has special significance which will be explained subsequently. If R_{pp} is constant, and the number of particles per unit volume,

* Meaning those monomers with low water solubility and those in which monomer and polymer are completely miscible over all ranges of composition.

N_p, is constant, then the overall rate of emulsion polymerization, R_p, is simply given by

$$R_p = N_p R_{pp} = N_p k_p \bar{n}[\text{M}] \qquad (4\text{-}44)$$

This result was obtained in a more circuitous fashion by W. V. Smith and R. H. Ewart in 1948.* The crux of their analysis was that the growing monomer-polymer particles contained a free radical only half the time or that the particle grew in alternating periods of activity and inactivity. This can be seen in the bar-diagram of Figure 4–8, where growth within a particle

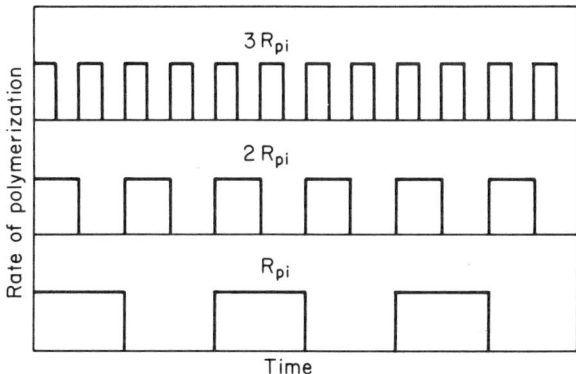

Figure 4–8. Effect of initiation rate on R_{pp}.

for various rates of free radical entry is shown. The important feature to observe is that the area under each of the curves is the same; consequently, the rate of polymerization per particle is independent of the rate of radical entry or particle-size. Thus, in calculating R_{pp}, $\bar{n}$ can be set equal to $(\frac{1}{2})/(6 \times 10^{23})^{-1}$ moles/particle. The concepts embodied in the foregoing discussion are known collectively as the *Smith-Ewart Theory, Case* II.

Experimental verification for $\bar{n} = \frac{1}{2}$ and constant is achieved in initiator perturbation studies in which the initiator concentration during Stage II is suddenly increased. For example according to Eq. (4–3), doubling the initiator concentration approximately doubles the rate at which free radicals are generated; if there are no other changes in the system, the rate of radical entry should also double. However, according to the rationale associated with the Smith-Ewart Theory and the bar diagram in Figure 4–8, the rate of polymerization should remain constant. Such behavior has been observed

* W. V. Smith and R. W. Ewart, *J. Chem. Phys.*, **16**, 592 (1948).

up to about 60 percent conversion for styrene emulsion polymerizations of the sort being considered here.

We must now consider the role that the emulsified monomer droplets and the monomer concentration [M] within the growing particles play in describing the kinetics. In the previous section we saw that the sole purpose of the emulsified monomer droplets was to serve as reservoirs, and supply monomer to the growing monomer-polymer particles. According to numerous

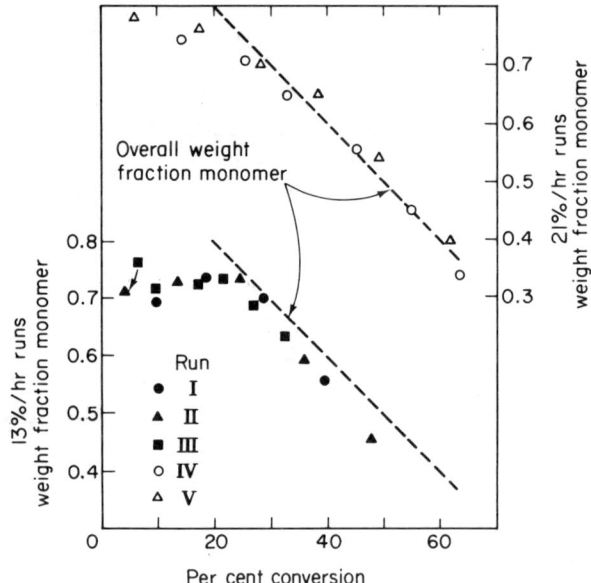

Figure 4–9. Particle monomer concentration vs. percent conversion for styrene emulsion polymerization of 13%/hr and 21%/hr, runs: constant rate up to 60% conversion. [M. R. Grancio and D. J. Williams, *J. Polymer Sci.* **8**, 2617 (1970).]

experimental observations, these reservoirs disappear at 25 to 30 percent conversion. At this point in the reaction the monomer-polymer particles are about 75 percent monomer. The behavior of [M] below this point is still a matter of controversy: one school of thought maintains that the particles are essentially saturated and [M] remains approximately constant, but there is experimental evidence that [M] may decrease. Shown in Figure 4–9 are data for styrene systems which support both schools. These data were obtained for specially formulated runs in which the particles were nucleated instantaneously; as a consequence, they do not exhibit a Stage I period, and the rate curves for these data exhibit linear behavior from zero to about 60 percent conversion. The lower curve in Figure 4–9 is for a conversion rate of 13 percent/hour and a final (uniform) particle diameter of 1950 Å, whereas the upper curve is for a rate of 21 percent/hour and a final (uniform) particle diameter of 2300 Å. The most important point is that [M] decreases continuously along the overall weight fraction monomer curve (as it must) after about 30 percent conversion. And herein lies the rub.

Consider Eq. (4–43) and (4–44) and the fact that R_p, N_p, k_p, and $\bar{n}$ are all constant, but that [M] decreases after 30 percent conversion. For this reason, as well as the fact that there is experimental evidence for [M] $\simeq$ 0.75 and constant at less than 25 to 30 percent conversion, the Smith-Ewart Theory has long been thought to be valid only at low conversions (< 30 percent) and for small particle sizes (< 1200 Å). The "apparent" constant-rate behavior between 30 and 60 percent conversions is thought by many to be the result of a fortuitous balance between decreasing [M] and increasing radical concentration within the particles. However, there has never been any direct experimental evidence to support this viewpoint for the types of systems under consideration here, namely, those which follow the behavior described in the previous section and Figure 4–7.

420 A Core-Shell Morphology

A rather straightforward scheme has recently been proposed by M. R. Grancio and the author* to resolve the foregoing dilemma. The experimental results to be described are for styrene emulsion polymerization. The extension of these results to other styrene-like systems and nonstyrene-like systems remains to be established. From Eq. (4–43) and (4–44), since R_{pp}, $\bar{n}$ and k_p are constant, we concluded that [M] *must* be constant at the actual site of polymerization. It follows that a zone of constant monomer concentration must exist within the particle during constant rate polymerization and that this zone must be the major locus of polymerization. On the basis of this argument, we suggested that the growing particle is actually heterogeneous rather than homogeneous, and that it consists of an expanding polymer-rich (monomer-starved) core surrounded by a monomer-rich (polymer-starved) spherical shell. In this *core-shell model* the outer shell serves as the major locus of polymerization while virtually no polymerization occurs in the core because of its monomer-starved condition. A Smith-Ewart (on-off) mechanism necessarily prevails within the monomer-rich shell. The core is pictured as

---

* M. R. Grancio and D. J. Williams, *J. Polymer Sci.*, A-1, **8**, 2617 (1970).

growing outward somewhat as a ball of string constructed from many single strands. With such a morphology the changing overall particle monomer concentration is consistent with constant particle rates because reaction takes place in an essentially pure monomer environment.

The proposed core-shell morphology can be further supported by kinetic evidence if we consider R_{pp} calculated via Eq. (4-44) and measurement of N_p ad R_p. For the 13 percent/hour and 21 percent/hour formulations

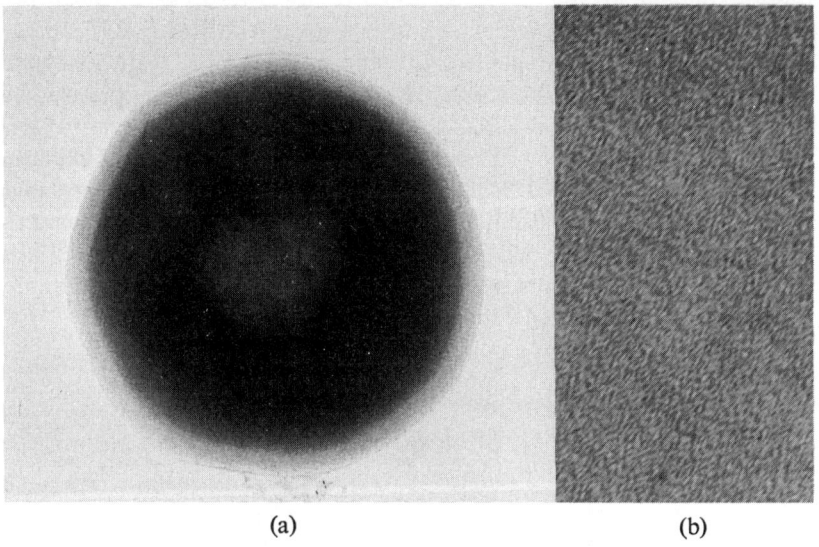

(a)   (b)

Figure 4–10. (a) Electron micrograph showing core-shell morphology. (250,000 ×). (b) Electron micrograph of thin film cast from test particles. (500,000 ×). [M. R. Grancio and D. J. Williams, *J. Polymer Sci*, **8**, 2617, (1970).]

previously described, these calculations yield R_{pp} values of 2.38×10^{-19} and 2.42×10^{-19} g/particle sec., respectively. The fact that the R_{pp} values are identical and constant, independent of particle size or overall particle monomer concentration is consistent with the proposed morphology.

In addition, we substituted appropriate values for $\bar{n}$, [M], and k_p into Eq. (4-44); namely, $\bar{n} = (\frac{1}{2})/(1/6.02 \times 10^{23})$ moles/particle, M = 905 g./l (the density for pure styrene), and k_p = 282.1/mole sec. (a value calculated from bulk experiments). We thus calculated the rate of polymerization for a single active species growing half the time in pure monomer just as the model suggests. We calculated $R_{pp} = 2.12 \times 10^{-19}$ g./particle sec. for rather good agreement with the experimental value of 2.40×10^{-19}.

The following experiment was performed to provide additional corroborating evidence for the core-shell morphology: To a 13 percent/hour styrene run at 20 percent conversion, a portion of butadiene was charged. We expected that copolymerization should result, and if the model were correct the final latex particle should consist of a pure polystyrene core—corresponding to the core which existed when the butadiene was charged—surrounded by a spherical styrene-butadiene copolymer shell. Since the unsaturated butadiene repeat unit could be stained with osmium tetroxide, an ultra-thin section taken through the center of the particle should appear as a doughnut shape under electron microscope observation.

Figure 4-10(a) is an electron micrograph of a thin section obtained from such a particle. It clearly shows the expected unstained core and the stained ring. The very thin light outer ring could have arisen from one of two sources: interaction of the embedding material with the particle surface; or, since butadiene polymerizes at a faster rate than styrene, all the butadiene may have been consumed before the styrene, and the outer ring would consist of pure polystyrene.

It was also necessary to ascertain how the inherent incompatibility of the styrene homopolymer and butadiene-styrene copolymer molecules might affect any conclusions we intended to draw from the proposed study. Thus, we prepared for electron microscope observation thin films of the latex polymer, formed by casting from an initially homogeneous solution. As shown in Figure 4-10(b), the structures which result from actual phase separation because of the inherent incompatibility of the component molecules are orders of magnitude smaller than the expected characteristic structure of the latex particle. This result indicates that compatibility considerations should not interfere with the stated objectives.

What is remarkable about the core-shell morphology is the fact that in macroscopic solution styrene and polystyrene are completely miscible one in the other over all ranges of concentration. On the basis of macroscopic solution behavior, the core-shell separation is completely unexpected. The important factor to consider here is the size of the latex particle relative to the space a polymer chain would occupy in its random coil conformation in a macroscopic solution. Consider the following example: a particle 2,000 Å in diameter with [M] = 0.5, containing polymer with $\bar{M}_n = 500{,}000$. Here the contour chain length of the polymer is over six times greater than the particle diameter, and the radius of gyration of one of these molecules in an ideal solvent would be about one-sixth the particle diameter. Thus, a portion of the elements comprising the particle would be, in macroscopic solution, on the same order of size as the particle itself. Obviously, the polymer chains cannot realize their favored conformations in the crowded and confined habit of a latex particle. It is evident that in the case of a latex particle these conformational restrictions must give rise to a large contribution to the free energy of mixing by an entropic term.

While one aspect of the applicability of the Smith-Ewart Theory has been clarified, another remains to be resolved. Thus, we are led to the topic of our next section.

421 Molecular Weight Development

We are again concerned with Stage II polymerization of styrene. We will utilize the results of measurements made with the 13 percent/hour formulation previously referred to.

Since this formulation exhibits Case II, Smith-Ewart kinetics, the instantaneous number average molecular weight, $\bar{M}_{ni}$, can be predicted from the kinetic chain length. For steady state, termination by combination, and 100 percent radical efficiency with first order decomposition, the result is simply

$$\bar{M}_{ni} = [(R_p/k_d[I]_o) \exp (k_d t)](M_o) \qquad (4\text{-}45)$$

where M_o is the monomer molecular weight. Thus for any given rate R_p, with k_d evaluated independently, and M_o and $[I]_0$ given, $\bar{M}_{ni}$ can be evaluated numerically. For the system of interest, viscosity and number average molecular weights were measured as a function of time and conversion.* These molecular weights were converted to instantaneous values by material balance. The results of these measurements and computations as well as $\bar{M}_{ni}$ predicted from Eq. (4–45) are all shown in Figure 4–11. The divergence between the experimental results and those predicted by the Smith-Ewart Theory for zero to 60 percent conversion is striking. Of further interest is the fact that the ratio $\bar{M}_v/\bar{M}_n$ remains essentially constant, ranging from 3 to 4, which indicates that little or no chain branching is occurring. The most straight-forward way to interpret these findings is that free radicals are not utilized with 100 percent efficiency (even though they are generated in the aqueous phase with 100 percent efficiency). This interpretation remains to be verified, and the means by which radicals are rendered inactive is yet to be established.

Thus, in conclusion, we see how the Smith-Ewart Theory seems to represent the rate behavior of styrene emulsion polymerization rather well; but that it must be amended in yet some unknown way to successfully account for molecular weight development. The direct applicability of these findings

---

* M. R. Grancio and D. J. Williams, *J. Polymer Sci*, A–1, **8**, 2733 (1970).

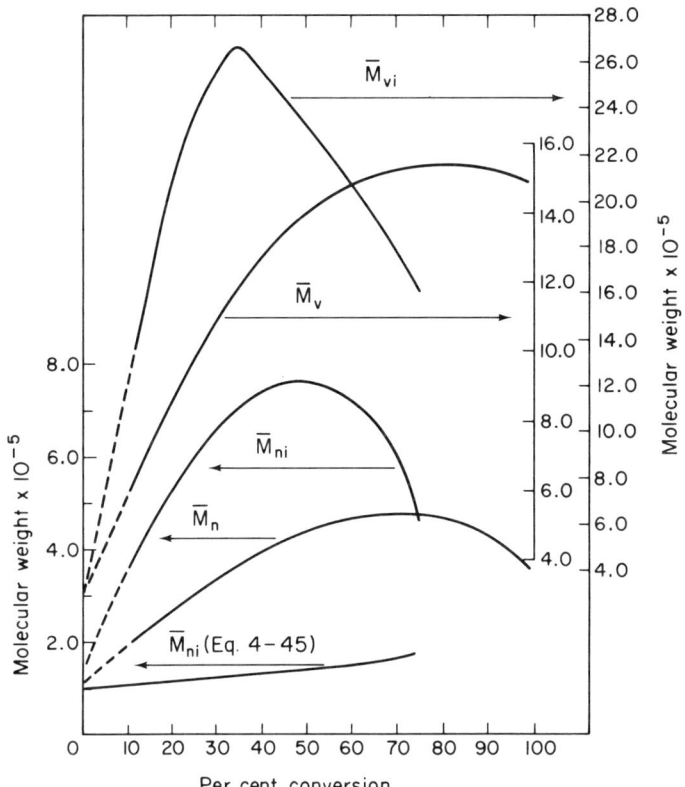

Figure 4–11. Molecular weight development for a 13%/hr styrene emulsion polymerization. [M. R. Grancio and D. J. Williams, *J. Polymer Sci.*, **8**, 2733 (1970).]

to other emulsion polymerization systems remains to be established. These studies point up in a very clear fashion that in polymerization studies proper attention must be paid to the nature of the physical environment. This is especially true in coordination polymerizations. Another case in point is the work undertaken by Bobalek et al. cited in conjunction with Figure 3–7 and gel point predictions in polyfunctional condensation systems.

5
Addition Copolymerization

As already indicated, by far the vast majority of commercially important synthetic polymers are copolymers. Copolymerization may be employed in a variety of ways to achieve a variety of purposes. Both monomers may comprise a significant portion of the reaction mixture, or one of the monomers may be present in small amounts—for example, 2 to 3 percent. In the former case, the properties of the copolymer will vary significantly from either of the homopolymers. In the latter case, the gross physical characteristics of the copolymer would probably be quite like those of a homopolymer of the major monomer component; the minor constituent may be a diene introduced to provide sites for postpolymerization crosslinking, or it may be a carboxyl containing monomer introduced to improve solubility, dieability, or postpolymerization reactivity. More will be said about the use of copolymerization to modify polymer properties in the ensuing chapters. It represents one of our most powerful tools in tailoring polymer systems to meet specified engineering needs, and it greatly extends the range of utility of many monomers that would otherwise be of limited value.

We will be concerned here with those copolymerization processes that involve only two monomers, centering around the addition mechanism. We first look at the formation of random and alternating-type copolymers and then at the formation of block and graft types. Copolymers involving three monomers, and perhaps four or more, are also important, but we cannot consider these more complex reactions here. An important type of terpolymer system is encountered where two components are present in major amounts, but a third is added to make slight modifications in what would otherwise be the copolymer—for example, diene or carboxyl containing monomer. The generation of the various types of copolymers via condensation polymerization is straightforward, and the discussion of these processes was undertaken in Chapter 3.

Even though we will discuss the properties of vinyl copolymers in detail in later chapters (7 and 11), it is appropriate to consider briefly some of their outstanding property differences and characteristics here. First, we should point out that the properties of random and alternating copolymers are normally quite different from those of block and graft copolymers. Generally random and alternating copolymers comprise homogeneous, single-phase systems in the solid state with properties that are intermediate to those of the parent homopolymers. On the other hand, block and graft copolymers form heterogeneous, multiphase systems with properties that are not simply intermediate to the parent homopolymers. In fact, they possess unusual properties which are directly attributable to their multiphase character. This multiphase character arises from the inherent incompatibility of two polymer chain sequences of different composition. A simple and common, but important, example will serve to illustrate the case in point. Styrene and butadiene readily copolymerize to form random, single-phase systems. If styrene monomer and polybutadiene are reacted, styrene chain sequences become grafted on the butadiene chain sequences, phase separation results, and the product assumes a multiphase character. Systems of this sort exhibit glass-transition behavior quite different from single-phase homopolymer or copolymer systems. With appropriate composition and design, multiphase block and graft copolymers systems exhibit far superior impact resistance than do many single-phase systems.

Both systems find important engineering application. Random and alternating, single-phase systems are the oldest and most studied. Block and graft, multiphase systems are experiencing a period of expanding development and application.

RANDOM AND ALTERNATING COPOLYMERIZATION

501 The Copolymer Equation

Contrary to our approach in Chapters 3 and 4, we begin here with a discussion of the kinetics of the process and then use these results as a basis to discuss the chemistry. Two types of active centers are denoted and identified according to the last monomer unit in the chain—that is, the one containing the active center. Thus polymerization of monomers M_1 and M_2 leads to active species A_1 and A_2, as illustrated below for the four possible reactions.

$$\sim\sim\sim\sim M_1^* + M_1 \longrightarrow \sim\sim\sim\sim M_1M_1^* \rightleftharpoons A_1$$

$$\sim\sim\sim\sim M_1^* + M_2 \longrightarrow \sim\sim\sim\sim M_1M_2^* \rightleftharpoons A_2$$

$$\sim\sim\sim\sim M_2^* + M_1 \longrightarrow \sim\sim\sim\sim M_2M_1^* \rightleftharpoons A_1$$

$$\sim\sim\sim\sim M_2^* + M_2 \longrightarrow \sim\sim\sim\sim M_2M_2^* \rightleftharpoons A_2$$

The kinetic problem is first approached by assuming that the rate of addition of monomer depends only on the nature of the end group. Experimental and theoretical analysis shows that the effect of the penultimate unit is usually negligible. The four possible reactions give rise to four propagation constants and four kinetic equations.

Reaction Rate

$$A_1 + M_1 \xrightarrow{k_{11}} A_1 \qquad -\frac{d[M_1]}{dt} = k_{11}[A_1][M_1]$$

$$A_1 + M_2 \xrightarrow{k_{12}} A_2 \qquad -\frac{d[M_2]}{dt} = k_{12}[A_1][M_2]$$

$$A_2 + M_1 \xrightarrow{k_{21}} A_2 \qquad -\frac{d[M_1]}{dt} = k_{21}[A_2][M_1]$$

$$A_2 + M_2 \xrightarrow{k_{22}} A_2 \qquad -\frac{d[M_2]}{dt} = k_{22}[A_2][M_2]$$

(5-1)

where the first subscript on the rate constant designates the active center and the second designates the monomer.

We further assume a steady-state mechanism in which the concentrations of A_1 and A_2 remain constant. This assumption requires that the rate of conversion of A_1 to A_2 must equal that of A_2 to A_1, or in mathematical terms,

$$k_{21}[A_2][M_1] = k_{12}[A_1][M_2] \qquad (5\text{-}2)$$

The rate of disappearance of each monomer is given by

$$-\frac{d[M_1]}{dt} = k_{11}[A_1][M_1] + k_{21}[A_2][M_1] \qquad (5\text{-}3)$$

$$-\frac{d[M_2]}{dt} = k_{22}[A_2][M_2] + k_{12}[A_1][M_2]$$

The combination of Eq. (5–2) and (5–3) yields an expression for the composition of the copolymer being formed at any instant. This is the so-called *copolymer equation*:

$$\frac{d[M_1]}{d[M_2]} = \frac{[M_1]}{[M_2]} \frac{r_1[M_1]+[M_2]}{[M_1]+r_2[M_2]} \tag{5-4}$$

where r_1 and r_2, *the monomer reactivity ratios*, are defined as

$$r_1 = \frac{k_{11}}{k_{12}} \quad \text{and} \quad r_2 = \frac{k_{22}}{k_{21}}$$

The copolymer equation was originally derived for radical copolymerization, but as one can see, it also suitably describes any steady-state addition copolymerization.

502 The Monomer Reactivity Ratios

These are the important parameters that evolve from the foregoing analysis. They compare the propensity for a given species to add its own monomer to that to add the other monomer. Thus $r_1 > 1$ means that species A_1 prefers to add M_1, whereas $r_1 < 1$ means that it prefers to add M_2. For example, in the radical copolymerization of styrene (M_1) and methyl methacrylate (M_2), at 60 °C, $r_1 = 0.52$ and $r_2 = 0.46$; and each radical tends to add the other monomer about twice as fast as its own. Their actual relative rates of addition will, of course, also depend on the relative concentration of the monomers.

Three special cases of copolymerization can be distinguished, depending on the specific values of the reactivity ratios:

1. A copolymer system is said to be *ideal* if both A_1 and A_2 show the same relative preference toward adding M_1 and M_2. This equal preference can be expressed in terms of the rate constants as $k_{11}/k_{12} = k_{21}/k_{22}$ or as $r_1 r_2 = 1$. To amplify, we are comparing the relative preference for each of the following combinations to occur:

$$(A_1 + M_1) \text{ compared to } (A_1 + M_2) \text{ represented by } \frac{k_{11}}{k_{12}}$$

and

$$(A_2 + M_1) \text{ compared to } (A_2 + M_2) \text{ represented by } \frac{k_{21}}{k_{22}}$$

If these relative preferences are the same, $k_{11}/k_{12} = k_{21}/k_{22}$ as stated. As an important corollary, we observe that the nature of the end group has no specific influence on which monomer adds, so that *random copolymers* are generated in ideal copolymerization. The copolymer equation reduces to

$$\frac{d[M_1]}{d[M_2]} = r_1 \frac{[M_1]}{[M_2]} \qquad (5\text{-}5)$$

which shows that once the rate constants are specified, the instantaneous copolymer composition depends only on the relative monomer feed composition.

2. An *alternating copolymer* results if each active species prefers to react exclusively with the other monomer, in which case $r_1 = 0$ and $r_2 = 0$. The copolymer equation then reduces to

$$\frac{d[M_1]}{d[M_2]} = 1 \qquad (5\text{-}6)$$

3. *Block* copolymers would result if r_1 and r_2 were both somewhat greater than unity. In this case, an A_1 species would rather add its own monomer, M_1, and many units of M_1 would successively add to a particular growing chain until a unit M_2 happened to add, changing the species to an A_2. At this point, M_2 units would preferentially add until an M_1 happened to add. The alternation of several such events would thus give rise to a block copolymer. In the limit, as both r_1 and r_2 became very large, the two units would homopolymerize in each other's presence. However, no copolymer systems are known where r_1 and r_2 are both greater than one, and one must resort to specialized techniques to make block copolymers.

The values of these ratios are markedly dependent on the specific mechanism of polymerization, as illustrated in Figure 5–1. This figure shows the results of studies with styrene and methyl methacrylate, using cationic, anionic, and free radical processes. Further amplification of this dependence will be taken up subsequently.

It should be noted that most copolymer systems do not fit the three special cases cited even though many of commercial importance do. Numerous free radical systems exist where r_1 and r_2 are both less than one, such that $0 < r_1 r_2 < 1$, in which case we will speak of a *random-alternating*-type system. For many purposes, copolymer systems of the random, alternating, and random-alternating type have the most useful properties. The tendency of free-radical copolymerization systems to produce these copolymers accounts for the commercial significance of radical polymerizations.

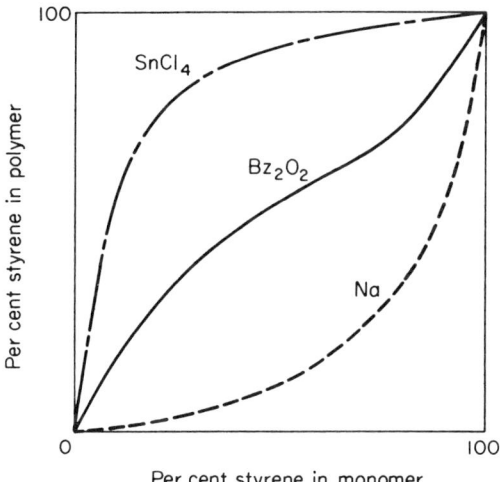

Figure 5-1. Instantaneous composition of copolymer as a function of monomer composition for the system styrene-methylmethacrylate polymerized by cationic (SnCl$_4$), free radical (benzoyl peroxide, Bz$_2$O$_2$) and anionic (Na) mechanisms. [Y. Landler, Compt. rend., *230*, 539 (1950).]

Still other systems occur with $r_1 < 1$ and $r_2 > 1$, but $r_1 r_2 > 1$; such systems are particularly troublesome in that special polymerization techniques must be applied to obtain random-type copolymers. The problems encountered in the last case are left as a student exercise as is a discussion of the methods of solution.

At this juncture, we should recognize that it is not always possible to make copolymers by combining any two monomers. A wide variety can be made, but we can only operate within the limits specified by nature in terms of the reactivity ratios.

503 The Distribution of Sequence Lengths

In this section we again apply elementary probability theory to gain some insight into the detailed structure of polymer chains—in this case, the distribution of sequence lengths in *random copolymers*. Let P_{11} be the probability that A_1 will add M_1 rather than M_2, where the first subscript again designates the active center and the second the monomer. The probabilities P_{12}, P_{21}, and P_{22} are similarly defined. Now, A_1 can either add M_1 or M_2, or it can terminate. Since termination occurs so rarely in the formation of high polymer, we can neglect it for the purposes of this analysis. This means that $P_{11} + P_{12} = 1$, and $P_{22} + P_{21} = 1$. Also the probabilities are simply given by comparing the rates of the possible propagation processes as

$$P_{11} = \frac{k_{11}[A_1][M_1]}{k_{11}[A_1][M_1]+k_{12}[A_1][M_2]} = \frac{r_1[M_1]}{r_1[M_1]+[M_2]} \qquad (5\text{-}7)$$

$$P_{12} = \frac{[M_2]}{r_1[M_1][M_2]}$$

$$P_{21} = \frac{[M_1]}{[M_1]+r_2[M_2]}$$

$$P_{22} = \frac{r_2[M_2]}{[M_1]+r_2[M_2]}$$

The probability, $P(M_1, n_1)$ that a sequence of M_1 monomer units will form of length n_1 is given by

$$P_N(M_1, n_1) = P_{11}^{n-1}(1-P_{11}) \qquad (5\text{-}8)$$

The other pertinent probabilities are similarly defined. Note that Eq. (5-8) is of the same form as the most probable distribution for linear-condensation polymerization. For any particular value of P_{11} (or P_{12}, P_{21}, P_{22}), the tendency is always to produce a very short sequence length. For example, when the molar composition is 50/50, with $r_1 = 1$, 50 percent of the sequences contain only one unit. This tendency to produce short-sequence lengths will, be even more pronounced in systems exhibiting strong alternating tendencies, with both r_1 and $r_2 < 1$ so that $r_1 r_2 < 1$.

To get an idea of the detailed structure of a random copolymer, one needs only to visualize a chain made up by alternately "drawing" chains comprised of M_1 and M_2 units from separate boxes, containing chains with the sequence distributions given by Eq. (5-8).

504 Instantaneous Feed and Polymer Composition

From a consideration of the relative rates of the possible growth reactions one sees that the composition of the feed and polymer will not be equal or constant throughout the polymerization. This situation finds its analog in batch-distillation operations. To amplify, the first copolymer to form from a comonomer mixture will not, except in special cases, be of the same composition as the comonomer mixture. Furthermore, the composition of the copolymer being formed will continue to change as the reaction proceeds. At 100 percent conversion, the overall, average copolymer composition must match that of the original comonomer charge. This drift in copolymer composition with conversion is the predominant state of affairs; hence most copolymers are *heterogenous in composition*. In special cases where composition does not drift, analogous to azeotropic distillation, copolymers that are *homogenous in composition* are generated. We shall see in Chapter 11 that homogenous and heterogenous copolymers, even of the same overall composition and molecular weight, will display different mechanical properties.

To proceed with the analytic development, let F_1 and F_2 be the mole fractions of monomers M_1 and M_2 in the polymer being formed at any instant, and then

$$F_1 = 1 - F_2 = \frac{d[M_1]}{d([M_1]+[M_2])} \qquad (5\text{-}9)$$

Furthermore, let f_1 and f_2 represent the mole fractions of monomer present, and then

$$f_1 = 1 - f_2 = \frac{[M_1]}{([M_1]+[M_2])} \qquad (5\text{-}10)$$

Utilizing Eqs. (5–9) and (5–10), the copolymer equation can be written as

$$F_1 = \frac{(r_1 f_1^2 + f_1 f_2)}{(r_1 f_1^2 + 2 f_1 f_2 + r_2 f_2^2)} \qquad (5\text{-}11)$$

Equation (5–11) has been used to calculate curves of monomer composition for various reactivity ratios. A series of curves are shown in Figure 5–2 for ideal copolymerization. Only a small range of feed compositions gives copolymers containing appreciable amounts of both components unless the monomers have very similar reactivities.

Suppose, for example, that it is desired to synthesize an ideal copolymer 60 percent in M_1 in a system where $r_1 = 0.5$. From Figure 5–2, one sees that a feed composition of about 75 percent M_1 is required. At this composition, M_1 will be consumed 1.5 times faster than M_2, and f_1 will increase as it must to yield a final product with an overall composition of 75 percent. To obtain a constant copolymer composition of 60 percent M_1, it would be necessary to maintain a constant feed composition of 75 percent. In certain copolymerization processes, especially with gaseous monomers, this can be accomplished in a continuous manner by recycling the unreacted monomer and adding fresh M_1 feed.

In practice, when strict control of the composition is not critical and where the reactivity ratios do not differ appreciably from unity, the com-

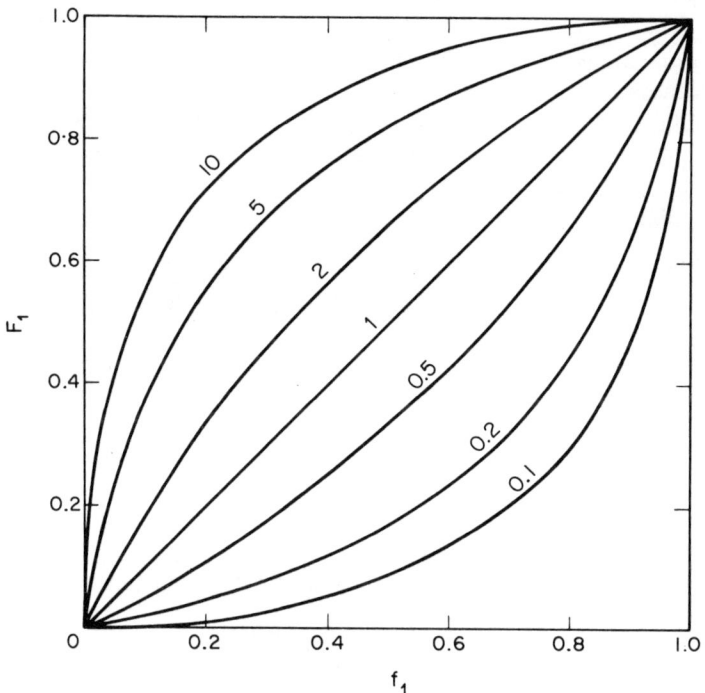

Figure 5-2. Instantaneous composition of copolymer (mole fraction F_1) as a function of monomer composition (mole fraction f_1) for ideal copolymers with the values of $r_1 = 1/r_2$ indicated. [F. W. Billmeyer, op. cit.]

positions are allowed to drift. When r_1, $r_2 = 1$, and $f_1 = F_1$, there is no drift in composition. Other means of composition control include stopping the reaction short of completion or incrementally adding the faster-reacting monomer to a reaction mixture that is initially richer in the slower reacting component.

Curves for several nonideal cases are shown in Figure 5-3. These curves show the effect of increasing tendency toward alternation, since a wider range of monomer compositions yields copolymers containing significant amounts of each monomer. This factor allows for the practical preparation of numerous copolymers, in that considerable drift in feed compositions can be tolerated.

In situations where r_1 and r_2 are both less than or both greater than unity, the curves of Figure 5-3 cross the line $F_1 = f_1$. *Azeotropic copolymerization* is said to occur at these points of intersection. Here reaction occurs without

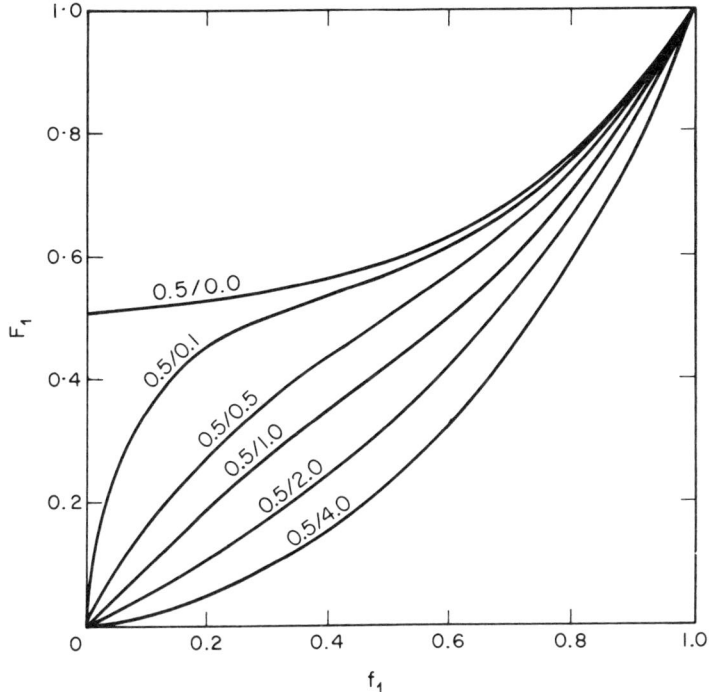

Figure 5-3. Instantaneous composition of copolymer F_1 as a function of monomer composition f_1 for the values of the reactivity ratios r_1/r_2 indicated. [F. W. Billmeyer, op. cit.]

change in either monomer or copolymer composition. Solution of the copolymer equation with $d[M_1]/d[M_2] = [M_1]/[M_2]$ gives the critical composition, $(f_1)_c$, for the azeotrope:

$$\frac{[M_1]}{[M_2]} = \frac{(1-r_2)}{(1-r_1)} \tag{5-12}$$

and

$$(f_1)_c = \frac{(1-r_2)}{(2-r_1-r_2)} \tag{5-13}$$

505 Integration of the Copolymer Equation

Following the composition drift requires integration of the copolymer equation. This problem is rather complex, and the result is not particularly convenient. The most convenient way utilizes a numerical or graphical method for which Eq. (5–11) forms the basis; M_1 is chosen as the monomer in which $F_1 > f_1$ (i.e., the polymer being formed contains more M_1 than the feed); and there are a total of M moles of monomers present. When dM moles have polymerized, the polymer will contain $F_1 d$M moles of M_1, while the feed content of M_1 will be reduced to $(M - d M)(f_1 - df_1)$ moles. The material balance for M_1 gives

$$f_1 M - (M - dM)(f_1 - df_1) = F_1\, dM \tag{5-14}$$

which becomes

$$\frac{dM}{M} = \frac{df_1}{F_1 - f_1}$$

Integration gives

$$\ln\frac{M}{M(0)} = \int_{f_1(0)}^{f_1(t)} \frac{df_1}{F_1 - f_1} \tag{5-15}$$

Next the quantities F_1 and $1/(F_1 - f_1)$ are computed from Eq. (5–11) at suitable intervals for $0 < f_1 < 1$. The fractional conversion $[1 - M/M(0)]$ is computed from Eq. (5–15). The average overall copolymer composition for any conversion can then be determined by graphical integration of a plot of F_1 versus M_1 or as the difference between the composition and amount of residual monomers and those present initially.

The results of these computations can be plotted as F_1 versus the initial feed composition versus percent conversion or as f_1 versus percent conversion versus average overall M_1 copolymer content versus initial feed composition. Such plots are illustrated in Figures 5–4 and 5–5 for the styrene-2-vinyl thiophene ($r_1 = 0.35$, $r_2 = 3.10$) copolymer system, which is nearly ideal. With the aid of these plots, one can follow either F_1 or the overall M_1 copolymer content with conversions for any initial feed composition.

One can also compute for any initial feed composition the average composition formed during successive intervals of conversion and obtain copolymer composition distributions. The results of such computations are illustrated in Figure 5–6 for the styrene-2-vinyl thiophene system. Observe that quite different distributions are obtained for different feed compositions and that the copolymers are quite heterogeneous in composition.

Shown in Figures 5–7, 8, and 9, are analogous plots for the alternating system styrene-diethyl fumarate, with $r_1 = 0.30$, $r_2 = 0.07$ and an azeotrope at $f_1 = 0.57$. Notice that at low conversions all feeds near the azeotropic composition remain quite constant, until, at high conversion, feeds with $f_1 > 0.57$ drift toward pure styrene whereas those with $f_1 < 0.57$ drift toward

Sec. 505 Integration of the Copolymer Equation

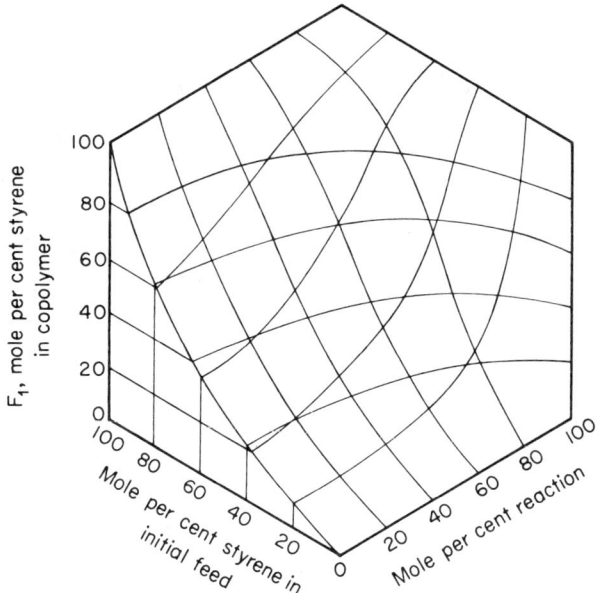

Figure 5–4. Variation in instantaneous composition of copolymer with initial feed and conversion for the system styrene 2-vinylthiophene. [F. R. Mayo and C. Walling, Chem. Revs., 46, 191 (1950).]

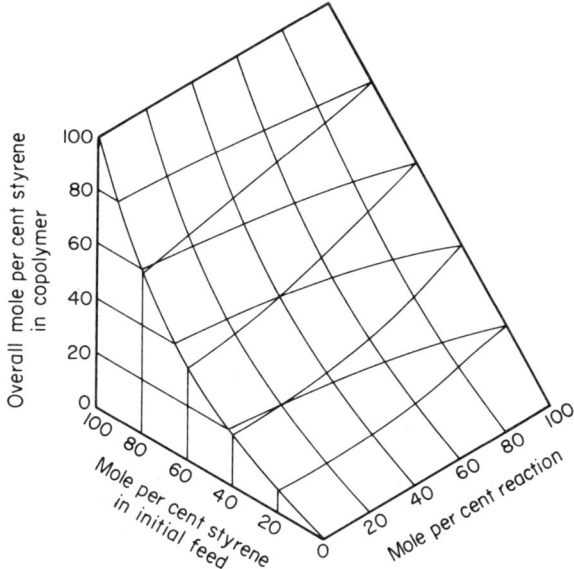

Figure 5–5. Variation of total polymer composition with initial feed and conversion for the system styrene 2-vinylthiophene. [F. R. Mayo and C. Walling, op. cit.]

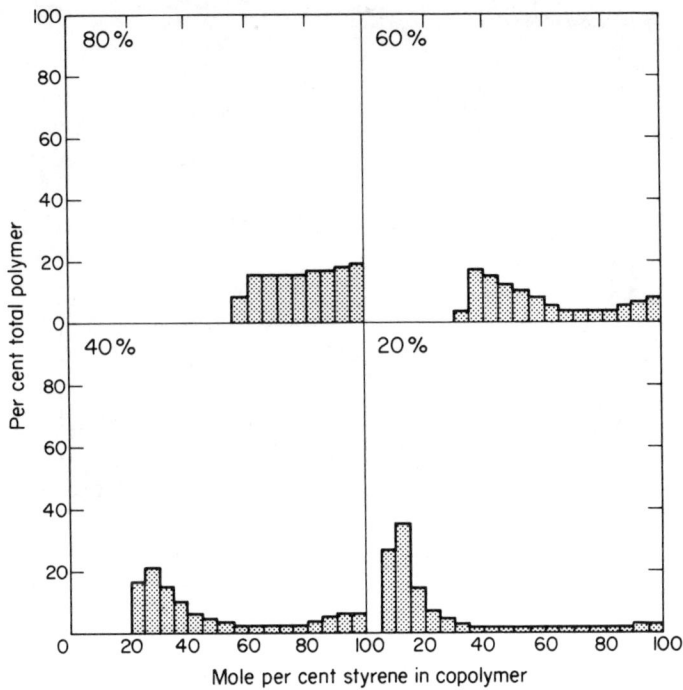

Figure 5–6. Distribution of copolymer composition at 100% conversion for the indicated values of initial mole percent styrene in the feed, for the system styrene-2-vinylthiophene. [F. R. Mayo and C. Walling, op. cit.]

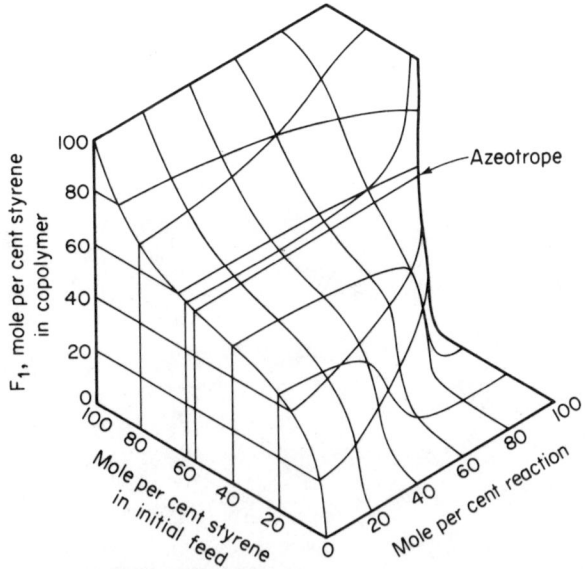

Figure 5–7. Variation in instantaneous composition of copolymer with initial feed and conversion for the system styrene-diethyl fumarate. [F. R. Mayo and C. Walling, op. cit]

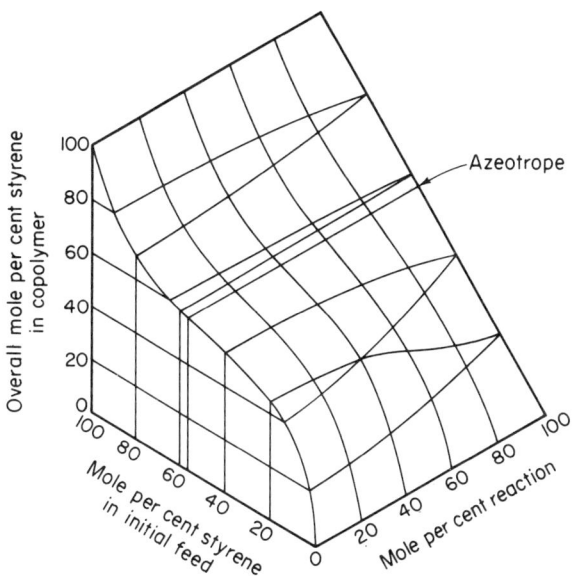

Figure 5–8. Variation of total polymer composition with initial feed and conversion for the system styrene-diethyl fumarate. [F. R. Mayo and C. Walling, op. cit.]

pure diethyl fumarate. In order to maintain a homogeneous composition in these cases, one would merely shortstop the reaction at about 80 percent. The distribution diagrams show that feeds with $f_1 > 0.57$ give only polymer containing more styrene than the azeotrope, and vice versa.

506 Rate of Copolymerization

The elementary approach to representing the overall rate in the copolymerization process is dealt with in a manner similar to that used in treating ordinary addition polymerization. The treatment here is thus subject to the same limitations, plus a few more, because of the additional complications imposed by the nature of the process. That is, we must consider two rates of initiation, three rates of termination, and four rates of propagation. It is first assumed that the initiator efficiency is high for both monomer types, and, therefore, that the two rates can be considered simultaneously as an overall rate, v_i. The steady state condition is applied to the overall active species concentration as well as the separate concentrations, and the overall rate of initiation must equal the overall rate of termination. For bimolecular termination, this condition becomes

$$v_i = 2k_{t11}[A_1]^2 + 2k_{t12}[A_1][A_2] + 2k_{t22}[A_2]^2 \tag{5-16}$$

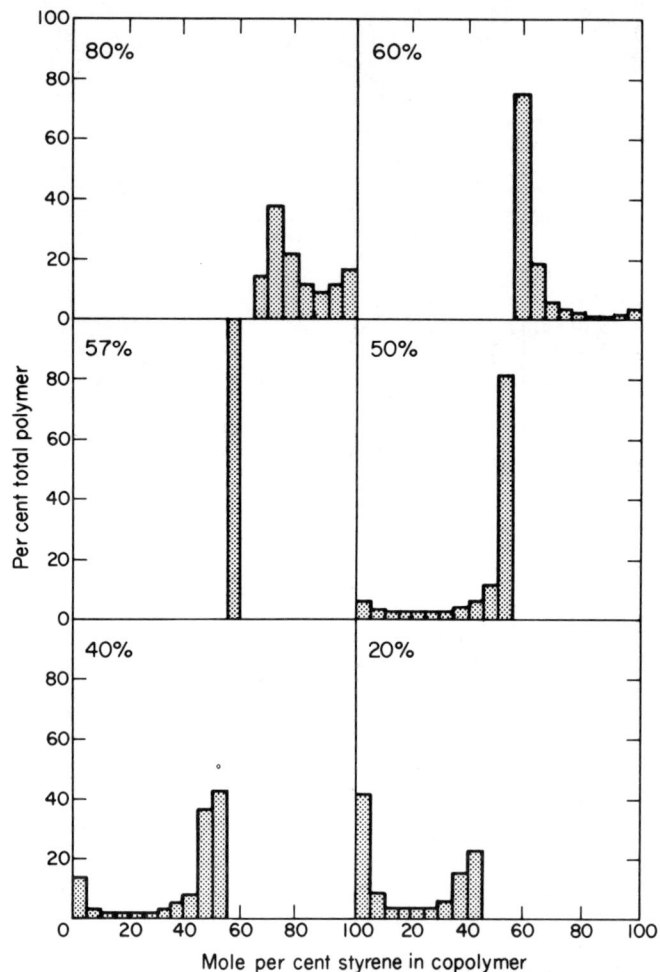

Figure 5-9. Distribution of copolymer composition at 100% conversion for the indicated values of initial mole percent styrene in the feed, for the system styrene-diethyl fumarate. [F. R. Mayo and C. Walling, op. cit.]

where k_{t11} and k_{t22} are the termination constants for like species and k_{t12} is the constant for cross termination.

The overall rate of propagation, v_p, is obtained by combining Eq. (5-2), (5-3) and (5-16) to obtain

$$v_p = \frac{r_1[M_1]^2 + 2[M_1][M_2] + r_2[M_2]^2 (v_i^{\frac{1}{2}}/\delta_1)}{\{r_1[M_1]^2 + 2(\phi r_1 r_2 \delta_2/\delta_1)[M_1][M_2] + (r_2 \delta_2/\delta_1)[M_2]^2\}^{\frac{1}{2}}} \quad (5\text{-}17)$$

where

$$\delta_1 = \left(\frac{2k_{t11}}{k_{11}^2}\right)^{\frac{1}{2}} \quad \delta_2 = \left(\frac{2k_{t22}}{k_{22}^2}\right)^{\frac{1}{2}}$$

$$\phi = \frac{k_{t12}}{2\{k_{t11}k_{t22}\}^{\frac{1}{2}}}$$

Equation (5–17) is difficult to verify, and to use it, it is necessary to determine ϕ as well as r_1 and r_2 experimentally. Figure 5–10 shows the application of Eq. (5–17) to the radical polymerized (AIBN) styrene-methyl methacrylate system. The rate data are, however, only for low conversions. A value of ϕ greater than unity—here equal to 13—is typical and shows that cross termination is favored.

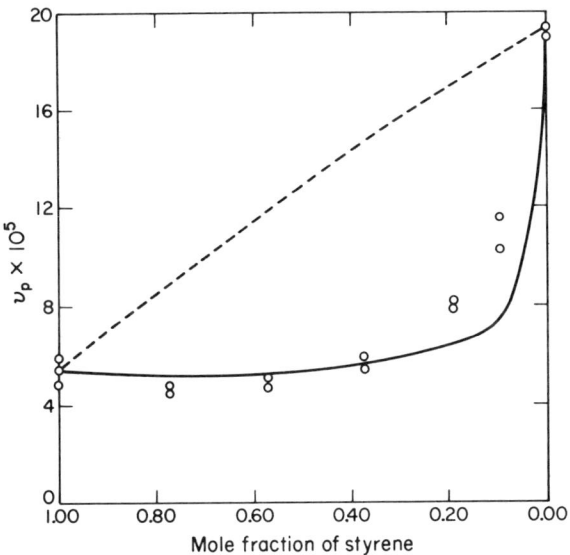

Figure 5–10. Copolymerization rate data for the system styrene-methyl methacrylate at 60°C. The broken line assumes $\phi = 13$. [C. Walling, J. Am. Chem. Soc., *71*, 1930 (1949). Reprinted with permission of the copyright owner. The American Chemical Society.]

CHEMISTRY OF ADDITION COPOLYMERIZATION: RANDOM AND ALTERNATING

The chemistry of radical copolymerization is the best understood. The chemistry of the other types is in the early stages of investigation; knowledge of cationic copolymerization is more advanced than either the coordination or anionic types. Our discussion will be based on the effect of the reactivities of monomer and active species as well as the effect of reaction environment on the reactivity ratios.

507 Radical Copolymerization

Here the reactivity ratios simply depend on monomer and radical reactivity. Because of the neutral-charge character of radical reactions, the reactivity ratios are generally unaffected by such environmental factors as the presence of inhibitors, chain-transfer agents, or solvent type. They also remain unchanged in heterogeneous polymerization processes, provided monomer availability is not affected. The foregoing characteristics allow a careful analysis on the basic effect of the structure and reactivity of monomer and radicals on the reactivity ratios. Although always isolating the influence of any single factor is difficult, considerable insight into their nature is gained by such an analysis. This analysis will be based on the effect of (1) steric hindrance, (2) conjugation, and (3) polarity.

Steric factors

We have already noted that 1,2–disubstituted vinyl monomers exhibit a reluctance to homopolymerize; they do, however, add quite readily to monosubstituted, and perhaps 1,1–disubstituted monomers. For example, styrene (M_1) and maleic anhydride (M_2) copolymerize with $r_1 = 0.01$ and $r_2 = 0$ at 60 °C, and a 50/50 alternating copolymer is obtained over a wide range of monomer feed compositions. This behavior seems to be more a consequence of steric hindrance rather than a result of cancelling of the polar or inductive influences by one substituent over the other, since both symmetrical and unsymmetrical 1,2–disubstituted compounds show the same trend.

We observe that the most common 1,1–disubstituted monomers—for example, isobutylene, methyl methacrylate, and methacrylonitrile—react quite readily in both homo- and copolymerization. Thus 1,1–disubstitution does not introduce nearly as serious a steric effect as the 1,2-type. In α-methyl styrene, steric hindrance begins to manifest an effect because of the increasing bulk, and thus stiffness, imparted to the chain.

Conjugation

The most important factor in governing the reactivity of monomers and radicals seems to be determined by the extent of resonance stabilization that can be achieved. First consider monomer reactivity: a conjugated monomer exhibits a strong tendency to add to a given radical because the resulting adduct radical will be strongly stabilized by resonance. On the other hand, an unconjugated monomer will exhibit a much weaker tendency to add to the same given radical because the resulting adduct will be much less stabilized. For example, styrene monomer is found to be 30 to 50 times more reactive toward radicals than vinyl acetate. Studies of relative monomer reactivities indicate that substituents tend to enhance monomer reactivity according to the following order:

$$-\langle\bigcirc\rangle > -CH=CH_2 > -COCH_3 > -CN > -COOR > -Cl > -OCOCH_3 > -OR$$

$\underbrace{\qquad\qquad\qquad\qquad\qquad}_{\text{Conjugated monomers}} \underbrace{\qquad\qquad\qquad\qquad}_{\text{Unconjugated monomers}}$

The placement of a second substituent in the α-position lends an additive effect, provided steric hindrance does not dominate. The styryl radical has a resonance energy of about 20 kcal/mole, whereas unconjugated systems have resonance energies only on the order of 1 to 4 kcal/mole, for only polar forces contribute.

Consider next radical reactivity. As should be anticipated, the reactivity of radicals toward a given monomer will follow the reverse order of monomer reactivity, for the resonance stabilization that enhanced monomer reactivity will, in turn, suppress radical reactivity. As it turns out, resonance stabilization is more successful in suppressing radical reactivity than in enhancing monomer reactivity.

Consider, for example, styrene (M_1)–vinyl acetate (M_2) copolymerization, which represents an extreme in the resonance stabilization order cited above. We would expect the addition of the vinyl acetate radical to styrene monomer to be the most favored reaction (k_{21} large) and the addition of the styryl radical to vinyl acetate to be the least favored (k_{12} small). For this system at 60 °C, $r_1 = 55$ and $r_2 = 0.01$. Absolute rate constant studies give k_p for styrene at 176 liters mole^{-1} sec^{-1} and k_p for vinyl acetate as 3700 liter mole^{-1} sec^{-1}. In general, data from homopolymerization studies should be used with caution in copolymerization studies; with this point in mind, we let $k_{11} \simeq 176$ and $k_{22} \simeq 3700$. These data immediately show the overwhelming effect of the importance of stabilization in suppressing the radical reactivity of styrene compared to the unstabilized vinyl acetate in that $k_{11} \ll k_{22}$. If these effects were mutually compensating, we might expect k_{11} and k_{22} to be somewhat closer in value although not equal because of steric and polar factors. To proceed with our analysis, since $k_{12} = k_{11}/r_1$ and $k_{21} = k_{22}/r_2$, we see that $k_{12} = 3.2$ and $k_{21} = 370,000$; and $k_{12} \ll k_{21}$ as predicted.

TABLE 5–1

PRODUCT OF REACTIVITY RATIOS WITH MONOMERS ARRANGED IN ORDER OF ALTERNATING TENDENCY[a]

	Vinyl Acetate	Butadiene	Styrene	Vinyl chloride	Methyl methacrylate	Vinylidene chloride	Methyl acrylate	Methacrylonitrile	Acrylonitrile	Diethyl fumarate
Butadiene	~0.5									
Styrene	0.6	1.0								
Vinyl chloride	0.3	0.2	0.35							
Methyl methacrylate	0.1	0.1	0.25	~0						
Vinylidene chloride	~1	0.04	0.15	~0.35	0.6					
Methyl acrylate	0.25	0.01	0.15	0.75	0.8					
Methacrylonitrile	0.25	0.05		0.4						
Acrylonitrile	0.25	0.02	0.02	0.1	0.35			1.1		
Diethyl fumarate	0.004	0.02	0.02	0.06	0.55				~0	

[a] F. W. Billmeyer, *Textbook of Polymer Science*, op. cit.

It is interesting to note that most commercial polymers with molar compositions on the order of 50/50 consist either of two conjugated monomers (e.g., styrene-butadiene and acrylonitrile-butadiene), or of two unconjugated monomers (e.g., vinyl chloride-vinyl acetate). Mixed conjugated-unconjugated combinations are difficult to obtain on the order of 50/50 because of the differences in copolymerization reactivities brought about by resonance stabilization. Mixed combinations can be achieved quite readily, however, to the extent of about 10 percent. For example, acrylonitile is copolymerized with vinyl acetate to the extent of 8 to 12 percent to improve the solubility of poly(vinyl acetate) in conventional hydrocarbon solvents. Also, acrylonitile can be copolymerized with vinyl chloride or vinylidene chloride to the extent of 5 to 10 percent to improve the adhesive properties and raise the glass transitions of the homopolymers.

Polarity

A substituent that lends a positive or negative character to the double bond of the monomer has a similar effect on its corresponding free radical. Since opposite charges attract, there will be a distinct tendency toward alternation in those systems where the commoner pairs have different electronegativities or especially in those systems where one substituent is electropositive and the other is electronegative. This tendency will be partly reflected in the product $r_1 r_2$ as shown in Table 5-1, which arranges sets of monomers in the order of the electronegativity of their substituents. The farther apart any two monomer pairs appear in the table, the lower the $r_1 r_2$ product will tend to be. Exceptions arise in instances where resonance or steric factors exert a dominating, competing influence. For example, vinyl chloride and styrene alternate more readily than vinyl chloride and vinyl acetate, although the former are closer together than the latter in the table.

508 Ionic Copolymerization

In contrast to radical copolymerization, the reactivity ratios in ionic copolymerizations depend strongly on the reaction environment, as embodied in the solvating power and dielectric constant of the solvent. This is in addition to their dependence on resonance, polar, and steric factors. This additional dependence introduces far more difficulty in independently assessing the influence of the various factors on the reactivity ratios. For this reason, the various mechanisms of the ionic copolymerizations are not nearly as well delineated as they are in radical copolymerization.

If the solvent had no effect on the ion-pair character of the active centers, one would expect random copolymers to form with the value of the reactivity ratios being determined primarily by polar considerations. A number of ideal copolymer systems do exist in cationic copolymerization, but relatively few are known in anionic copolymerization. In most cases of ionic copolymerizations, however, the product $r_1 r_2 > 1$ with $r_1 < 1$ and $r_2 > 1$, so that copolymers with a good sequence distribution are difficult to obtain.

The relative order of monomer reactivity toward active species has been well established for carbonium ions. This order correlates with carbonium ion stabilities, and according to the substituents the order is as follows:

$$p\text{---}CH_3OC_6H_4\text{---}> p\text{---}RC_6H_4\text{---}> \begin{Bmatrix} C_6H_5\text{---} \\ CH_3\text{---} \end{Bmatrix} > (CH_3)_2\text{---} >$$
(α-methyl styrene)

$$C_6H_5\text{---}> \underset{\underset{CH_3}{|}}{CH_2=C\text{---}} > p\text{---}XC_6H_5\text{---}> ROOC\text{---}$$
(X ⇌ halogen)

Unexplained inversions of order are often observed in anionic types when comparing data from different solvent systems.

At present, the most important copolymer obtained by ionic polymerization is butyl rubber, composed generally of 97 percent isobutene and 3 percent isoprene. The reaction is carried out at about $-100\ °C$, using aluminium chloride as catalyst in a solution of methyl chloride. The reactivity ratios for the isobutene-isoprene pair are about 2.5 and 0.4, respectively, so that $r_1 r_2 = 1$, and a random copolymer results.

509 Copolymerization of α-Olefins

Mixtures of α-olefins have been copolymerized by Ziegler-type catalyst systems. Most of the available literature is concerned with the copolymerization of ethylene and propylene. The reason is because of the low cost and inherent chemical simplicity of the combination, as well as the anticipated commercial significance of the copolymer products. Two of the most important copolymer products are ethylene-propylene and ethylene-propylene terpolymer which include a minor amount of reactive monomer for cross-linking. Our discussion will center around ethylene-propylene copolymers, but the general principles developed here should be applicable to other α-olefin combinations as well.

When mixtures of ethylene and propylene are brought into contact with ordinary solid, Ziegler-type catalyst systems, the copolymer product is quite heterogeneous with respect to its composition and molecular weight—in spite of attempted control of monomer composition as dictated by predetermined reactivity ratios. This heterogeneity is due to a gel effect in which the active catalyst sites become coated with polymer. The reaction thus becomes monomer diffusion controlled, and differences in monomer diffusion rates control the copolymer composition.

To form a uniform copolymer product, with a narrow molecular-weight distribution, one must use a soluble catalyst system as well as control the monomer composition. One such homogeneous system is the combination of $(C_2H_5)_2AlCl$ and pentavelent vanadium esters, $VO(OR)_3$, in chlorobenzene. Since the catalyst has a short active life, on the order of 5 to 10 minutes at 30 °C, one portion of the catalyst is charged initially, and the remaining portion is charged during the reaction so that the active species is formed slowly and continuously, *in situ*. When the catalyst is formed in this way, the entire reaction mixture remains homogeneous, and the rate is not allowed to be so high that heat buildup is a serious problem. The viscosity increases substantially, but even in the final stages of reaction, the mixture is free from gel or precipitation.

For most Ziegler-type catalyst systems, the product $r_1 r_2$ is close to unity so that ideal copolymerization prevails. For ethylene (M_1) and propylene (M_2) copolymerization at 30 °C, $r_1 = 5$ and $r_2 = 0.2$ when measured by the liquid-phase composition and $r_1 = 26$ and $r_2 = 0.0392$ when measured by the liquid-phase composition. For both, $r_1 r_2 = 1$. Unlike systems where both monomers are liquids, the ethylene-propylene system can be easily regulated to produce constant composition copolymer by continuous regulation of the gas-phase composition.

510 The Q—e Scheme

The *Q-e* scheme is an intuitive and empirical treatment of free-radical copolymerization data that attempts to separate reactivity ratio data, for various monomer pairs, into copolymerization parameters characteristic of the individual monomers. The following equations were proposed to relate the individual rate constants with three parameters characteristic only of the individual monomers.

$$k_{11} = P_1 Q_1 \exp(-e_1 e_1)$$
$$k_{12} = P_1 Q_2 \exp(-e_1 e_2)$$

(5-18)

Here P is considered to be solely a function of the radical reactivity and Q is supposed to be solely a function of the monomer reactivity. Both P and Q are determined by the resonance properties of the species. The constant e, on the other hand, is considered to represent the polar characteristics of both the radical and monomer. The polarity of any species is independent of whether it is in monomer or radical form, for the free radical is a neutral entity. Steric effects are not accounted for in this scheme, and it cannot be expected to apply when steric hindrance manifests a dominating effect. Equation (5–18) combined yields

$$r_1 = \frac{k_{11}}{k_{12}} = \frac{P_1 Q_1 \exp(+e_1 e_1)}{P_1 Q_2 \exp(-e_1 e_2)} = \frac{Q_1}{Q_2} \exp[e_1(e_2 - e_1)] \quad (5\text{-}19)$$

Thus the effects of radical activity cancels, and r_1 can be expressed analytically in terms of parameters independent of the paired interdependence of M_1 and M_2.

Some selected Q-e values are shown in Table 5–2. Numerous tabulations exist in the literature. Negative e values indicate electron-rich monomers, whereas positive e values indicate electron-poor monomers. Unfortunately, the listed Q-e values are not necessarily unique, for they may vary according to the monomer with which they were paired in the experimental parameter evaluation. This lack of consistency may be caused by errors in the measurement of reactivity ratios or by inadequacies inherent in the scheme itself. Table 5–3 shows some monomer-pair systems to illustrate this variation.

BLOCK AND GRAFT COPOLYMERIZATION

Block and graft copolymers are prepared by taking advantage of special chemical or kinetic situations. Our ability to synthesize a wide variety of these materials is even more restricted than it is in the synthesis of random, alternating, or random–alternating-type copolymers. Attempts to generalize concepts are less fruitful here than elsewhere because of the special conditions

TABLE 5–2

SOME REPRESENTATIVE Q AND e VALUES[a]

Monomer	Q	e
Acrylonitrile	0.60	1.20
n-butyl methacrylate	0.50	1.06
Methyl acrylate	0.42	0.60
Methacrylonitrile	1.12	0.81
Vinyl chloride	0.044	0.20
Methyl methacrylate	0.74	0.40
Vinyl acetate	0.026	−0.22
Styrene	1.0	−0.80
1,3-Butadiene	2.39	−1.05
Isoprene	3.33	−1.22

[a] J. Bandrup and E. H. Immergut, *Polymer Handbook* (New York: Wiley, 1966).

TABLE 5-3

Q AND e VALUES FOR A GIVEN MONOMER
AND DIFFERENT COMONOMERS[a]

Monomer (M_1)	Comonomer	$Q(M_1)$	$e(M_1)$
α-methyl styrene	Methyl methacrylate	0.70	−1.2
	Acrylonitrile	0.55	−1.1
	Methacrylontrile	0.50	−0.8
Acrylonitrile	Vinyl chloride	0.37	1.3
	Styrene	0.44	1.2
	Vinyl acetate	0.67	1.0
Vinyl acetate	Vinyl chloride	0.01	−0.5
	Methyl acrylate	0.028	−0.3
	Methyl methacrylate	0.026	—0.4

[a] C. C. Price, *J. Polymer Sci.*, 3, 772 (1948).

required for successful synthesis. Ingenious techniques have been devised to synthesize block and graft copolymers, some of wider applicability and importance than others. After a brief review of some general concepts, we will discuss certain examples of the more widely applied techniques.

511 Generalized Concepts

The routes leading to block and graft copolymers generally involve the reaction of fresh monomer(s) with a previously prepared parent homopolymer or copolymer.* The various sequences may be either homopolymer or copolymer in nature. In the special case of generating block copolymers via the anionic living polymer technique, the parent homopolymer is still active. For the most part, block and graft copolymerizations involve initiating polymerization reactions through active sites bound on the parent polymer molecule. Block copolymerization involves terminal active sites, whereas graft copolymerization involves active sites attached either to the backbone or to pendant side groups. Polymerization is usually conducted in a mixture of the parent polymer, the monomer(s) to be grown on the parent polymer, and fresh initiator.

If polymerization is conducted in a mixture of a monomer, initiator, and polymer, a mixture of products may result: (1) homopolymer of fresh monomer, (2) homopolymer of the parent molecules that did not take part in the copolymerization, (3) crosslinks between the parent polymers in graft copolymerization or branched structures in block copolymerization, and (4) the

---

* For this reason, these reactions are often considered under the topic heading of "reactions of polymers."

desired copolymer. For any system having the potential to produce a high yield of the desired product, reaction conditions must be controlled so as to suppress those leading to the first three types of products and to enhance those leading to the desired copolymer. In some instances, it may be possible to apply separation processes to obtain a pure product, but this is a troublesome route for industrial purposes.

Because reaction so often involves a medium of high viscosity, steady-state reaction kinetics may not apply to the rate of block or graft growth. Such diffusion control can introduce important solvent effects.

512 Block Copolymerization by the Living Polymer Technique

At this point, we simply resume our discussion begun in Section 405 on the living polymer technique. It is only necessary to add here that the ability of active species to add another type of monomer is highly selective; or, in other words, not all active species will add all monomers. For instance, although polystyrene anions will readily add methyl methacrylate monomer, poly(methyl methacrylate) anions cannot add styrene. Thus many copolymers prepared by this route will consist only of three sequences as

$$B\sim//\sim BA\sim//\sim AB\sim//\sim B$$

In the following examples of this sort of block structure, the first monomer designated corresponds to the inner sequence, while the second corresponds to the outer: styrene-alkyl acrylates and alkyl methacrylates; styrene-acrylonitrile and methacrylonitrile; methyl methacrylate-alkyl acrylates and alkyl methacrylates; and methyl methacrylate-acrylonitrile. On the other hand, styrene and isoprene form multiblock copolymers as well as random copolymers. This selectivity is believed to be due to the fact that the anions of the internal sequences must be sufficiently basic to promote addition of another type of monomer. When the anion of the primary sequence is not sufficiently basic, further addition cannot occur.

513 Block Copolymerization by Free Radical Processes

Block copolymerization by this process calls for the regeneration of terminal group radicals. Most such block copolymerizations involve initiation by decomposition of peroxide groups that are introduced into the polymer by special means. These peroxide groups may form an internal part of the chain backbone, or they may be incorporated as stable end groups. When these peroxide-containing polymers are mixed with fresh monomer and the proper conditions are introduced to decompose the peroxides, block copolymerization results. Terminal radicals leading to block copolymerization may also be introduced by mechanically cleaving polymer chains.

Two examples of means for introducing internal peroxide sequences involve copolymerization of small amounts of oxygen with olefinic monomers and initiation of the parent polymer with polymeric phthaloyl peroxide. Oxygen copolymerizes with styrene, methyl methacrylate, and vinyl acetate to produce peroxide linkages as

$$n\text{CH}_2=\underset{X}{\text{CH}}+\text{O}_2 \xrightarrow{A_o} \sim\sim\text{CH}_2-\underset{X}{\text{CH}}-\text{O}-\text{O}-\text{CH}_2-\underset{X}{\text{CH}}\sim\sim$$

Phthaloyl polyperoxide has the structure

$$\text{HOOC}-\text{C}_6\text{H}_4+\text{OOC}-\text{C}_6\text{H}_4-\text{COO}\}_n\,\text{C}_6\text{H}_4-\text{COOH}$$

It decomposes by random cleavage to form shorter diradical species, which become incorporated into the parent polymer backbone. Terminal radicals form hydroperoxide end group units. With fresh monomer and additional heating, the remaining peroxide linkages decompose to form terminal radicals and polymerization resumes.

Isolable polymers with hydroperoxide or peroxide end groups are primarily prepared by converting other types of end groups into peroxide groups in a separate reaction step. Two examples of such processes are (1) initiation with m-diisopropyl benzene monohydroperoxide in the presence of cumyl mercaptan as a transfer agent to form the parent polymer, followed by oxidation conversion of the two types of cumyl end groups to hydroperoxides, and (2) initiation with azobiscyanopentanoic acid to form the

$$\left(\text{HO}-\overset{\overset{\text{O}}{\|}}{\text{C}}-\text{CH}_2-\text{CH}_2-\underset{\underset{\text{CH}_3}{|}}{\overset{\overset{\text{CH}_3}{|}}{\text{C}}}-\right)_2 \text{N}_2$$

parent polymer, followed by conversion of the carbooxylic acid end group to the acid chloride and then reaction of these with t-butyl hydroperoxide to form t-butyl perester end groups. Reactivation of growth leads to the production of homopolymer as well as copolymer. With hydroperoxide end groups, homopolymer formation can be suppressed, but not eliminated, through initiation with redox decomposition

$$\text{\footnotesize\textasciitilde\textasciitilde\textasciitilde}CH_2-CH-ROOH \xrightarrow{\text{heat}} CH_2-CH-RO\cdot + \cdot OH$$

$$\xrightarrow{Fe^{++}} CH_2CH-RO\cdot + HO{:}^- + Fe^{+3}$$

The amount of thermal decomposition can be further suppressed by reducing the temperature since redox decomposition occurs at reasonable rates at lower temperatures.

Terminal free radicals are generated when a polymer chain is mechanically ruptured. This action frequently occurs in such polymer-processing operations as cold milling or mastication. This method of free radical generation can be used to produce block copolymers by milling under nitrogen either two different polymers together or a polymer in the presence of a second monomer. When two polymers are milled together, a complex mixture of homopolymers and copolymers results. Highest efficiencies with negligible grafting can be achieved by milling monomer and polymer. A wide variety of block copolymers can be produced in this way.

The efficiency of bond scission by mechanical working depends on the rheological properties of the polymerizing medium. Highest efficiencies are achieved between the viscous and glassy states. If the medium is too fluid,

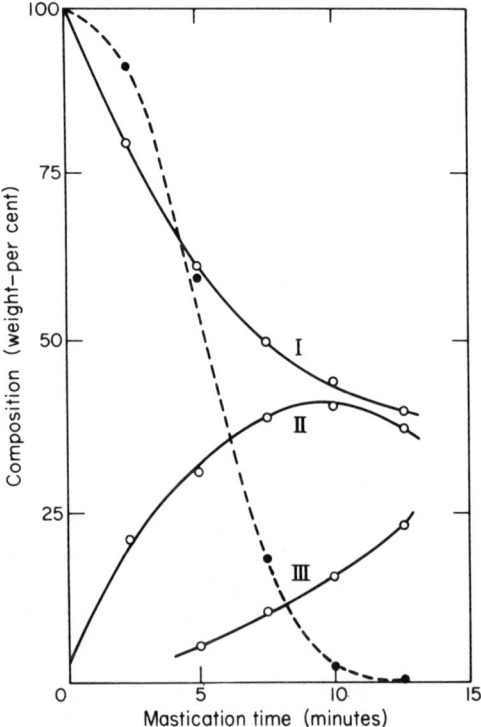

Figure 5–11. Change in composition of the polymer mixture during the mastication of the methyl methacrylate-polystyrene system (24% methyl methacrylate initially in polystyrene): curve I—free polystyrene; curve II—block copolymer, curve III—free polymethyl methacrylate; dashed curve—percentage of unpolymerized monomer. [D. J. Angier, R. J. Ceresa, and W. F. Watson, J. Polymer Sci., **34**, 699 (1959).]

flow will be favored over bond rupture. On the other hand, a glassy substance will not flow in the process equipment. The variables controlling efficiency are directly related to the effect they have on the rheological state of the polymerizing medium. Generally, efficiencies increase with (1) increase in the initial molecular weight of the polymer, (2) decrease in the temperature, (3) decrease in the monomer concentration, and (4) decrease in the viscosity of the reacting medium as the copolymer forms. Efficiency may also depend on the design of the mastication equipment.

The process conditions for a wide variety of monomer-polymer combinations follows. The polymer is first swollen with 80 to 100 percent of its weight with monomer in an inert atmosphere. The resulting swelled mixtures are cold-masticated for periods generally under one hour at 15 to 50 °C. Reaction seems to proceed in two stages. In the first stage, the backbone degrades with the subsequent formation of short-chain blocks on the ruptured chain ends. Negligible homopolymer is formed and efficiencies are high. In the second stage, the molecular weight begins to rise with increasing block length: monomer is continuously depleted; and consequently the viscosity of the reaction medium increases. If the viscosity is high enough, the result may be a rapid increase in block formation due to the associated gel effect. Homopolymer may form at the very end, perhaps because of scissions occurring in the higher-molecular-weight blocks. These effects are illustrated in Figure 5-11 for a methyl methacrylate-polystyrene system.

514 Graft Copolymerization by Free Radical Processes

There are three main techniques for preparing graft copolymers via a free radical mechanism. All, of course, involve the generation of active sites along the chain backbone. These approaches are (1) chain transfer to both saturated and unsaturated backbone or pendant groups, (2) radiative or photochemical activation, and (3) activation of pendant peroxide groups.

The success of the *chain transfer* method depends on the backbone's ability to compete with the primary or growing radicals for monomer. Reasonable efficiencies require rather high chain-transfer coefficients. Estimates of the value required can be obtained from low-molecular-weight analogs of the repeat-unit structures. Examples of appropriate model compounds would be ethyl benzene for polystyrene and ethylene chloride for poly(vinyl chloride). Table 5-4 lists some representative transfer coefficients, from which we can see that polymers containing carbon-halogen or sulfur-hydrogen bonds might be susceptible to transfer and graft polymerization. The carbon-hydrogen bonds in the α-position to a carbonyl group, as in poly(vinyl acetate), and the carbon-hydrogen bond adjacent to the double

TABLE 5-4

SOME REPRESENTATIVE CHAIN-TRANSFER CONSTANTS OF GROWING
FREE RADICALS AT 60°C[a]

Chain-Transfer Agent	Poly-styrene	Poly(methyl Metha-crylate)	Polyacrylo-nitrile	Poly(methyl acrylate)	Poly (vinyl acetate)
Benzene	0.023	0.4	2.46	0.326 (80°C)	1.07
Toluene	0.105	0.17	2.63	2.7	20.75
Carbon tetrachloride	87	0.925	0.85	1.25 (80°C)	7300
Carbon tetrabromide	17800	2700	500	4100	7.39×10^6
n-butyl mercaptan	210000	6600	—	6900	4.8×10^5
Triethylamine	1.4	1900	1900	400	370

[a] J. Bandrup and E. H. Immergut, *Polymer Handbook*, op. cit.

bond in unsaturated polymers, such as polyisoprene, are also suitable for transfer growth.

Especially reactive groups may be introduced through prior treatment of a parent polymer or copolymer that has been prepared for this purpose. For example, a copolymer of methyl methacrylate and glycidyl methacrylate can be reacted with H_2S or thioglycolic acid to incorporate —SH groups along the backbone:

$$\begin{array}{c}
CH_3 CH_3 \\
\sim\sim CH_2-\underset{|}{C}-CH_2-\underset{|}{C}\sim\sim \\
COOH_3 CO \\
 | \\
 O \\
 | \\
 CH_2 \\
 | \\
 CH \\
 \diagdown \\
 O \\
 CH_2 \diagup
\end{array}
\quad \xrightarrow{HS\text{-}CH_2COOH} \quad
\begin{array}{c}
CH_3 CH_3 \\
\sim\sim CH_2-\underset{|}{C}-CH_2-\underset{|}{C}\sim\sim \\
COOCH_3 CO \\
 | \\
 O \\
 | \\
 CH_2 \\
HS-CH_2COO-\underset{|}{CH} \\
HS-CH_2COO-CH_2
\end{array}$$

$\downarrow H_2S$

$$\begin{array}{c}
CH_3 CH_3 \\
\sim\sim CH_2-\underset{|}{C}-CH_2-\underset{|}{C}\sim\sim \\
COOCH_3 CO \\
 | \\
 O \\
 | \\
 CH_2 \\
 | \\
 CHOH \\
 | \\
 CH_2-SH
\end{array}$$

Monomers like styrene, acrylates, and methacrylates have been grafted onto these materials with high efficiencies.

The unsaturated rubber polymers are especially important grafting vehicles when the graft chains are glassy polymers in that the resulting materials have superior impact properties over their normal linear amorphous polymer counterparts. We use styrene as an example. A 5 percent solution of rubber (e.g., butadiene) in styrene monomer is first prepared. Activation with a suitable catalyst or initiator leads to activation of the backbone either by addition to a double bond or by removal of an atom of hydrogen from an α-methylenic group. Polymerization proceeds in homogeneous solution up to about 5 to 10 percent conversion of styrene. Then phase separation occurs, as the grafted chains agglomerate in the styrene-rubber solution matrix. At a later stage of reaction, phase inversion occurs, and the rubber chains, to which polystyrene chains have become grafted, now become the discrete phase in a polystyrene-styrene monomer matrix. This condition persists to complete conversion. Phase separation and phase inversion result because of the inherent incompatibility of polymer chains that are even slightly different in composition. The product, characterized by a continuous glasslike matrix in which there are dispersed microparticles of rubber, exhibits remarkably improved impact resistance over ordinary polystyrene. This material is known as "impact resistant" polystyrene.

Grafting efficiency also depends on the initiator. For example, benzoyl peroxide generally serves as a very efficient initiator for grafting, whereas AIBN generally produces no graft copolymer. This *initiator effect* may be attributed to the superior resonance stability of the $(CH_3)_2$—C(CN)· radical compared to C_6H_5· or C_6H_5COO· radicals. Although the former is sufficiently reactive to polymerize monomers, it does not react with polymers.

Grafting efficiencies can also be increased by increasing temperature. This increased efficiency is in accord with our analysis in Chapter 4 for situations where competing kinetic processes are operating. The process with the highest activation energy, and generally lower growth constant, can become relatively more important as temperature is raised. Efficiencies also increase with increasing initiator concentration but decrease in the presence of low-molecular-weight compounds with high-transfer coefficients.

Both ultraviolet light and high-energy irradiation can be used to initiate graft copolymerization. When monomer and polymer are present together initially, the process in known as *mutual irradiation*. In a *preirradiation* process, the polymer is first exposed to the radiation to produce *long-lived* or *trapped radicals*. On exposure to monomer, grafting occurs. Although high-grafting efficiencies can be achieved in this latter process, it is difficult to generate and maintain an adequate concentration of radicals in solid polymer, using tolerable dose rates. Only mutual irradiation processes will be considered further here.

Photolytic activation of polymer growth relies on the ability of certain functional groups to be efficient absorbers of radiation. The absorption of sufficient energy causes bond cleavage and the formation of free radicals. Most ultraviolet initiations for grafting involve the photolysis of polymers having either pendent carbonyl groups or pendent halogen atoms, for these are easliy activated by ultraviolet radiation. Examples include vinyl acetate, acrylonitrile, or methyl methacrylate grafted on poly(methyl vinyl ketone) and styrene or methyl methacrylate grafted on brominated polystyrene. Homopolymer is also produced in these processes, since bond cleavage of pendent group results in an unattached low-molecular-weight radical as well as one that is bound to the polymer chain.

When high-energy radiation is used, the most important factor affecting grafting efficiency is the radiation sensitivity of the graft monomer relative to the parent polymer. Since high-energy radiation is less selective than ultraviolet radiation in breaking bonds, efficient grafting can be obtained only if the monomer is less sensitive than the polymer. In this way, active sites will be primarily generated on the polymer backbone. Comparison of the so-called *G values*, which is the number of radicals formed per 100 ev absorbed, for the monomer and polymer provides an indication of the efficiencies to be expected. According to the data shown in Table 5-5, the process of grafting styrene onto poly(vinyl chloride) should be relatively efficient. Such is, indeed, the case, and pure polystyrene is virtually absent in such graftings.

In general, the total radiation dose determines the number of grafted chains, while the dose rate determines their length. That is, the dose rate controls the rate of initiation, which, in turn, determines the kinetic chain length and thus the molecular weight. The longer the radiation is applied for any specified dose rate, the more sites are initiated. Care must be taken to avoid crosslinking or degradation.

Chemical grafting involves grafting through preformed labile groups on the polymer chain. These groups may be introduced in the initial polymerization or through subsequent chemical reaction of the polymer. This is a particularly versatile approach for both radical and ionic grafting reactions.

Peroxide or hydroperoxide groups for radical grafting may be introduced by (1) irradiation of the polymer in the presence of oxygen, (2) ozonation of such polymers as polyethylene, polystyrene, and poly(vinyl chloride), (3) autoxidation of such polymers as poly-*p*-isopropylstyrene and poly (styrene-co-4-vinyl-cyclohexane), (4) reaction of polymers containing pendent acid chloride groups with hydroperoxides or peracids, and (5) copolymerization with peroxide-containing monomers such as *t*-butyl peracrylate.* Reactions (1), (2), (3), and (4) (with hydroperoxide) lead to the generation of hydroperoxide groups. As indicated in Section 512, homo-

---

* For specific references, see R. W. Lenz, *Organic Chemistry of Synthetic High Polymers* (New York: John Wiley, 1967), pp. 716–719.

polymer formation is suppressed by using redox decomposition with hydroperoxide initiator groups.

TABLE 5-5

RELATIVE SENSITIVITIES OF MONOMERS AND POLYMERS TO IONIZING RADIATION[a]

Material	G-value[b]
Monomers	
Styrene	0.7
Acrylonitrile	5–5.6
Methyl methacrylate	5.5–11.5
Vinyl acetate	9.6–12
Vinyl chloride	~10
Polymers	
Polystyrene	1.5–3
Polyisoprene	2–4
Polyethylene	6–8
Poly(methyl methacrylate)	6–12
Poly(vinyl acetate)	6–12
(Poly(vinyl chloride)	10–15

[a] A. S. Hoffman and R. Bacskai, in G. E. Ham (Ed.), *Copolymerization* (New York: Interscience, 1964), p. 377.
[b] Number of radicals formed per 100 ev absorbed.

515 Graft Copolymer Formation by Ionic Mechanisms

Virtually all procedures for inducing graft growth by the anionic mechanism involve initiator groups bound to the polymer backbone; consequently, high-grafting efficiencies are achieved. Grafting by cationic growth is much less efficient because of extensive chain transfer, and much less work has been devoted to grafting by this mechanism. One example of an ionic grafting reaction involves first the preparation of poly(*p*-lithiostyrene) from poly(*p*-iodostyrene) by reaction with *n*-butyllithium followed by reaction with acrylonitrile. Other anionic grafting reactions with various other pendant groups have also been carried out.*

In summary, some important reactions and procedures for preparing graft and block copolymers have been presented. Although several general principles are available for our guidance, numerous processes rely on specific chemical and kinetic situations for their successful execution. The number of practical preparation routes is likely to continue expanding. Very few detailed kinetic investigations have been conducted to date, but the need for such studies is acute.

*For specific references, see R. W. Lenz, *Organic Chemistry of Synthetic High Polymers* New York: John Wiley, 1967), pp. 716–719.

PART III

Physics of the Solid State

The ultimate goal of all science and engineering is to derive the fundamental relationships that describe the physical, chemical, or mechanical behavior of the system under consideration in terms of parameters whose values can be determined independently, with minimum recourse to experimentation on the actual system of interest. This goal pertains to any system ranging from a collection of molecules, to a distillation column or an entire plant operation. On a molecular level this endeavor falls into the realm of statistical mechanics. Although some success has been realized with this approach, more often than not, we must rely on direct observation and measurement to develop empirical correlations to describe the systems of interest. The state-of-the-art in polymer science and engineering offers no exception to this rule; nevertheless, some important breakthroughs have been made. Even though statistical mechanics cannot yield a final answer to many problems, it has succeeded in providing considerable insight into molecular mechanisms in many situations. Such partial results aid in the compilation and judicious application of empirical rules of behavior.

Part III probes into some of the molecular aspects associated with describing polymeric behavior in the solid state. First, Chapters 6 and 8 are devoted to describing the spatial placement of molecules within a particular physical state, i.e., the manner in which the chains are packed relative to one another. In this regard we consider two extremes: Chapter 6 is devoted to highly ordered systems, and Chapter 8 to completely random systems. The material in Chapter 6 on order and morphology in crystalline systems represents one of the major developments of modern polymer science. In Chapter 9, which uses Chapter 8 as a basis, statistical mechanics is applied with considerable success to a practical random system—elastomers—to gain both fundamental and practical insight into their behavior. The theoretical treatment of rubber elasticity represents one of the salient successes of statistical mechanics.

The approach in Chapter 7 on transitional phenomena is more phenomenological than is the rest of Part III. In this chapter we attempt to gain insight into the modes of molecular motion and the available states of molecular aggregation as a function of temperature. This chapter is important from a practical point of view, and it provides prerequisite reading for Part IV on polymer rheology.

6

Morphology and Order in Crystalline Polymers

In the 1920s it was discovered that some polymers yielded x-ray diffraction patterns. In contrast to metals and inorganic salts, which exhibit well-defined patterns, these polymers produced only a few broad Bragg diffraction peaks superimposed on a diffuse, liquidlike scattering pattern. The interpretation of these patterns, which held for many years, was that small, relatively perfect crystallites existed in an otherwise amorphous matrix. *The fringed micelle model* that evolved is pictured in Figure 6-1. It was thought that the molecules passed successively through a number of these crystalline and intervening amorphous regions. The crystallites were pictured as sheaves of chains aligned in a parallel fashion; the x-ray diffraction patterns showed their dimensions to be on the order of several hundred angstroms. The crystallties were thought to serve as mechanical crosslinks and to affect the physical properties in much the same way as chemical crosslinks in vulcanized rubber. The development of the fringed micelle model would seem to be a natural result of the concept that polymers crystallize from a melt in which the chains are intimately entangled.

This model has now been all but abandoned in favor of a more ordered and more complex model. The modern concept of polymer morphology has evolved since 1957, and its development represents one of the major advances in polymer science of that decade. One of the major factors in the evolution of this concept was developments in the field of electron microscopy and

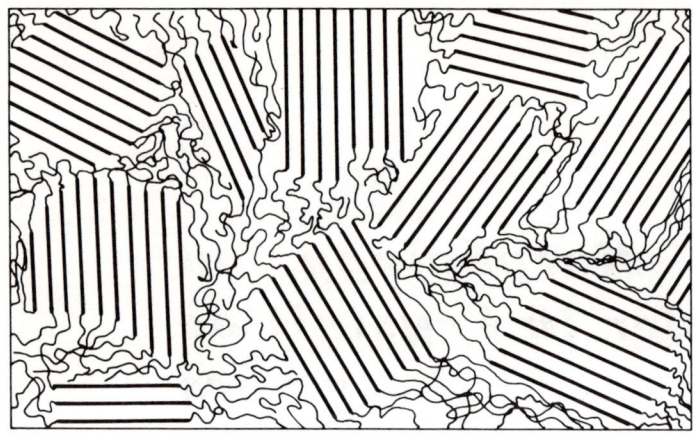

Figure 6–1. The fringed micelle model.

electron diffraction in the 1950s. That is, the interesting features of polymer crystals are of such a size that magnifications on the order of several tens of thousand times with resolutions on the order of 10 Å are required for their observation and measurement.

Our discussion of the morphology and order prevailing in polymeric-crystalline materials will be developed in several distinct stages: (1) Discussion will open with a description of their crystallographic features. Here we will be concerned with how neighboring chain atoms and chain sections fit into the most fundamental element of crystal structure—the unit cell. (2) Attention will then focus on the general morphological features of polymer single crystals. We will show how a polymer's crystallographic and long-range conformational habits mesh to establish its overall single crystal morphology. (3) Interest will then turn to bulk crystallized material. Such material is actually compositelike, being comprised of stacks of crystallites with intervening noncrystalline regions. We will see that many of the features of polymer single crystals are preserved in the crystallites of bulk crystallized material. (4) Final attention will focus on the morphological changes arising through orientation by drawing.

601 Polymeric Crystal Structures

In the development of concepts of polymer crystal structure, there has been a natural tendency to apply crystallographic concepts and nomenclature already familiar to the metallurgist and inorganic chemist. However, the long-chain, often helical, nature of polymers and the presence of crystal structures not ordinarily found in other substances make it difficult to apply, in all instances, concepts such as the unit cell, symmetry, and space groups.

The crystallographic structure of polymers can be described in terms of three factors: (1) configuration, (2) local conformation, and (3) molecular packing.

The *configuration* of a polymer molecule is defined in terms of its chemical repeat unit and a statement concerning its molecular architecture. The requirements for crystallizability are embodied in the configurational properties of the polymer system. Typical crystalline polymers are those with a high degree of structural regularity. Occasional irregularities, such as those introduced in copolymerization or branching, limit the extent of crystallinity but do not necessarily prevent its occurrence. Typical noncrystalline polymers exhibit little structural regularity. Other configurational properties affecting polymeric crystallizability are the size and structure of backbone and side-chain substituents, plus to a lesser extent molecular weight.

The *local conformation* of a molecule refers to geometrical arrangements of neighboring groups in the molecule, which can be altered only by rotation about primary valence bonds. The planar zigzag or trans-conformation is the minimum energy form for an isolated section of a hydrocarbon chain. The energy of the trans form is about 800 cal/mole less than that of the gauche form. The trans form is therefore favored in polymer crystal structures unless steric hindrance causes the chain to assume other minimum energy conformations—for example, the helical conformation. Typical polymers exhibiting planar zigzag conformations in the crystalline state are polyethylene, poly(vinyl alcohol), most polyamides, cellulose, and the syndiotactic forms of poly(vinyl chloride) and 1,2-polybutadiene. Most isotactic polymers and 1,1-disubstituted vinyl polymers exhibit the helical conformation in the crystalline state. Some typical helical structures are shown in Figure 2–6.

The *molecular packing* refers to the arrangement of the molecules in the crystal. Packing is described most completely in terms of the unit cell and its contents. To illustrate the concepts to be developed, we will frequently refer to polyethylene. Of all the organic polymers, the crystal structure, morphology, and their interrelationships are the simplest and the best understood for polyethylene.

The unit cell of polyethylene is shown in Figure 6–2. It has the shape of a parallelepiped with axes a, b, and c and angles α, β, and γ. The unit cell constants for polyethylene at 30 °C are given below. The spacing of a and b are temperature dependent.

$a = 7.41$ Å
$b = 4.94$ Å $\qquad \alpha = \beta = \gamma = 90°$
$c = 2.55$ Å (chain axis)

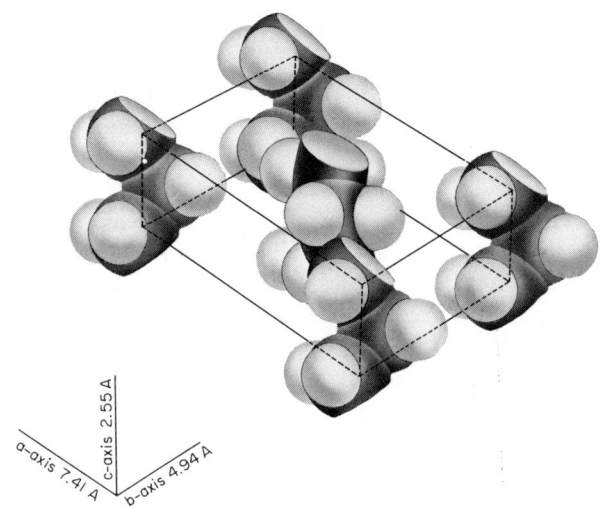

Figure 6-2. Arrangement of chains in the unit cell of polyethylene. [Courtesy of E. S. Clark, E. I. duPont deNemours.]

By convention, the c axis corresponds to the direction the polymer chains assume in passing through the cell. The length of the c axis is determined by the *crystallographic repeat unit*. This is the simplest arrangement of atoms capable of generating the structure of the crystallized molecule by translation and duplication. The crystallographic repeat unit will contain one or more chemical repeat units. To illustrate: in polyethylene, the crystallographic repeat unit consists of one —CH_2—CH_2— chemical repeat unit. In polypropylene, the crystallographic repeat unit consists of three chemical repeat units, for the helical conformation requires three units to make one turn and thus duplicate its structure. The *crystallographic repeat distance* is the axial length of the crystallographic repeat unit and corresponds to the length of the c axis.

The arrangement of the crystallographic repeat units in the unit cell is much more difficult to determine than the unit cell dimensions. For polyethylene, these are best observed as illustrated in Figure 6-3, which shows a top view of the unit cell. The planes of chain zigzag and the plane containing the consecutively folded chains are shown as they pass through the unit cell. How the polyethylene unit cell fits into a description of polymer morphology will be discussed subsequently.

A complete description of the unit cell is not always possible for two reasons. (1) A poorly developed lattice will not yield sufficient data to permit its determination. (2) Certain features of a cell may not exist due to special

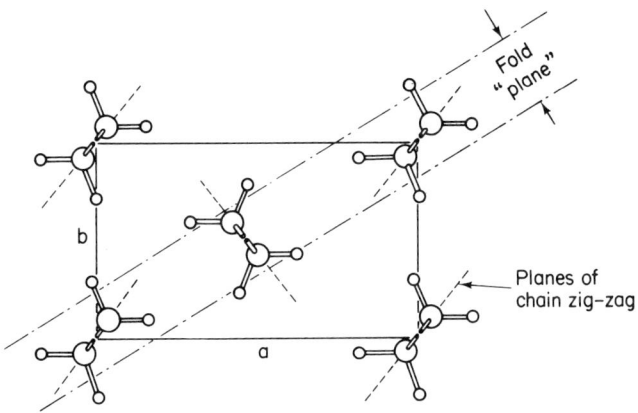

Figure 6-3. Packing in the crystal structure of polyethylene as viewed along the c-axis. [After G. Natta and P. Corradini, Rubber Chem. Tech., **33**, 703 (1960).]

types of disorder, such as the random linear translation of the molecules along their axes. The ideal crystal density can be calculated from the unit cell dimensions. This density may be used to determine percent crystallinity, x, on a density basis as

$$x = \left(\frac{\text{actual density—amorphous density}}{\text{ideal crystal density-amorphous density}}\right)(100 \text{ percent})$$

602 Some Representative Crystal Structures

In this section we will take note of some of the more interesting, complex crystallographic structures that may occur in polymer systems. Beyond this section, the relationship between crystallographic structure and crystal morphology will be discussed primarily in terms of polyethylene. In more complex situations, this relationship is not so clear, and a detailed description would be beyond the intended level of our discussion. The material given here is presented only to acquaint the reader with some of the intricate structures that may occur.

Poly(hexamethylene adipamide)

The configuration of this material is well known. Its crystallographic repeat unit corresponds to its chemical repeat unit which is linear and stereoregular. Its conformation is essentially planar zigzag. Poly(hexamethylene adipamide) or Nylon 66 is polymorphous, existing in three crystallographic

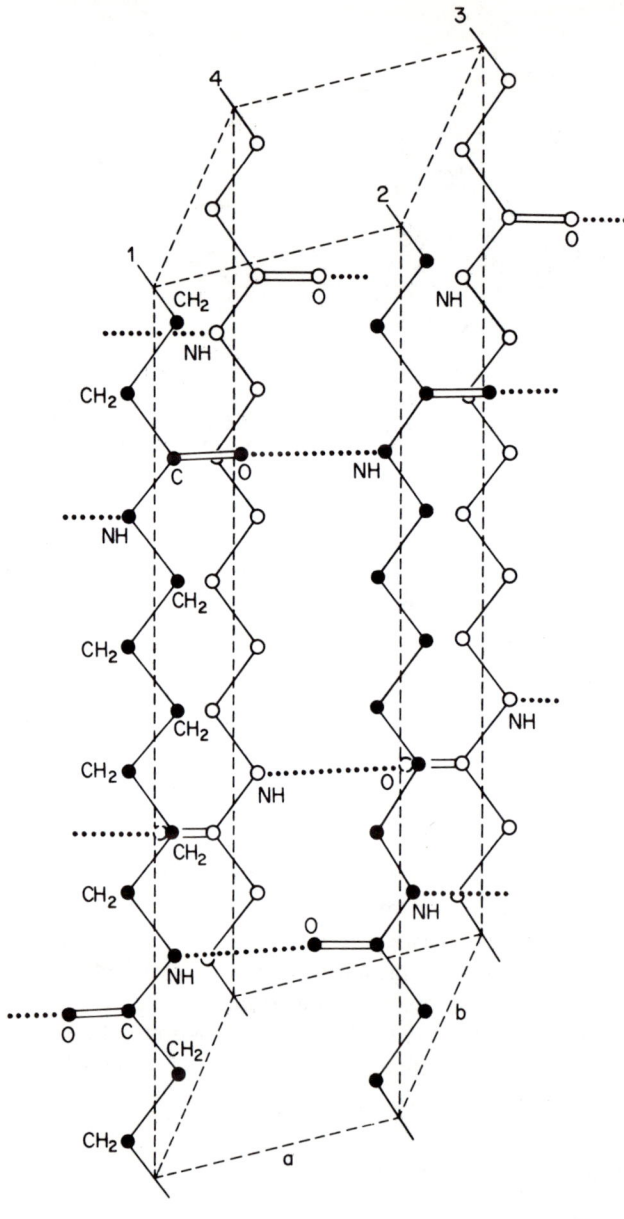

Figure 6-4. Packing of Nylon 66 molecules in the triclinic unit cell: α-form. [C. W. Bunn and E. V. Games, Proc. Roy. Soc. (London) **189A**, 39 (1947).]

forms, α, β, and γ, as well as a mixture of these. The lattice spacings vary with temperature: the α and β forms are triclinic and stable at room temperature, but the γ form—a high-temperature form—is not. The detailed structure of the γ forms is still a matter of conjecture. The cell constants for the α and β form are given as

$$\alpha \text{ form}$$
$$a = 4.9 \text{ Å} \quad \alpha = 48.5°$$
$$b = 5.4 \text{ Å} \quad \beta = 77°$$
$$c = 17.2 \text{ Å} \quad \gamma = 63.5°$$

$$\beta \text{ form}$$
$$a = 4.9 \text{ Å} \quad \alpha = 90°$$
$$b = 8.0 \text{ Å} \quad \beta = 77°$$
$$c = 17.2 \text{ Å} \quad \gamma = 67°$$

The packing of the α form is shown in Figure 6–4. The most interesting feature of this crystal structure, and the one that dominates the properties of this material, is the hydrogen bonding that occurs between —NH and —C=O groups on neighboring chains. The result is the formation of sheets of molecular segments where the intersegmental bonds are stronger within the sheets than between neighboring sheets. Infrared evidence indicates that more than 99 percent of the possible sites are so bonded. The crystallographic forms of other polyamides as well as some polyurethanes are similar to those of Nylon 66. The crystalline content of these materials ranges from 20 to 55 percent.

Polybutene

Isotactic polybutene can exist in either a threefold or fourfold helical conformation, and three crystallographic forms are known to exist. The threefold helix, with a rhombohedral unit cell, is the most stable. Its unit cell is shown in Figure 6–5 as viewed along the c axis. Note particularly how the right-handed and left-handed chains (as indicated by the arrows on the central axis of each chain) alternate in interlocking rows. The fractional numbers refer to the elevation of the carbon atoms above the basal plane.

Poly(ethylene terephthalate)

Its conformation is nearly a planar zigzag with the benzene ring in the plane of the zigzag. As shown in Figure 6–6, there is some distortion along the axis of the chain as the

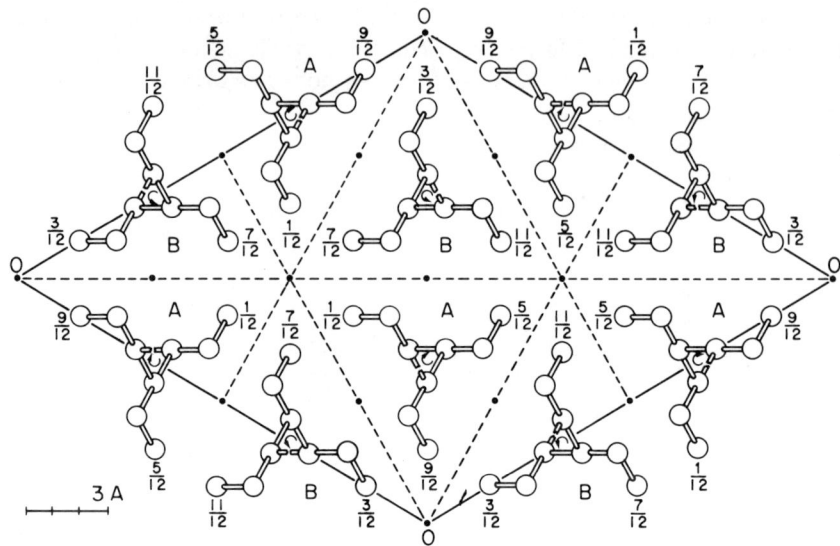

Figure 6–5. Projection of polybutene molecules of the unit cell along the chain-axis. [G. Natta, P. Corradini, and J. W. Bassi, Nuovo cimento, Suppl. to Vol. 15, **1**, 52 (1960).]

group makes a slight angle with the axis by rotation about the C—O bond in order to allow for close packing.

The unit cell is shown in Figure 6–7, and its constants are given below.

$$a = 4.56 \text{ Å} \quad \alpha = 98.5°$$
$$b = 5.94 \text{ Å} \quad \beta = 118°$$
$$c = 10.75 \text{ Å} \quad \gamma = 112°$$

Polypropylene

Isotactic polypropylene can crystallize under normal conditions in two forms, both with a conformation of three units in one turn. Although a hexagonal form has been observed, the monoclinic form is the more usual. The monoclinic unit cell constants are given as

$$a = 6.65 \pm 0.05 \text{ Å}$$
$$b = 20.96 \pm 0.15 \text{ Å} \quad \beta = 99°20' \pm 1°$$
$$c = 6.50 \pm 0.04 \text{ Å}$$

It is shown in Figure 6–8 as viewed along the c axis. Again the arrows show the direction of rotation of the helix, and the numbers give the height of the carbon atoms above the basal plane.

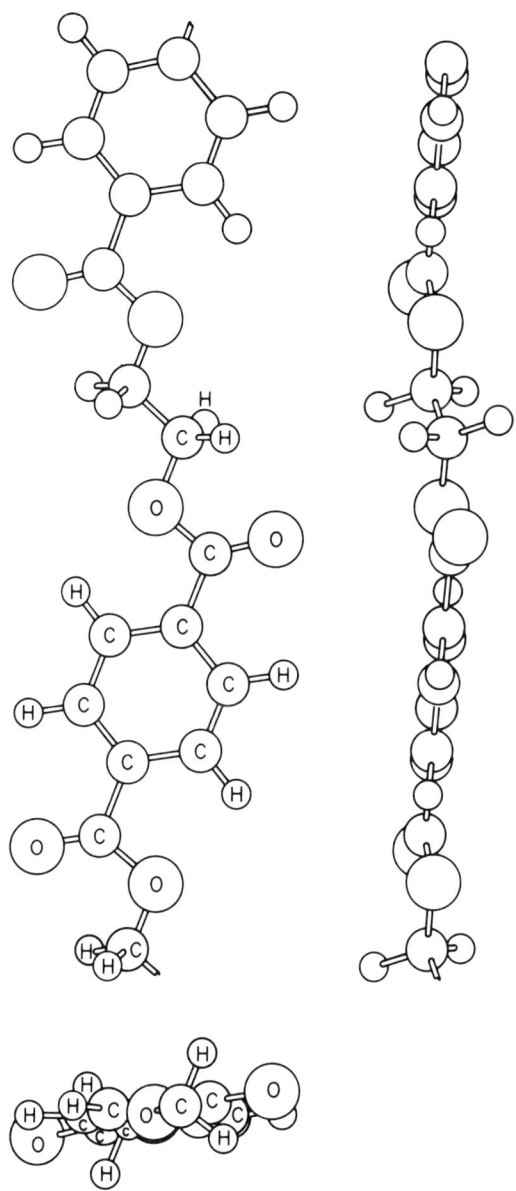

Figure 6-6. Molecular conformation of polyethylene terephthalate. R. deP. Daubeney, C. W. Bunn, and J. C. Brown, Proc. Roy. Soc. (London), **226A**, 531 (1954).]

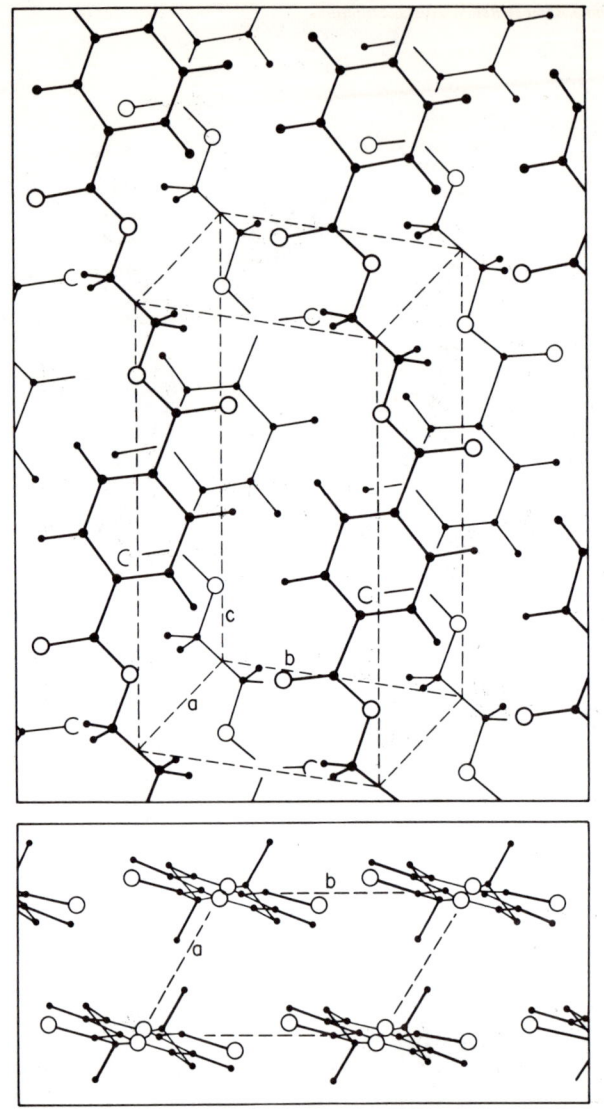

Figure 6–7. Molecular packing of polyethylene terephthalate. [R. deP. Daubeney, C. W. Bunn, and J. C. Brown, Proc. Roy. Soc. (London), **226A**, 531 (1954).]

603 Polymer Single Crystals Grown from Solution

The growth of polymer single crystals from dilute solution (0.1 percent or less) was first reported in 1953 and later in several laboratories in 1957. It was long believed that such crystals could not be produced because of complexities

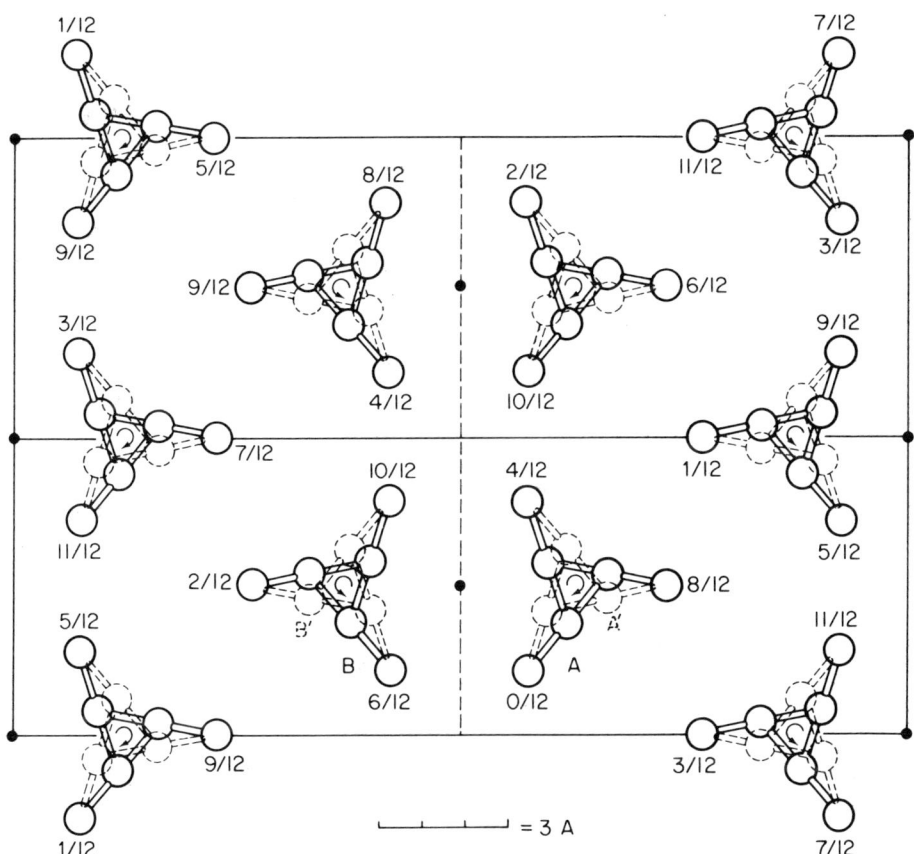

Figure 6-8. Projection of the monoclinic unit cell of polypropylene along the chain-axis. [G. Natta and P. Corradini, Nuovo cimento, Suppl. to Vol. 15, **1**, 40 (1960).]

introduced by the coiled nature of the chain in solution. This phenomenon has been reported for so many polymers that its occurrence is considered to be universal. These single crystals are grown by cooling a dilute solution from a temperature above the crystalline melting point to one below. For example, polyethylene single crystals can be grown isothermally from xylene at temperatures ranging from 70 to 90 °C (T_m = 135°C). A wide variety of morphologies result, the exact nature of which seem to depend on polymer type, solvent, concentration, temperature, and growth rate.

All polymer single crystals seem to have the same general appearance and structure. In their simplest form, they appear in the electron microscope as thin, flat platelets on the order of 100 to 200 Å thick and several microns in lateral dimensions. A diamond-shaped single crystal of polyethylene, as shown in Figure 6–9, will be used to illustrate the prominent features of polymer single crystals.

Electron-diffraction studies of these crystals indicate that the polymer chains are oriented perpendicular to the plane of the lamellae. Since the crystals are only 100 Å or so thick and since polymer chains are generally on the order of 1000 to 10,000 Å long, the chains must be folded back and forth on themselves. In this context, we shall speak of *chain folding*. Each molecule folds up and down in a regular fashion to establish a *fold plane*. As illustrated in Figure 6–10, a single fold plane may contain many polymer chains. In polyethylene, five carbon atoms are required for the fold—that is,

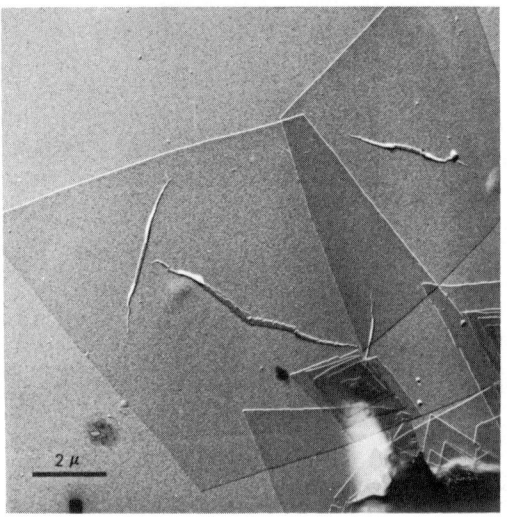

Figure 6–9. Single crystals or linear polyethylene. [D. H. Reneker and P. H. Geil, J. Appl. Phys., **31**, 1916 (1960).]

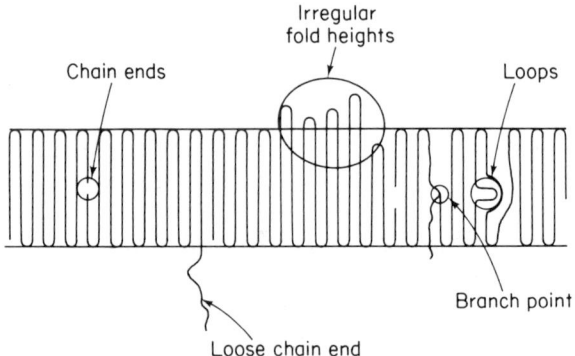

Figure 6–10. Model of fold plane illustrating chain folding with imperfections which may occur in the structure.

for the chain to reverse its direction. Also illustrated are some of the imperfections that may occur in polymer crystal structure—namely, chain ends, branches, loops, or irregular fold heights. The height of the fold plane is known as the *fold period*. As we shall see, it corresponds to the thickness of the crystal. Generally the fold period is quite uniform, as is chain folding.

We must now consider how the preceding conformational features fit into the overall morphological features of the polymer single crystal. Shown in Figure 6–11 is a schematic top view of an idealized model of a diamond-shaped polyethylene single crystal, showing its skeletal structure as viewed along the c axis. Just one of its many fold planes is outlined by the broken-

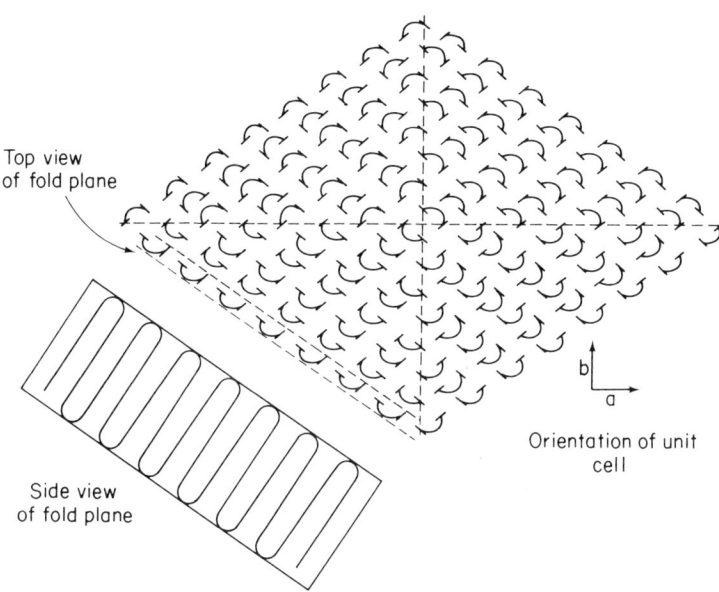

Figure 6–11. Fold packing in a polyethylene single crystal. [D. H. Reneker and P. H. Geil, J. Appl. Phys., **31**, 1916 (1960).]

line, rectangular envelope. A side view of the crystal, showing this fold plane, is shown to the bottom left. In the main diagram, the curved lines represent the fold regions, and the solid lines at each end of the curved lines represent the plane of the zigzag. Note that as a chain folds within its fold plane, it also twists, as represented by the alternating parallel orientation of the zigzag planes. The crystal is also divided into four quadrants by broken lines. Each quadrant contains fold planes that are parallel to the outside edge of that quadrant. Thus the entire crystal is comprised of four triangular quadrants, which contain parallel rows of fold planes. Also shown is the orientation of the polyethylene unit cell within the crystal. Compare this representation with Figure 6–3, where we can now see that fold planes actually pass diagonally through the unit cell. As we have indicated, many polymers possess more complicated crystal structures than polyethylene. Since the morphology is controlled by the detailed structure (and not vice versa), more complicated morphologies than that illustrated must be considered. Such morphologies will be the subject of the next sections.

604 Hollow Pyramids

Close inspection of the electron-diffraction patterns of many single crystals indicates that the chains are not oriented exactly at 90 deg to the crystal surface, but that they are slightly tilted within the lamellae comprising the crystal. Shown in Figures 6–9 and 6–12, a truncated diamond-shaped crystal, are polyethylene single crystals with central ridges, apparently formed by

Figure 6–12. Truncated single crystal of polyethylene crystallized and filtered from xylene solution at 90 °C. [D. C. Bassett and A. Keller, Phil. Mag., 6, 345 (1961).]

pleats of extra material. Other crystals, shown in Figure 6–13, have four sets of parallel corrugations, one set in each fold domain. On the basis of these observations, as well as others, it is evident that polyethylene, and perhaps other polymers, may exist in solution as hollow pyramids. Four types of pyramidal shapes are recognized, either with or without corrugations. The type that forms depends on the conditions of crystallization. For example, if the crystal is grown at relatively high temperatures and held there for a sufficient period of time, very regular structures develop; each quadrant is planar; and the lamellar surfaces in which the folds lie are crystallographic planes of well-defined slope. The development of this form requires not only a regular fold and fold period but a uniform displacement of adjacent parallel fold planes as well. At lower growth temperatures, the direction of displacement of the adjacent fold planes may reverse periodically,

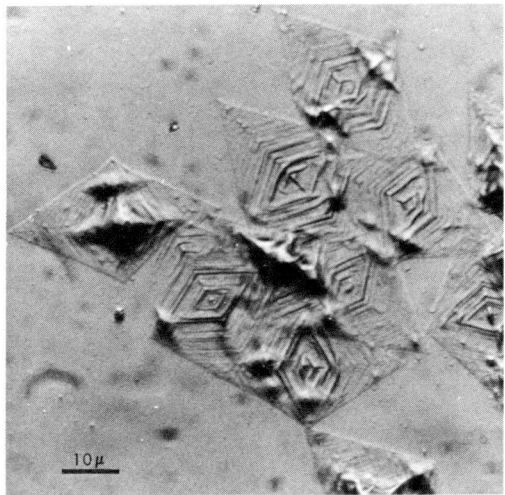

Figure 6-13. Optical micrograph of corrugated crystals of linear polyethylene precipitated from tetrachloroethylene. [D. H. Reneker and P. H. Geil, J. Appl. Phys., **31**, 1916 (1960).]

resulting in the formation of the corrugations. Displacement of one or two —CH_2— units between neighboring fold planes is thought to account for the formation of the various pyramidal forms. The plane containing the five carbon atoms of the fold may not be parallel to the molecular axis but they may be at an angle. This suggests a reason for the displacement of parallel fold planes. Figure 6-14 shows a sketch indicating how fold planes with this asymmetric folding might be packed. Uniform asymmetric folding would

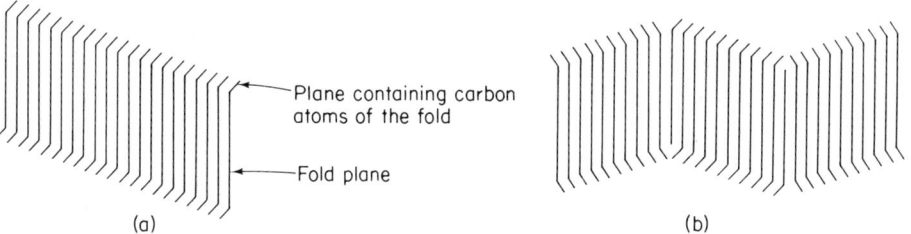

Figure 6-14. End-view of fold planes showing (a) regular and (b) irregular stacking required by assymetric fold which lead to formation of (a) hollow pyramids or (b) corrugated structures.

account for the formation of planar hollow pyramids, whereas a periodic alternation in direction of the asymmetric fold would lead to the formation of corrugated structures.

605 More Complex Morphologies

In addition to the diamond, truncated diamond, and hollow pyramid forms already discussed, more complex morphological forms may result by crystallization from solution. We will not elaborate on this aspect of polymer morphology here, beyond illustrating some of the varieties of structures that have been observed. The morphology that actually results from the crystalization of a particular polymer will depend on a number of factors interacting

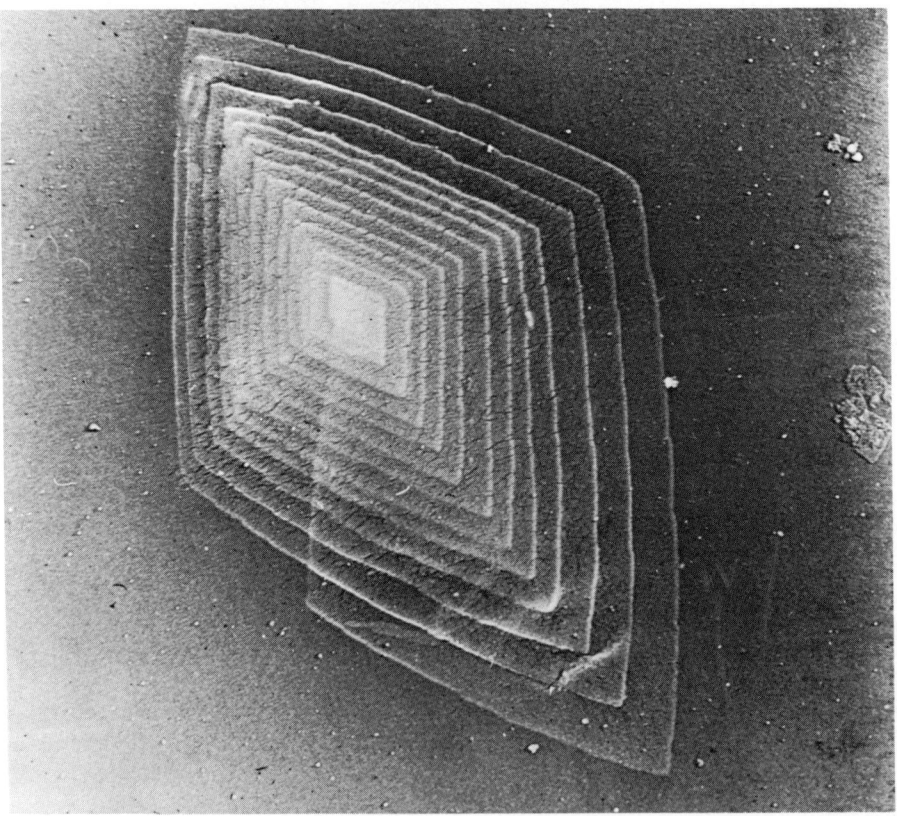

Figure 6-15. Electron micrograph of multilayered single crystal of linear polyethylene grown from dilute solution (0.1% from xylene). [B. G. Ranby, F. F. Moorehead, and N. M. Walter, J. Polymer Sci., **44**, 349 (1968).]

in a very complex, and not yet understood, fashion. Among these factors are solvent, temperature, molecular weight, and solution concentration.

Figures 6-15, 6-16, and 6-17 show spiral-growth morphologies for polyethylene and polyoxymethylene.

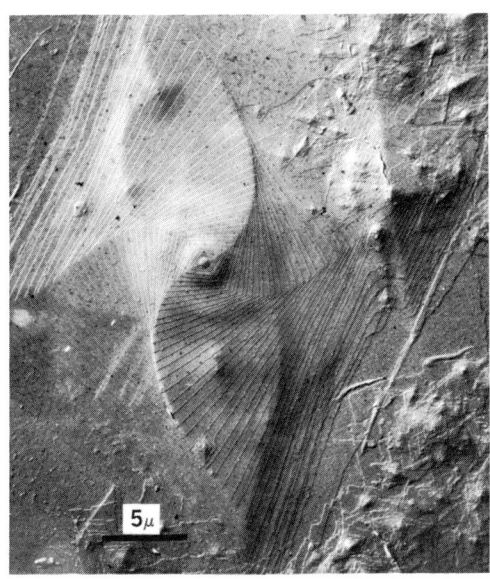

Figure 6-16. Crystal of polyethylene resulting from a spiral growth in which successive lamellae have developed with a regular rotation. [A. Keller, Polymer, 3, 393 (1962).]

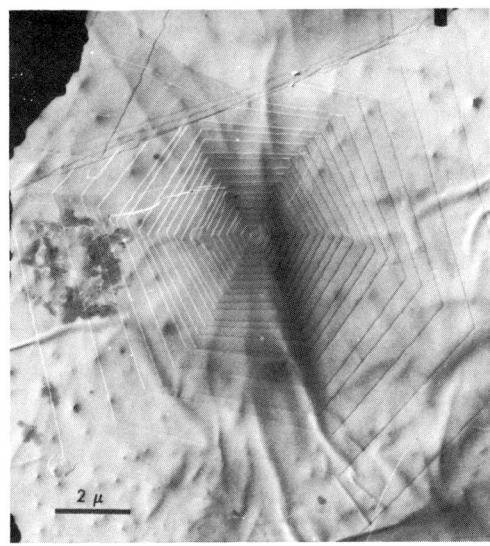

Figure 6-17. Spiral growth on a polyoxymethylene crystal precipitated from a dimethyl phthalate solution. The rotation of the lamellae is generally in the same direction with some irregularities. [D. H. Reneker and P. H. Gail, J. Appl. Phys., 31, 1916 (1960).]

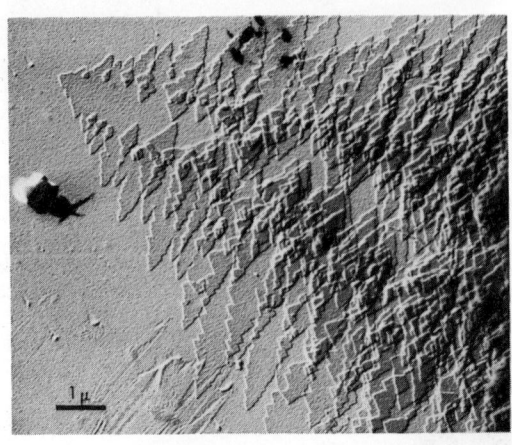

Figure 6–18. Electron micrograph of a portion of a dendrite crystal. [P. H. Geil and D. H. Reneker, J. Polymer Sci., **51**, 569.]

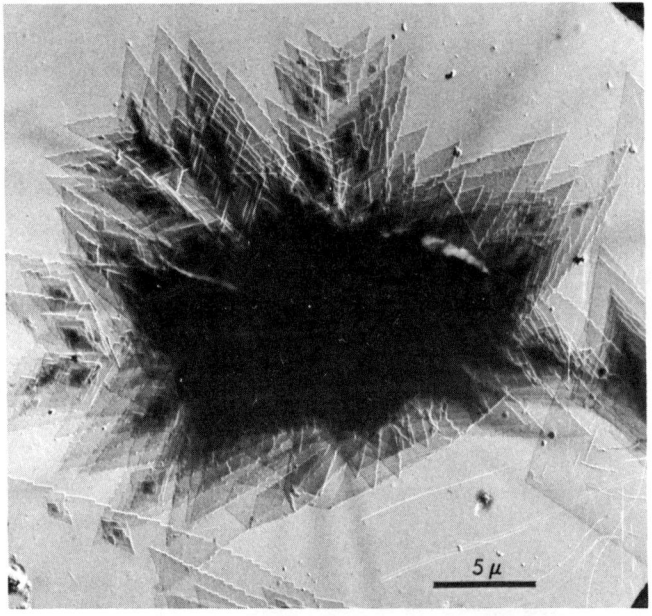

Figure 6–19. Complex growth twin crystal of polyethylene, [P. H. Geil, Polymer Single Crystals, Interscience Publishers, 1963, p. 147.]

Polyethylene crystals grown at high degrees of supercooling from a good solvent, are, in general, dendritic—that is, they are of a branched form. Shown in Figure 6–18 is an electron micrograph of such a crystal, grown from 0.1 percent xylene solution.

Figure 6–19 shows a complex "growth twin" crystal of polyethylene.

Hedrites are intermediate in character to the single crystals we have been discussing and the spherulites crystallized from the melt we will be discussing in the next sections. They may be grown from concentrated polymer solutions or from polymer melts by extremely slow crystallization. Hedrites are structures having a polygonal appearance when viewed from at least one direction. They are composed of numerous lamellae, to form a composite structure that is usually thicker than one micron. The lamellae are usually joined together along a common plane that is perpendicular to the lamellar surface. The structure is quite similar to two books glued together at their bindings, each page representing a lamelae and the joined bindings representing the common plane. Figure 6–20 shows several views of a hedrite in suspension.

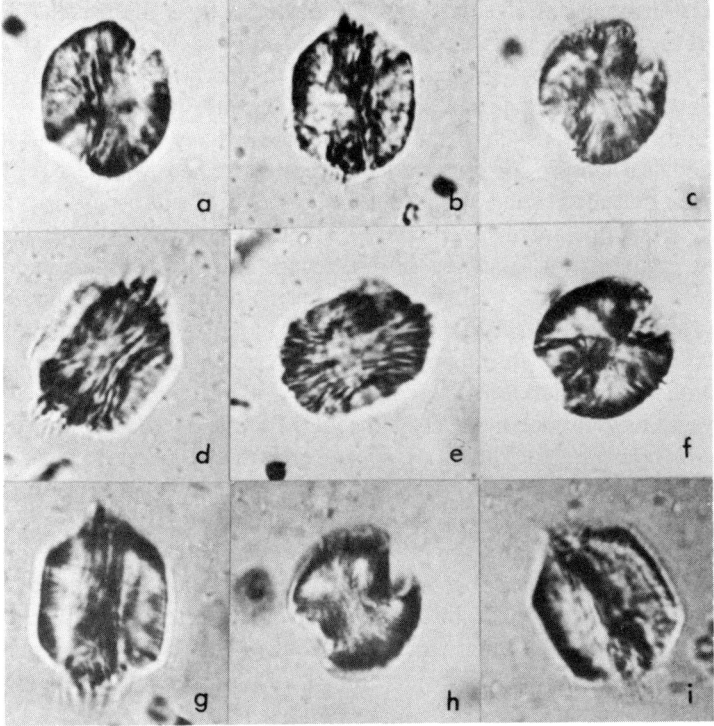

Figure 6–20. Several view of a hedrite in suspension. [D. C. Bassett, A. Keller, S. Mitsuhashi, J. Polymer Sci., 1A, 763 (1963).]

To date, polymer single crystals have no value as engineering materials. Aside from their inherent interest, the major use of these crystals is as model systems in experiments aimed at a basic elucidation of the properties of crystalline polymers and in studies of the deformation and orientation that occur when a polymer is processed. It now remains to extend our knowledge of single crystals to a discussion of polymer morphology obtained during crystallization from the melt. Most crystalline polymers of engineering value are obtained by crystallizing from the melt.

606 The Morphology of Polymers Crystallized from the Melt

Polymers crystallized from the melt seem to maintain the two most prominent structural features of single polymer crystals: aggregates of 100 Å-thick lamellae of different degrees of perfection are observed, and the chains are oriented perpendicular to the face of the lamellae so that chain folding must also be inherent in melt-crystallized materials. The degree of crystallinity of polymeric materials may range from 0 percent for noncrystallizable polymers through intermediate crystallinities, such as 20 percent for unoriented poly(vinyl chloride), 50 percent for branched polyethylene, 70 percent for isotactic polypropylene, and up to nearly 100 percent for polytetrafluoroethylene and linear polyethylene. Because it is difficult to detect crystallinity as low as 20 percent and because it is not clear as to the morphological features of such material, discussion here will pertain only to those materials displaying well-developed x-ray diffraction patterns. In this context, crystalline polymers are more correctly referred to as *semicrystalline*, but the prefix "semi" may be dropped with the same understanding we employ in dropping the prefix "high" from the term "high polymers." The detailed aspects of bulk crystal morphology will vary somewhat with percent crystallinity, but the ensuing discussion will refer to materials in which crystallinity is well developed.

Crystallinity in polymers crystallized from the melt develops through *spherulitic growth*. Nucleation of crystal growth occurs at various heterogeneous nuclei, and crystal growth proceeds in a radial fashion from each nucleus until the growth fronts from the neighboring structures impinge. These spherulitic structures are referred to as *spherulites*. They have different sizes and degrees of perfection, and they completely fill the volume of all well-crystallized material. Spherulites have somewhat the same role as the grain structures in polycrystalline metals. Control of their properties and morphology, is therefore, of primary importance in the engineering design of polymers.

Further discussion will center around two views of the structural features of spherulites—a gross, supramolecular view versus a molecular view. In the latter case, we will be concerned with the details of elemental crystal structure, lamellar packing, and chain folding, while in the former case we will ignore these details. Also, our discussion will be based on the structural features of crystalline films—essentially material crystallized from a two-dimensional melt.

In the *gross, supramolecular view*, polymer films and crystal growth are observed in a light microscope through a cross-polarizer lens system at magnifications of only several hundred times. Figure 6–21 shows a polymer film observed in such a way. The characteristic *Maltese-cross pattern* results

Figure 6–21. Spherulites of a low density polyethylene crystallized from a molten thin film as observed between crossed polarizers, showing the typical Maltese-cross patterns. [P. H. Geil, *Polymer Single Crystal*. Interscience Publishers, 1963, p. 224.]

from the birefringent nature of the polymer film. Crystal growth can be observed if a temperature-controlled mounting stage is used. The sample is first heated above its melting point and then supercooled by 10 or 15 C°. The development of spherulitic growth is shown in Figure 6–22. Rates of crystallization can be measured from such micrographs by comparing the area occupied by the spherulites to the total area.

The *detailed structural features* of spherulites are revealed by a combination of electron microscopy and electron diffraction. These techniques show that the spherulites are comprised of ribbonlike lamellae which grow radially from a central nucleus. The lamellae are parallel at the nucleus, but as they begin to grow outward, they diverge, twist, and branch to form an overall structure that is the radially symmetric spherulite. The branched, ribbonlike lamellae of which the spherulite is comprised are on the order of 100 Å thick, and chain folding is evidenced. Figure 6–23 is an electron micrograph of a section of a spherulite showing the twisted and branched growth of the ribbonlike lamellae.

It is well recognized that the degree of crystal imperfection is much higher in bulk-crystallized material than in material crystallized from dilute solution.

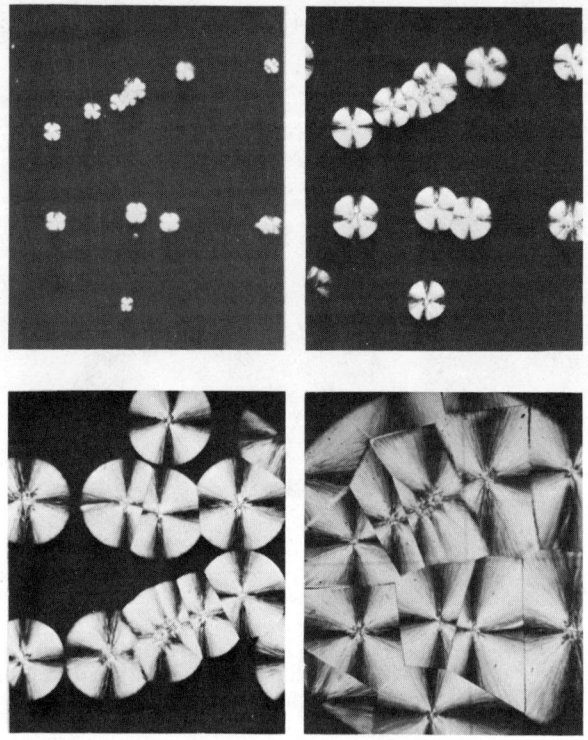

Figure 6–22. Sequence of growth of spherulites in polypropylene. [B. Maxwell, "Modifying Polymer Properties Mechanically" in *Polymer Processing*, Chemical Engineering Progress Symposium Series, **60**, No. 49, 10 (1964), J. V. D. Fear, ed.]

Even so, given sufficient time at temperatures at which the molecules are sufficiently mobile, the level of perfection can be quite high. In this regard, three models are considered for chain folding within the lamellae; these are shown in Figure 6–24. Observe in (a) adjacent reentry with sharp, regular folding and a uniform fold period—features associated with polymer single crystals; in (b), adjacent reentry with an irregular fold period; and in (c), nonadjacent reentry (switchboard model) in which the molecules meander through an amorphous surface layer on the lamella before reentering the lamella or a neighboring lamella. Models (a) and (b) are thought to be more nearly correct, with higher degrees of perfection being attained by the lamellae at high-crystallization temperatures and prolonged periods of time at this temperature.

It seems that the noncrystallizable material diffuses ahead of the crystal growth front. The resulting spherulites consist of radial arms of lamellae

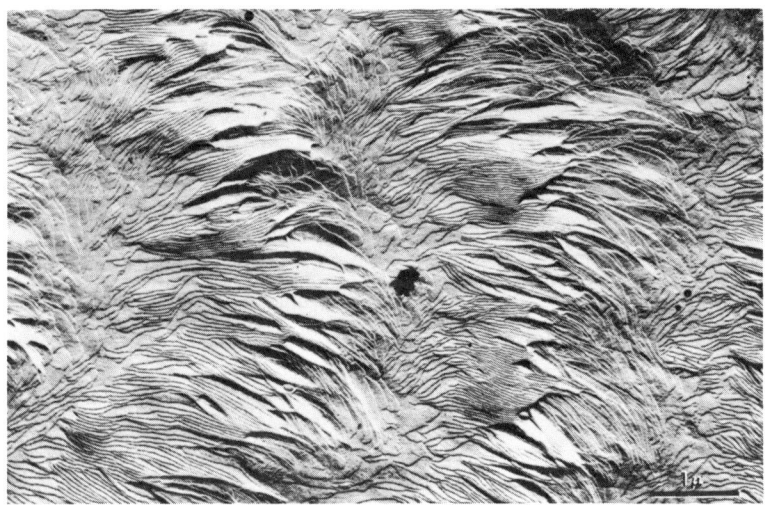

Figure 6-23. Surface replica of a portion of a linear polyethylene spherulite. [R. Eppe, E. W. Fischer, and H. A. Stuart, J. Polymer Sci., **34**, 721 (1959).]

separated by the segregrated, noncrystallizable material. That is, no noncrystallizable material is contained within the lamellae per se, but it collects between them. The properties of bulk-crystallized polymers depend on the type and amount of impurities and how they are dispersed within the sample, as well as on the properties of the crystalline material.

607 Interlamellar Ties

We must now consider how the individual, neighboring lamellae are held together. It is evident from the ductility and strength of polymers that the ties between lamellae must be stronger than the van der Waal's forces holding neighboring, parallel fold planes together, even in the case of lamellae whose surfaces are in intimate contact.

Evidently some molecules (tie molecules) must participate in the growth of two or more adjacent lamellae, thereby providing relatively short molecular links, which reinforce the structure. Such links were actually revealed by cocrystallizing fractionated samples of linear polyethylene with n-$C_{32}H_{66}$ and then removing the paraffin diluent by dissolution in toluene at room temperature. The exposed skeletal structure of a 50–50 sample ($\bar{M}w$ of the polymer = 726,000) grown at 95 °C is shown in the micrograph of Figure 6-25. The fibrilar links, which bridge radial arms of the various spherulites, are clearly evident. These connecting fibrils may measure up to 15,000 Å

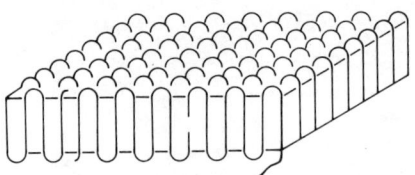

Regular, adjacent re-entry folds similar to those postulated as present in pyramidal crystals that have been grown from solution

(a)

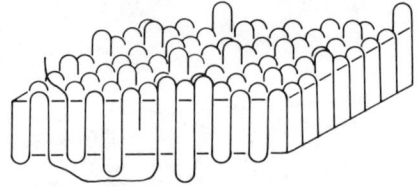

Irregular, adjacent re-entry folds in which the extent or thickness of the irregular layer is suggested to be proportional to the temperature

(b)

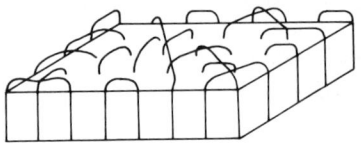

Switchboard, or nonadjacent re-entry model in which an even more nonordered amorphous layer is present on both sides of the lamellae than in the irregular model

(c)

Figure 6-24. Lamellar Crystals. Three models for the lamellar crystals observed in polymers crystallized from the melt have been proposed: (a) Regular, adjacent re-entry folds similar to those postulated as present in pyramidal crystals that have been grown from solution. (b) Irregular adjacent re-entry folds in which the extent or thickness of the irregular layer is suggested to be proportional to the temperature. (c) Switchboard, or nonadjacent re-entry model in which an even more nonordered amorphous layer is present on both sides of the lamellae than in the irregular model.

in length and may be between 30 and 300 Å in diameter. Selected area electron diffraction indicates that the chain axes of the molecules lie parallel to the long axes of the links, which leads to the contention that each link must be an extended-chain single crystal. Relatively few links were formed in samples where the molecular weight was 27,500, whereas the links became more profuse with increasing molecular weight. It is further postulated that even more linkages should be found in crystal samples formed from undiluted polymer.

The following mechanism has been postulated for the formation of these links: On crystallizing from the melt, chain molecules occasionally come into contact with, and begin to crystallize upon, the surfaces of different and often

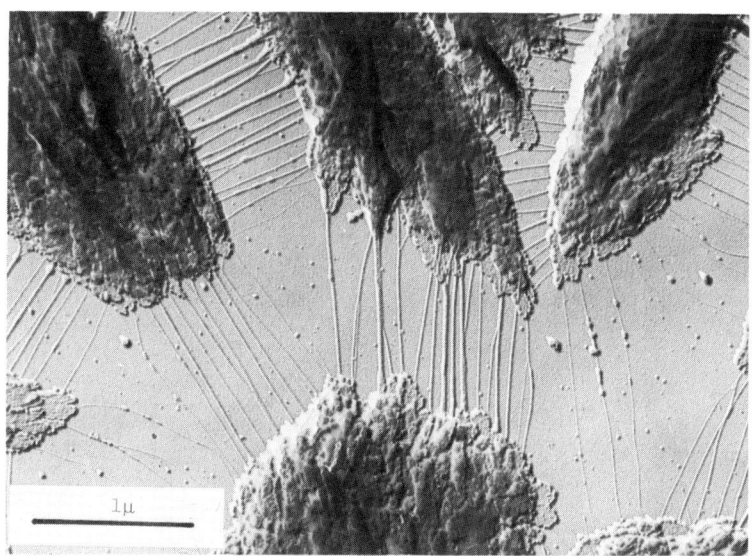

Figure 6–25. Boundary region between spherulite grown at 95 °C. in a polyethylene fraction, $\bar{M}w = 726.000$. [H. D. Keith, F. J. Padden and R. G. Vadimsky, J. Polymer Sci., **4**, A-2, 267 (1966).]

widely separated lamellae. These molecules crystallize until they are pulled taut. Additional molecules then condense upon the nuclei provided by the original tie molecule, thereby establishing the intercrystalline links observed. Such a mechanism accounts for the fact that the links are highly oriented and strongly anchored to the neighboring lamellae. These links should possess considerable strength.

608 Polyethylene Crystallized under Pressure

The morphological features of polyethylene crystallized from the melt under pressure are especially intriguing. At about 5000 atm and a crystallization temperature of 236 °C, the density may be as high as 0.997. This high a value cannot be explained for a perfectly folded chain even if one accounts for the folds, molecular ends, and branches. Electron micrographs, such as the one shown in Figure 6–26, indicates that all the molecules crystallize, not in a folded fashion, but in paraffinlike crystals in which the chains are, for the most part, fully extended. The molecules are parallel to the striations, and crystals up to 3 microns in thickness have been observed. In order for such crystals to form in mixtures of high and low molecular weight fractions, it has been shown that most of the molecules within the same crystallite must be nearly the same size. Present evidence indicates that the molecules, while

Figure 6-26. Fracture surface of a polyethylene spherulite center, crystallized at 300 atm and 12°C of supercooling. The sample contained 25w% high and 75w% low molecular weight linear polyethylene. [Courtesy of J. L. Kardos, Washington University.]

crystallizing under pressure, sort themselves out according to size (i.e., fractionate) and then combine to form crystals. The distribution of crystal thickness is in reasonably good agreement with the known molecular-weight distribution. Research in relating these observations to physical properties is in its infancy.

609 Annealing Polymer Crystals

Annealing a formed crystal is accomplished by maintaining the crystal at a temperature above its crystallization temperature. Evidence indicates that, upon annealing, extensive remelting and recrystallization take place and lead rapidly to a severalfold increase in the fold period of the lamellae. Increase in the thickness of the fold period (or lamellae) leads to the development of holes in the lamellae, as illustrated in Figure 6–27 for a single crystal of polyethylene.

Some polymers [e.g., poly(ethylene terephthalate) and polycarbonate] crystallize so slowly that they can be quenched to temperatures below T_g and solidified as a glass. In doing so, the molecules maintain the amorphous arrangement they had in the melt. When these materials are annealed for a sufficient period of time at temperatures where the molecules can diffuse $(T > T_g)$, well-formed, lamellae like crystals form, which exhibit chain folding. This formation is one of the stronger evidences for the lower free energy of the folded-chain crystal relative to the fringed-micelle crystal.

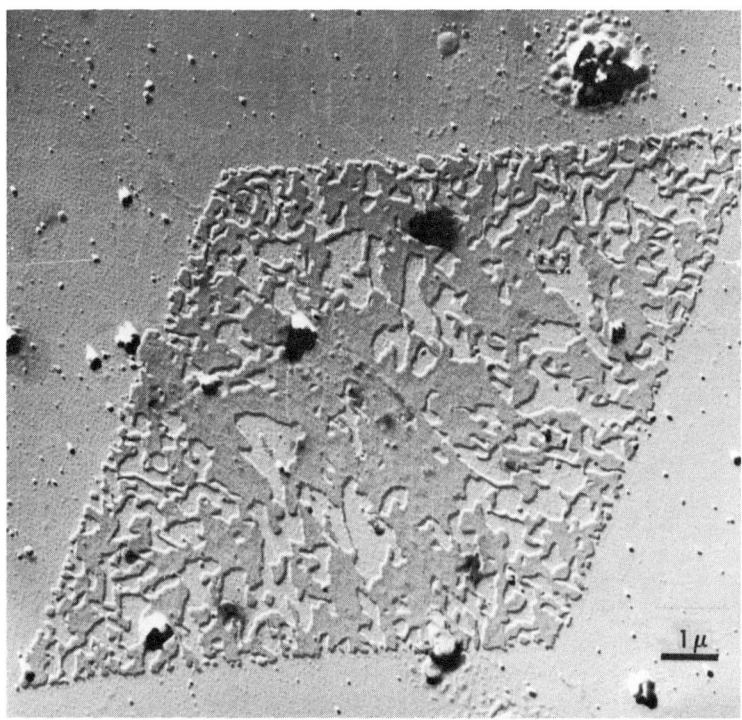

Figure 6–27. A single crystal of linear polyethylene crystallized from perchloroethylene solution, then annealed for 30 min. at 125°C, 10°below Tm. The fold period increased from about 100A to almost 200A during annealing. Compare to Figure 6-10. [W. O. Statton and P. H. Geil, J. Appl. Pol. Sc., **3**, 357 (1960).]

Several aspects of polymer crystal structure and morphology imply a high degree of molecular mobility in both the melt and solid state. The phenomenon of chain folding and the formation of structures with a high degree of order and perfection are strong evidence for prefolding and local orientation in the melt. Also, the molecules must undergo extensive changes in conformation during crystallization. Ample additional evidence for mobility in the solid state is provided by the examples of annealing cited above.

610 Orientation and Drawing

When a bulk polymer is crystallized in the absence of external forces, there is no preferred orientation of crystallites or molecules. If such a specimen is subjected to an external force, such as mechanical drawing, the crystallites and molecules become oriented. Figure 6–28(a) shows the powder pattern of concentric rings for an unoriented specimen. As the specimen becomes oriented, Figure 6–28(b), the rings break into arcs, and at high orientation this pattern is reminiscent of that for single crystals. The polymer is stronger in the draw direction than in any other direction, and it will be weakest in directions 90 deg to the draw direction. The increase in strength in the draw

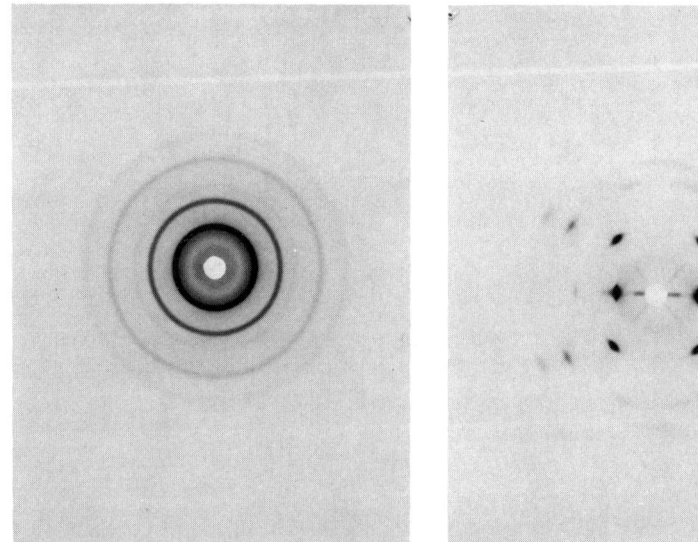

Figure 6–28. X-ray diffraction patterns for (a) unoriented and (b) oriented polyoxymethylene. [Courtesy of E. S. Clark. E. I. duPont deNemours.]

direction may be as much as twenty-five times. This anisotropy of bonding forces is used to advantage in numerous engineering applications. Many crystalline polymers, such as fibers and films, are oriented. Films can be biaxially oriented—that is, oriented in two directions at 90 deg. Orientation can be accomplished by mechanical drawing or rolling a crystalline polymer and by crystallizing an oriented melt. If a polymer is highly crystalline, drawing will not affect the degree of crystallinity; but if the sample is of low crystallinity, drawing is liable to increase the percent crystallinity.

The drawing process is usually carried out below T_m but above T_g. As a rule, in drawing to produce a fiber, the sample does not become gradually thinner; but as shown in Figure 6–29, it suddenly becomes thinner at one point in a process known as *necking*. As the sample is stretched, the ratio of the diameters of necked and unnecked portions remains the same, and the neck section increases at the expense of the unnecked. The length of the drawn sample to the original length, known as the *draw ratio*, is about 4 or 5 to 1 for many polymers.

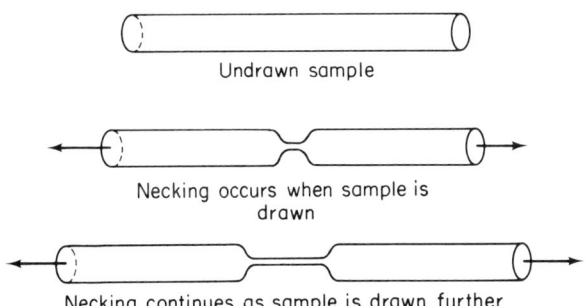

Figure 6–29. Schematic illustrating the necking process in a crystal line filament.

As yet, the detailed morphological and crystallographic changes taking place during orientation remain one of the unsolved problems of modern polymer science. From x-ray diffraction studies we do know that drawn fibers have a periodic structure with dimensions on the order of 100 Å along the fiber axis and that one polymer chain axis corresponds with the fiber axis. It is presumed that the lamellae-like character of the undrawn material is preserved. Two suggested models for an oriented polymer are presented in Figure 6–30; the draw direction is vertical. The oriented interlamellar amorphous model shown in (a) is thought to be appropriate for polymers that are heated during drawing (hot drawn) or annealed after drawing at room temperature (cold drawn). A high degree of lateral cohesiveness is preserved in both the lamellae and oriented amorphous regions. It is thought that cold drawing by itself causes the original lamellae to break up into

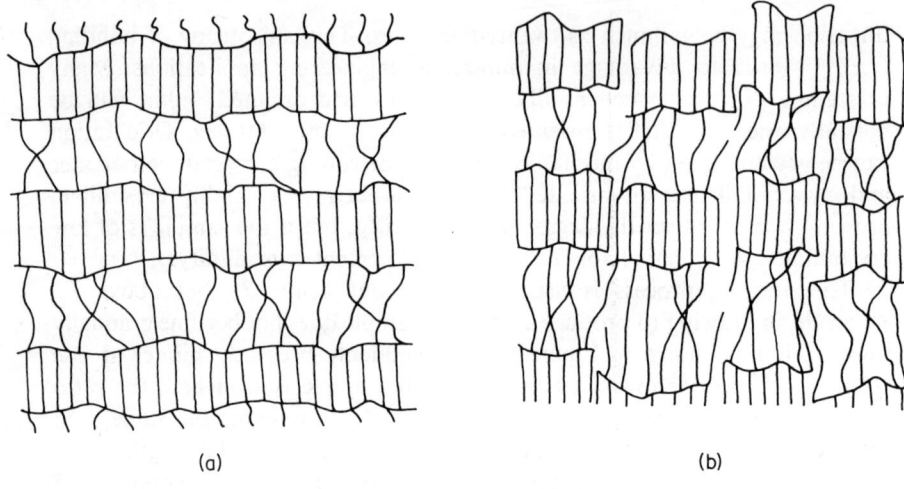

Figure 6-30. (a) Interlamellar amorphous model for an oriented polymer showing a considerable amount of lateral cohesiveness to the lamellae and the oriented amorphous regions. Stretch direction vertical. (b) Fibrils in drawn crystalline polymers formed during cold-drawing or hot-drawing to a very high draw ration. [G. G. Oppenlander, Science, **59**, No. 3821, 1311 (1968). Copyright 1968 by The American Association for the Advancement of Science.]

blocks accompanied by the formation of long fibrils, as shown in (b). The blocks are not in crystallographic register with one another, and the fibrils are 100 to 400 Å wide. Upon annealing, the fibrils disappear as the crystal blocks move back into crystallographic register to reform the lamellae. This mechanism allows for the dependence of the lateral coherency of the fiber on the drawing and annealing conditions.

7

Transitional Phenomena

Synthetic organic high polymers exhibit a variety of changes in state that drastically alter their physical and mechanical properties. The principal determinants in these transitions are temperature, external stress, and the time scale or rate of the experiment used to measure the transition under consideration. The temperatures at which these transitions occur for any particular time scale represent important engineering design parameters. After a brief introduction to the nature of transitional phenomena, we will discuss transitional behavior from the standpoint of following mechanical property changes as a function of temperature and the effect of polymer composition on these properties.

701   First-Order Transitions

When a low-molecular-weight crystalline solid melts or when a liquid boils, changes in volume and enthalpy, as well as other primary thermodynamic properties, take place at constant temperature. Such phase changes are termed *first-order transitions* and are true thermodynamic changes of state.

The first-order transitions can also be discussed on the molecular level. Recall that ordinary low-molecular-weight molecules exhibit three types of motion—vibrational, rotational, and translational. For example, in crystalline materials, the molecules have fixed mean positions in a space lattice about which they may vibrate but not depart, for they are held firmly in place by symmetrically directed intermolecular bonding forces. As the temperature is increased, the amplitude of these vibrations increases until the intermolecular forces are strained to the breaking point and the structure breaks down. Melting occurs and a liquid forms, in which the molecules may translate and rotate as well as vibrate. In boiling, molecules at the surface of the liquid must acquire sufficient thermal energy to overcome the intermolecular bonding and thus escape into the vapor phase.

Of the several ways available to follow the transitions of interest in this chapter, one of the simplest and most satisfactory is to trace the change in specific volume, $\bar{V}$, with temperature. Shown schematically in Figure 7–1 is the volume–temperature behavior of glycerine. Of immediate interest is the

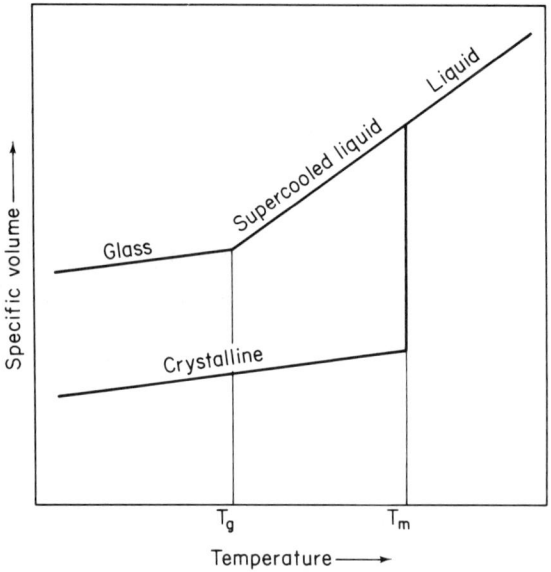

Figure 7–1. Schematic specific volume–temperature curves for glycerine.

lower curve for the crystalline solid. Observe that at temperatures below the *equilibrium crystalline melting point*, T_m, the volume expands slowly as the temperature is raised. At T_m the volume increases under isothermal conditions After melting is complete, the volume continues to expand with increasing temperature but somewhat more rapidly than below T_m.

Polymers, of course, do not boil, but crystalline polymers undergo a transition similar to the melting transition of low-molecular-weight crystalline solids. There are a number of important differences. For one, polymers melt or fuse over a temperature range, generally on the order of 2 to 10 C°. The fusion behavior of some linear polymers is shown in Figure 7–2. *The equilibrium crystalline melting point for polymers* is defined as that temperature at which the last of the crystallites melt. Another important difference is that the actual value of T_m is subject to a strong hysteresis effect. That is, T_m depends on the melt history of the polymer as reflected in the percent crystallinity and the size distribution of the crystallites. Thus equilibrium crystalline

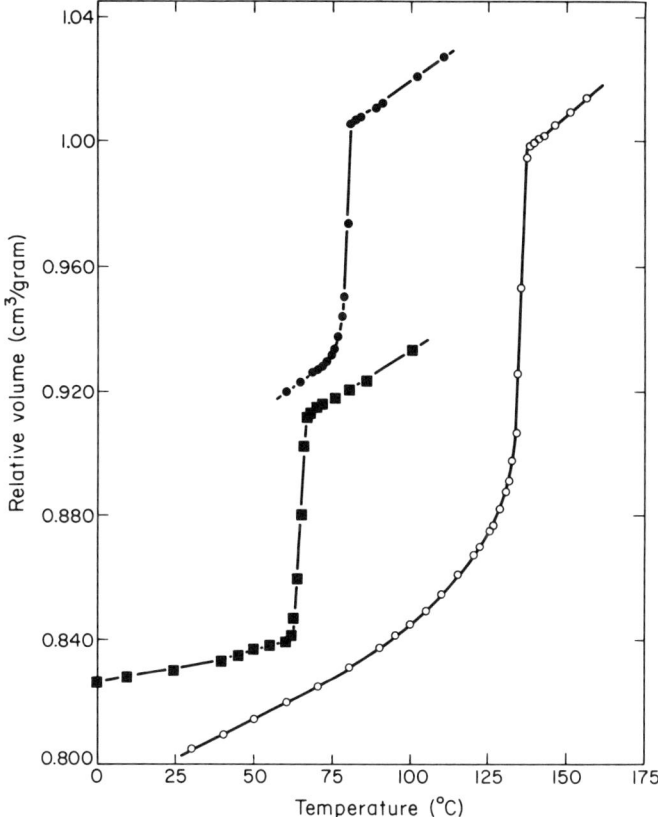

Figure 7–2. Plot of relative volume against temperature: open circles, polymethylene; solid squares, polyethylene oxide; solid circles, polydecamethylene adipate. [L. Mandelkern, Chem. Rev., **56**, 903 (1956).]

melting temperatures must be, and are, standardized by specifying the treatment of the melt prior to testing.* Thirdly, while low-molecular-weight materials become liquids on melting, polymers become very viscous (actually viscoelastic) fluids. These differences arise because the length of the polymer

* Unless indicated otherwise, all transition temperatures cited in this chapter were obtained from J. Bandup and E. H. Immergut, *The Polymer Handbook*, (New York: Interscience, 1966).

chains gives rise to long-range interactions and physical entanglements. Self-diffusion and flow of entire molecules past one another in the polymer melt are thus restricted and produce the observed behavior. Segments of the chain, however, may be quite mobile. For instance, a mobile segment may be formed by a loop in the polymer chain as it doubles back on itself. Considerable molecular flow can take place if the temperature or external stress is sufficiently high.

702 The Glass Transition

To develop the concept of the glass transition, consider the volume-expansion behavior of the *amorphous* (noncrystallizable) polymer illustrated in Figure 7-3. As such a material is heated from below a certain characteristic temperature, the specific volume increases at a fixed rate. At the *glass-transition temperature T_g*, this rate increases, and there is a discontinuity in the volume expansion curve. This phenomenon is termed a *glass transition* because amorphous polymers exhibit a change from glasslike behavior below T_g to soft, rubbery behavior as the temperature is raised above T_g. An excellent illustration of this remarkable change in properties has been observed by anyone who has ever attended a demonstration on the properties

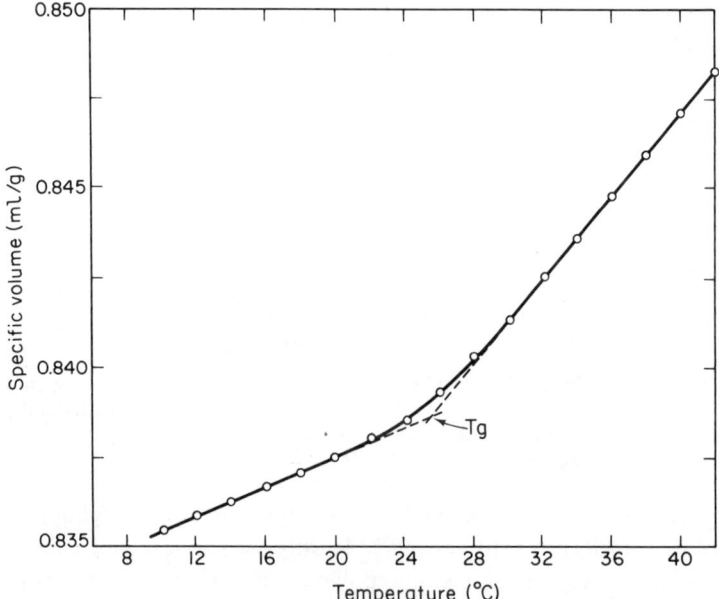

Figure 7-3. Specific volume of polyvinyl acetate as a function of temperature. [P. Meares, Trans. Faraday Soc., **53**, 31 (1957).]

of liquid air or nitrogen. The demonstrator at some point in his presentation will usually take a rubber ball (above the glass-transition temperature of the "rubber polymer" under ambient conditions), bounce it once or twice to prove that it is rubber, dip it into liquid air or nitrogen for a few moments (below the glass-transition temperature of the "rubber polymer"), and then hurl it against the nearest wall. As you would expect of a glassy object, the ball shatters.

There are two schools of thought regarding the fundamental nature of the glass transition. The most popular point of view, and the one we will follow in detail, treats the glass transition as a kinetic or relaxation phenomenon. The most important characteristic of this concept is that the value of T_g measured depends on the experimental time scale as well as on the method of measurement. The other school considers the transition to be an equilibrium phenomenon with T_g approaching the true transition temperature as the experiments are carried out more and more slowly. More will be said of this latter concept in Section 706 after we have considered the kinetic point of view in some detail.

Because the values of T_g depend on the experimental time scale and the means of measurement, values of T_g reported in the literature for any particular polymer may vary over a 10 to 15 C° range. For most practical purposes, this variation will not be important.

As we have implied, the phenomenological aspects of the glass transition are associated with the amorphous nature of polymers. In semicrystalline polymers, the rate of volume increase would be due principally to mechanisms operating within the amorphous zone, while the crystalline regions would remain relatively intact. The higher the degree of crystallinity, the lower the effect of the glass transition on the properties of the material. A highly crystalline polymer would not be expected to exhibit much of a glass transition. The lower curve of Figure 7-4 shows schematically the

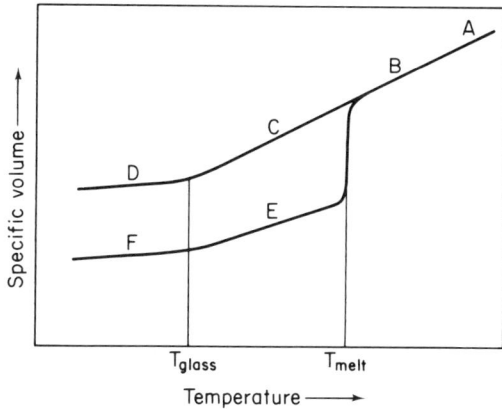

Figure 7–4. Specific volume–temperature curves for a semicrystalline polymer: (A) liquid region, (B) liquid with some elastic response, (C) rubbery region, (D) glassy region, (E) crystallites in a rubbery matrix, and (F) crystallites in a glassy matrix.

volume-expansion behavior of a typical semicrystalline polymer in which both the glass and melting transitions are observed.

The upper curves of Figures 7-1 and 7-4 represent volume expansion curves for crystalline materials that have been quench-cooled to a metastable supercooled liquid state. Only in rare instances is it possible to suppress crystallization completely in a crystallizable polymer; namely, those that are very stiff, such as isotactic polystyrene, or poly(ethylene terephthalate). Observe that they exhibit glass transitions but not crystalline transitions. Although the second-order transition is virtually universal with polymers, very few low-molecular-weight materials exhibit such a transition. In addition to the glass transition exhibited by supercooled glycerine, many paraffins exhibit a transition in which the cross section of the molecules changes from an elliptical to a circular form when the hydrocarbon chains become free to rotate about their long axes.

703 Mechanical Properties

In order to note the effect of the glass and crystalline melting transitions on the mechanical properties of some simple polymer systems, it is instructive to study the effect of temperature on a particular viscoelastic property for several types of polystyrene: namely, two linear atactic amorphous samples of different molecular weight ($\bar{M}_n$ = 140,000 and 217,000), a lightly cross-linked sample, and an isotactic, crystalline sample. The property chosen is the relaxation modulus, but the observed results are quite general in form for many mechanical properties and are not unique to this variable. A more detailed study of the mechanical properties as a function of temperature, time, and stress is undertaken in Chapter 11. In order to utilize these results, we must digress briefly to discuss the relaxation modulus.

The mechanical properties of a polymeric solid under conditions of low strain depend on time and temperature. If such a material is subjected to a sudden tensile strain and then held at that strain at constant temperature, we note that the stress required to maintain the strain in the sample decreases with time. This phenomenon is due to a process known as molecular relaxation. Analogous to Young's modulus for elastic bodies, we define a *relaxation tensile modulus*, $E_r(t, T)$, for viscoelastic bodies at low strains as

$$E_r(t, T) = \frac{S(t, T)}{\epsilon(0)} \tag{7-1}$$

where $S = S(t, T)$ is the stress, which is a function of time and temperature, and where $\epsilon(0)$ is the constant tensile strain applied at time zero. If either time or temperature is held constant, one can measure a modulus that is either a function of temperature or time, respectively. In relaxation experiments, one must hold temperature constant and monitor stress decay with time. At various temperatures, a series of curves like those shown in Figure 11-3 is obtained. To obtain data at a constant time scale, one merely cross-plots the modulus values at constant time for various temperatures.

Figure 7-5 shows such data for the polystyrene samples of interest: in this case, for a time scale of 10 seconds.* This result is extremely important

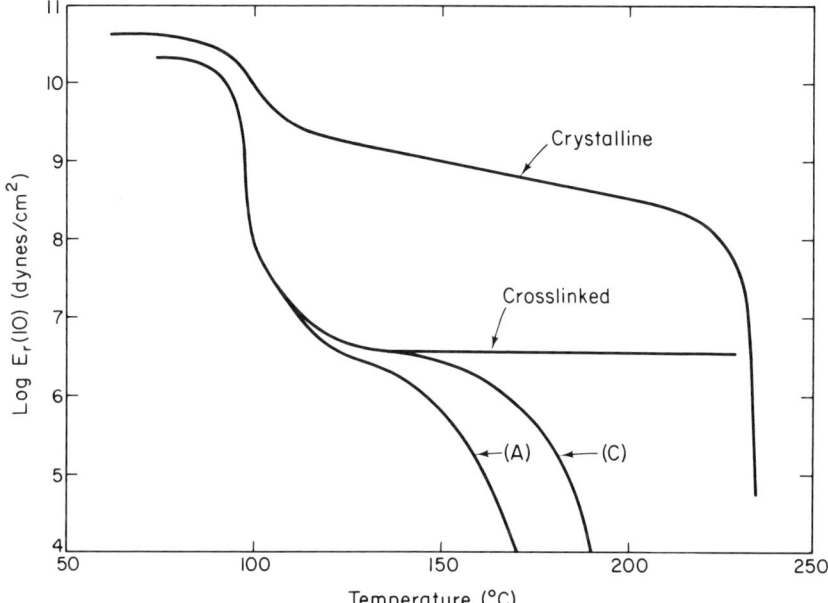

Figure 7-5. $E_r(10)$ versus temperature for crystalline isotactic polystyrene, for polystyrene samples A and C, and for lightly cross-linked atactic polystyrene. [A. V. Tobolsky, *Properties and Structure of Polymers*, John Wiley and Sons, New York, 1960.]

in that its general characteristics are shared (for small strains) by all linear amorphous polymers and their crystalline homologues. First consider the amorphous samples in which we observe *five regions of viscoelastic behavior*.† Below 90 °C is the *glassy region* where the modulus is on the order of 10^{10} dynes/cm^2. In the *transition* region between 90 and 120 °C, the modulus decays from 10^{10} to about $10^{6.5}$ dynes/cm^2. Above 120 °C, up to 150 °C for sample C and 135 °C for sample A, where the modulus remains essentially constant at $10^{6.5}$ dynes/cm^2, we have the *rubbery plateau region*. Between the modulus values of $10^{6.5}$ and $10^{5.5}$ dynes/cm^2, the samples behave as viscoelastic liquids, and this is the *rubbery flow region*. Below $10^{5.5}$ dynes/

---

*The approximate time scale for which many literature values of T_g are cited.
†After A. V. Tobolsky, *Properties and Structure of Polymers* (New York: Wiley, 1960).

cm^2, the material exhibits little elastic recovery to approach mainly viscous flow, and we have the *region of fluid flow*. For amorphous polymers in the transition region, the temperature at which the modulus value is halfway between the values in the glassy and rubbery plateau regions corresponds very closely to the glass transition temperature measured by volume-expansion techniques (for equivalent experimental time scales). At this midpoint, amorphous polymers exhibit tough, leatherlike behavior; how this property is utilized is of considerable interest and will be discussed subsequently.

The effect of molecular weight beyond the transition region is the result of molecular entanglements that serve as temporary physical crosslinks. The higher the molecular weight, the higher the degree of entanglement, and the higher the resistance to viscous flow. An important extension to the effect of molecular entanglement is that of light crosslinking. This light crosslinking makes entanglement permanent, and it extends the rubber plateau region (as shown in Figure 7-5) to the polymeric decomposition temperature. All elastomeric materials of practical value are crosslinked to some extent to extend the temperature (as well as time-scale) range of their utility.

The crystalline form of isotactic polystyrene exhibits a melting transition in the neighborhood of 250 °C as well as a glass transition at 105 °C. The modulus curve for this material is also shown in Figure 7-5. Notice that the modulus is higher than for the amorphous form even at temperatures below T_g. Moreover, the effectiveness of temperature in changing the modulus at the glass transition is not as great as in the amorphous form. This behavior is to be expected, since a crystalline polymer has relatively few amorphous regions. Between T_g and T_m, this material exhibits high impact resistance (toughness) and high rigidity. The exact shape of the modulus curves for crystalline polymers depends on the thermal history of the sample, particularly on how it was cooled from the melt.

704 The Principle of Corresponding Temperatures

To illustrate further the importance of the glass transition in terms of polymeric mechanical properties, we consider the *Principle of Corresponding Temperatures* (as first advanced by Tobolsky) for linear amorphous polymers. This important qualitative concept arises from the observation that under modest strain conditions and for equivalent time scales, all linear amorphous polymers have modulus values between 10^{10} to 10^{11} dynes/cm^2 in the glassy region and $10^{6.5}$ to $10^{7.5}$ dynes/cm^2 in the rubbery zone. Furthermore, the shapes of the modulus-temperature curves, as exemplified in Figure 7-5, are identical. This principle can be simply stated as: *two linear amorphous polymers of the same molecular weight distribution are approximately equivalent at corresponding temperatures.* If we use the glass-transition temperature as the critical temperature, we can define a *reduced temperature*, T_r, as $T_r = T/T_g$. A plot of modulus versus reduced temperature for all linear amorphous polymers at modest strains thus results in the reduction of all such data to a single band, as illustrated schematicaly in Figure 7-6. It is important that the polymers so correlated have equivalent molecular architectural properties and that the comparison be made for a fixed time scale. Large differences in such factors as molecular weights and their distributions, chain stiffness, and side-chain mobility will produce corresponding differences in mechanical properties. It is also important that strains remain moderate such that the materials remain completely disoriented. The principle can also be extended to amorphous crosslinked polymers of the same crosslinked density and concentration of terminal chains.

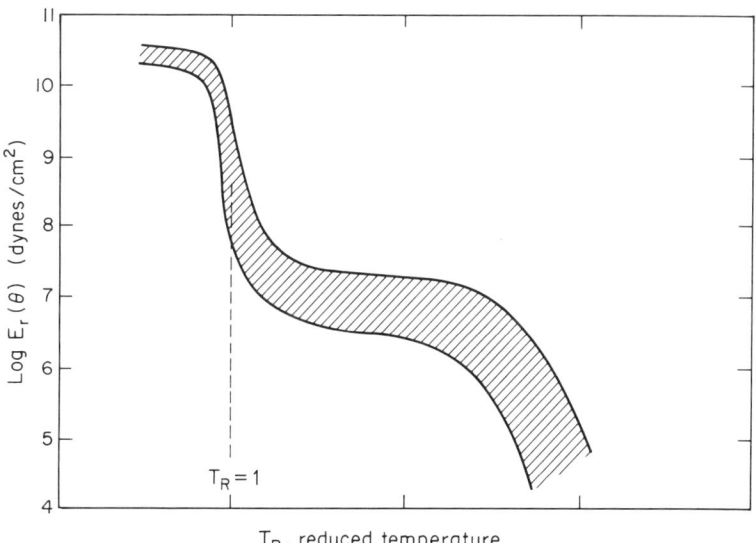

Figure 7-6. $E_r(\theta)$ vs. reduced temperature, T_r, for linear amorphous polymers of equivalent molecular architecture, and a fixed time-scale, θ.

705 Transition Temperatures and Engineering-Use Temperatures

To gain a further appreciation of the importance of transition temperatures in engineering design, we divide polymeric materials into five classes and discuss the temperature ranges in which they may be used in relation to their transition temperatures.

1. *Elastomers.* These materials must be used at temperatures well above T_g to maintain the high, local-segmental mobility required in such materials. The values of T_g for polymers typically used as elastomers range from $-57\,°C$ for styrene-butadiene (25/75) rubber to about $-70\,°C$ for polyisoprene and polyisobutylene.
2. *Amorphous structural polymers.* Such polymers rely on their glasslike rigidity for their utility, and they must be used at temperatures well below T_g. For example, styrene and methyl methacrylate-based polymers have T_g values of about 105 °C and are used mainly under ambient conditions.
3. *Tough, leatherlike polymers.* These polymers are limited to use in the immediate vicinity of their glass-transition temperatures. Such behavior is observed in the so-called vinyl plastics, which serve as substitutes for leather in automobile seat covers, ladies handbags, and travel luggage. (See Section 715.)
4. *Highly crystalline and oriented (fibrous) polymers.* These materials must be used at temperatures substantially below T_m (on the order of $100C°$), since changes in crystal structure can occur above T_g as T_m is approached. In some applications, the glass transition is not important, for it represents a minor transition for these highly crystalline materials. Typical fibrous materials like nylon and poly(1,4-ethylene terephthalate) with T_m on the order of 275 °C must be used at temperatures below 175 °C.
5. *Semicrystalline polymers.* At about 50 percent crystallinity, these polymers can be used at temperatures between T_g and T_m, where the material exhibits moderate rigidity and a high degree of toughness, somewhat analogous to a reinforced rubber. The classical example of this situation is branched (low-density) polyethylene, which is used under ambient conditions with $T_g = -120\,°C$ and $T_m = 115\,°C$.

706 Molecular Motion and Transitional Phenomena

It is instructive to consider the modes of molecular motion occurring in each region of viscoelastic behavior. First, consider heating a linear amorphous polymer from below its glass-transition temperature through the five regions of interest. Below T_g, there is relatively little molecular movement, and the chain segments are frozen in the fixed positions of a disordered quasilattice. The chain segments are able to vibrate about these fixed positions, but there it little opportunity for diffusional rearrangement of segmental position. As the temperature is raised, the amplitude of the vibrations becomes greater, imparting increased stresses on the secondary intermolecular bonding forces. As the temperature is increased through the transition zone, an increasing fraction of the chain segments acquire sufficient energy to overcome these intermolecular bonds, and they take on higher modes of molecular motion. These higher modes of motion involve rotational and translational motion of chain ends and chain segments or chain loops as depicted in Figure 7-7. These motions provide an important mechanism for energy absorption, thereby imparting toughness to the material. More will be said of this subject in Chapter 11 on dynamic mechanical testing. Molecular motion does not yet involve the motion of entire molecules, and in this context we refer to *segmental motion* or *short-range diffusional motion*. In the

Figure 7-7. Motion of chain ends, loops, and segments in the glass transition zone.

rubbery plateau region, the short-range segmental motions are very rapid, molecular motion is still restricted by chain entanglement, and there are no permanent changes in molecular conformation. In the rubbery flow region, molecular slip is becoming important as the degree of entanglement decreases, but some elasticity is still retained. In the liquid flow region, the slip of entire molecules is the dominant mode of motion, large changes in conformation occur, and little elasticity is exhibited.

In semicrystalline polymers, the same series of events occurs in the amorphous regions. In the crystalline regions, short-range vibration of the segments about their equilibrium lattice positions is the dominant mode of motion up to the onset of crystal melting. As the temperature is raised from below T_g, the vibrational amplitude increases, but the forces holding the segments in position are more directed than in the amorphous regions. Additional energy with higher temperatures is required to induce higher modes of segmental and chain-end motion in the crystalline regions. Hence these crystal segments maintain their equilibrium positions through the glass transition to the onset of crystal melting near T_m. Under conditions of stress, segmental motion may occur well below T_m.

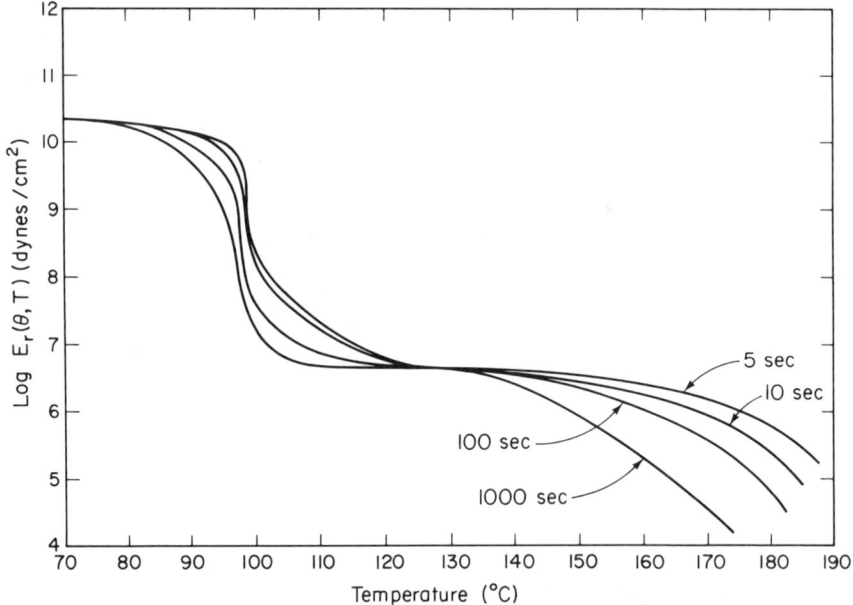

Figure 7–8. Modulus at different times versus temperature for atactic polystyrene sample C. [A. V. Tobolsky, op. cit., Figure 7–5.]

If we introduce the concept of a molecular time scale of motion, we can briefly discuss the effect of the experimental time scale on the value of T_g. More will be said of this facet of transitional phenomena in Chapter 11. A time scale of molecular motion is obviously related to the frequency of segmental motion. Figure 7–8 shows the modulus-temperature behavior of the 217,000 molecular weight, amorphous sample of Figure 7–5 for a variety of experimental time scales. If one asserts that a transition occurs when the molecular time scale corresponds to the experimental time scale, the dependence of T_g on the experimental time scale can be explained. This is left as a student exercise.

We now briefly compare the equilibrium point of view with the kinetic point of view of this transition. As already pointed out, the glass transition approaches a limiting equilibrium value as the experimental time scale is increased. According to the equilibrium concept, this limit occurs at the temperature at which the conformational entropy first becomes zero. In this context, the transition is considered to be a true *second-order thermodynamic transition* and T_g becomes a true thermodynamic property—the *second-order transition temperature.** The kinetic concept on the other hand, considers the rubbery and glassy states as a single thermodynamic phase. Unfortunately, little reliable data are available for transitions measured at extremely slow rates.

707 Secondary Transitions

Many polymers exhibit secondary relaxation transitions. The nature of these transitions is not clearly understood, but they are generally traced to motion of side groups or to motion of small segments of the main polymer chain rather than to motion of large backbone segments. The magnitude of these phenomena is sometimes too small to detect by methods based on transient techniques, including volume expansion, but they are easily measured in oscillating mechanical tests. These secondary relaxation phenomena will be discussed in Chapter 11 in the sections covering the dynamic mechanical properties of polymers.

THE EFFECT OF COMPOSITION ON POLYMERIC TRANSITIONAL PHENOMENA

The relationship between polymer properties and polymer composition was introduced in Chapter 2. We now wish to expand this discussion as it pertains to transitional phenomena. Because the transitional properties of a polymer are among the most important factors in determining its utility, it is imperative that we study the factors governing these properties. Their modification can then be wrought by judicious manipulation of polymeric composition. Discussion at this point falls into two parts: (1) homopolymer systems and (2) copolymer and polyblend systems.

---

* The term *second-order* is employed to distinguish a transition in which there is a sharp change in the derivative of the extensive property rather than in the property itself as in a first-order transition.

HOMOPOLYMER SYSTEMS

708 The Boyer-Beamen Rule

In linear, semicrystalline homopolymers, composition affects both the glass and crystal melting transitions in the same manner. R. F. Boyer and R. G. Beamen* inspected data for a large number of semicrystalline polymers, some of which are shown in Table 7-1 and Figure 7-9. They found that the ratio T_g/T_m (°K) ranged from 0.5 to 0.75. This ratio is closer to 0.5 for symmetrical polymers, such as polyethylene and polybutadiene, but closer to 0.75 for unsymmetrical polymers, such as polystyrene and polyisoprene.

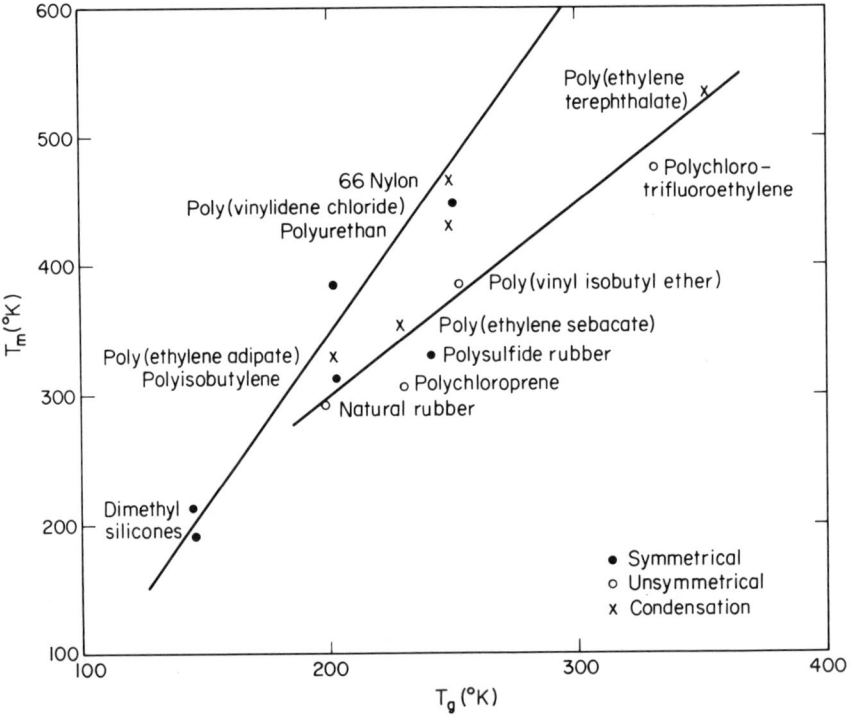

Figure 7-9. Relation between T_m and T_g for various polymers. [R. F. Boyer, J. Appl. Phys., **25**, 585 (1954).]

*R. F. Boyer, *J. Appl. Phys.*, **25**, 825 (1954), and R. G. Beamen, *J. Polymer Sci.*, **9**, 470 (1952).

TABLE 7-1

T_g, T_m, AND T_g/T_m FOR SOME SELECTED POLYMERS[a]

Polymer	T_g,°C	T_m,°C	T_g/T_m,°K
Silicone rubber	−123	−58	0.70
Polyisoprene	−70	28	0.67
Poly(vinylidene fluoride)	−39	210	0.48
Poly(vinyl chloride)	82	180	0.78
Polystyrene	100	230	0.75
Polyethylene	−68	135	0.50
Polypropylene	−18	176	0.57
Nylon	47	225	0.64

[a] *Polymer Handbook*, op. cit.

709 Intermolecular Bonding

Here we are concerned with intermolecular bond strengths ranging from the puny van der Waal forces through the relatively stronger hydrogen bonds

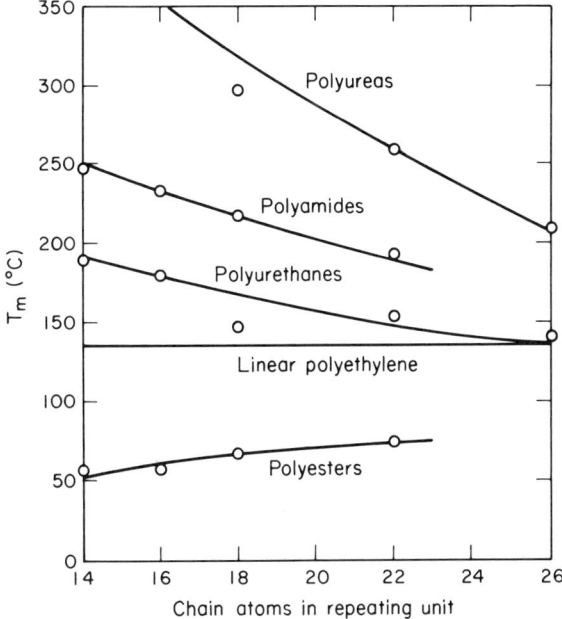

Figure 7–10. Trend of crystalline melting points in homologous series of aliphatic polymers. [R. H. Hill and E. E. Walker, J. Polymer Sci., 3, 609 (1948).]

to the strongest type of bonds provided by chemical crosslinks. Indeed, all intermolecular bonds can be considered as crosslinks, either physical or chemical. Thus we have a complete spectrum of bond strengths that can be controlled through manipulation of both the nature and density of the interchain connections. It can be safely stated that high-transition temperatures are favored by high-intermolecular bonding forces of attraction (and vice versa), since more thermal energy is required to separate strongly bonded chains than weakly bonded chains.

Figure 7-10 shows the trend of crystalline melting points for several homologous series of aliphatic polymers. Using linear polyethylene as a point of reference, note how the melting points approach that of polyethylene as the density of sites for intermolecular bonding decreases. Detailed examination of homologous series indicates that the melting points vary in a much more complex way. Figure 7-11 shows an alteration in T_m with odd and even numbers of carbon atoms in the repeat units. This behavior, which is typical, is superposed on the general trend and results from differences in crystal structure.

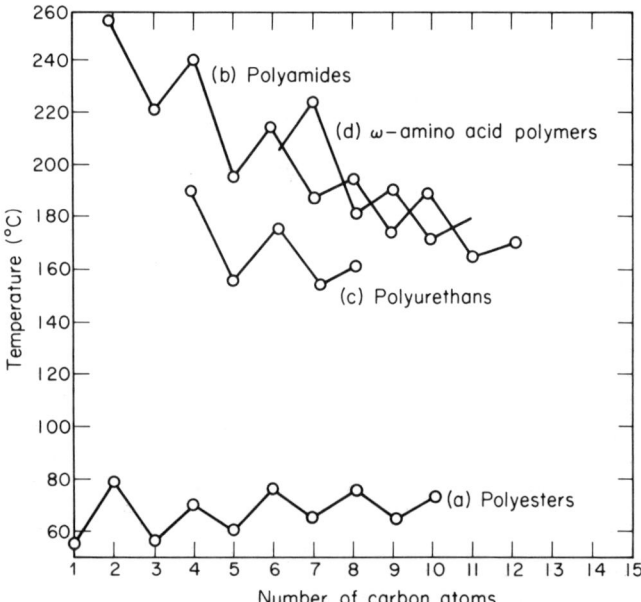

Figure 7-11. Dependence of crystalline melting point on spacing of polar groups. Number of carbon atoms refers to (a) acid for polyesters made with decamethylene glycol; (b) diamine for polyamides made with sebacic acid; (c) diamine for polyurethanes made with tetramethylene glycol; and (d) ω-amino acid polymers. [D. G. Bannerman and E. E. Magat, "Polyamides and Polyesters", Chap. VII in C. E. Schildknecht, ed., *Polymer Processes*, Interscience Publishers, New York, 1956.]

The melting points for polyesters fall below T_m for polyethylene because the weak polar bonding forces in the ester linkage is offset by the flexibility induced by the presence of oxygen in the main chain. In this particular case, note that T_m increases with increasing distance between interunit linkages.

As should be expected, T_g is raised and the manifestation of the glass transition is reduced by crosslinking. If the degree of crosslinking is sufficient, the glass transition is completely eliminated. Figure 7-12 shows plots of specific volume versus temperature for a series of styrene polymers crosslinked with various amounts of divinyl benzene. Observe that the transition points

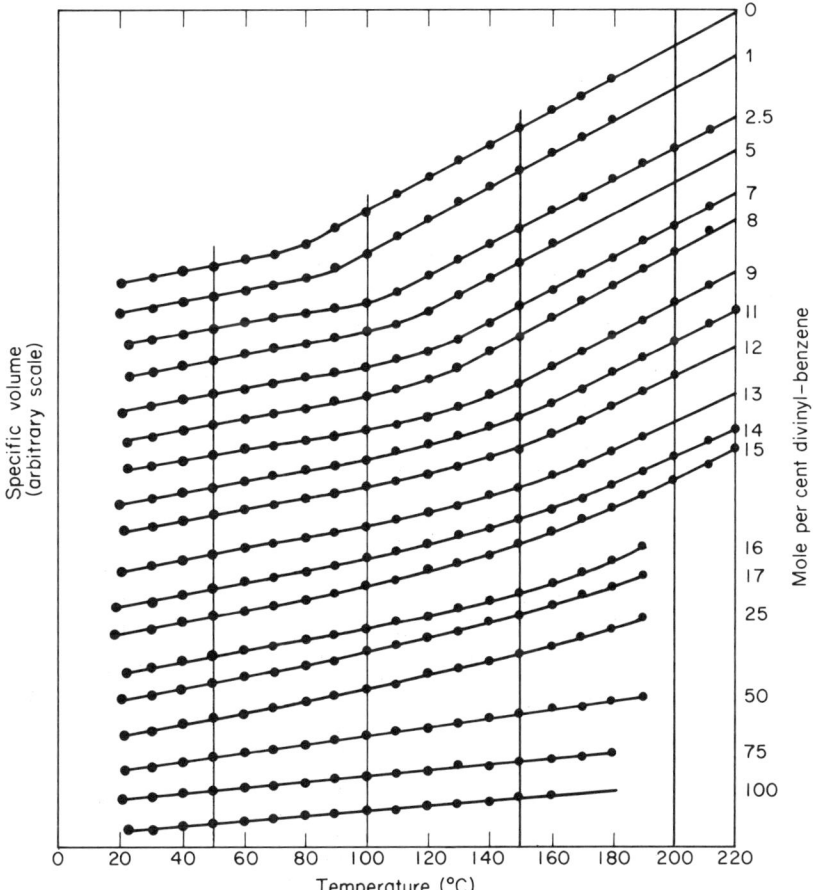

Figure 7-12. Effect of crosslinking on the glass transition of an amorphous polymer system: styrene with divinyl benzene. [K. Ueberreiter and G. Kanig, J. Chem. Phys., **18**, 399 (1950).]

shifts to higher temperatures and becomes more obscure as the divinyl benzene content is raised.

710 External Plasticization

The term *plasticization* refers to the process of making a material more susceptible to plastic flow. For example, in polymer molding and extrusion operations, plasticization is achieved by raising the temperature. Increased deformability can also be achieved by the addition of low-molecular-weight organic compounds, referred to as *external plasticizers*. These materials act to weaken intermolecular bonding by a solvating action, thus reducing T_g and T_m and increasing softness and flexibility. The process of adding a plasticizer is known as *external plasticization*.*

Plasticizers are generally organic, nonpolymeric, high boiling point,

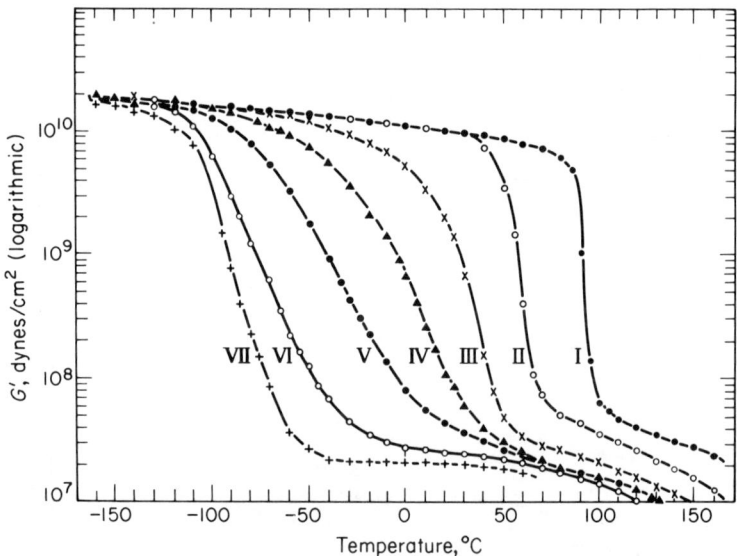

Figure 7–13. Shear modulus, G', vs. temperature, measured for a time-scale of approximately 1 sec., poly(vinyl chloride) plasticized with diethylhexyl succinate: I, 100% polymer; II, 91%; III, 79%; IV, 70.5%; V, 60.7%; VI, 51.8%; VII, 40.8%. [K. Schmeider and K. Wolf, Kolloid-Z., **127**, 65 (1952).]

* The term *external* is used to contrast this effect with a similar one achieved in certain copolymerization processes. In this context, copolymerization is sometimes referred to as *internal plasticization*. See Section 713.

weakly polar compounds with glass transitions on the order of −50 to −150 °C and boiling points on the order of 300 °C or more. The high boiling point is required to prevent loss by volatilization. The weakly polar esters make good plasticizers because they tend to be compatible with many polar and nonpolar polymers. Commonly used ester-based plasticizers are derived from phthalic acid and include the diethyl, dibutyl, and *n*-dioctyl phthalates. Water, as absorbed from the atmosphere, serves to plasticize several important polymers like nylon and the acrylates and methacrylates.

As previously indicated, a plasticizer performs its function by a solvating action. The individual plasticizer molecules distribute themselves through the polymer network and tend to separate the polymer chains. In addition to this physical action, the intermolecular bonding forces between polymer chains are further weakened because the bonding at plasticizer locations is shared by the plasticizer at that point.

Figure 7-13 shows, in a plot of elastic modulus versus temperature, how the transitional behavior of poly(vinyl chloride) varies with plasticizer (diethyl hexyl succinate) content. Observe that not only is T_g reduced by the addition of plasticizer but also that the temperature range of the transition region is broadened to a maximum width at an intermediate content. This

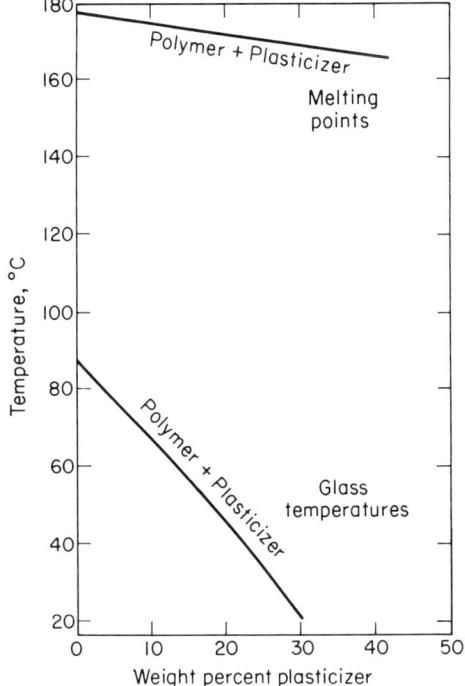

Figure 7-14. Relative lowering of melting point and glass-transition temperature by plasticization. [From *Mechanical Properties of Polymers* by L. E. Nielsen, Copyright 1962 by Reinhold Publishing Corp., by permission of Van Nostrand Reinhold Co.]

broadening of the transition region has practical implications that will be considered in detail in Chapter 11. Figure 7-14 shows how T_g and T_m vary with plasticizer content in poly(vinyl chloride). Notice how T_g drops more rapidly than does T_m. Plasticizer content is often limited to about 40 percent, as shown here, because of limited compatibility.

711 Steric Factors

Here we are concerned with chain flexibility and intermolecular packing distance. Chain flexibility is determined by the ease with which rotation can occur about primary valence bonds. Since rotation commonly involves an energy barrier that is of the same order as the molar cohesive forces (1 to 5 kcal/mole), chain flexibility is an important determinant of polymeric transitional behavior. As previously explained in Chapter 2, increasing steric hindrance, and thus decreasing chain flexibility, increases the transition temperatures. Steric hindrance is established by the constitution of backbone groups, as well as the size and shape of the substituent groups.

The most common backbone "stiffening agent" is the p-phenylene group. The substitution of this flexible group for six chain —CH_2— groups compared brings about a marked rise in the melting points of the polymers compared in Table 7-2. The effect is more pronounced when the p-phenylene group is connected to carbonyl groups as

$$-\underset{\underset{O}{\|}}{C}-\!\!\left\langle\bigcirc\right\rangle\!\!-\underset{\underset{O}{\|}}{C}-$$

instead of to CH_2 groups. This amplified effect occurs because the three groups resonate as a unit in a planar structure, thus producing an even larger stiffening group. Another effective "stiffening agent" incorporating p-phenylene is the

$$-\!\!\left\langle\bigcirc\right\rangle\!\!-\underset{\underset{CH_3}{|}}{\overset{\overset{CH_3}{|}}{C}}-\!\!\left\langle\bigcirc\right\rangle\!\!-$$

group found in epoxy resins (crosslinked) and polycarbonates (T_g = 150 °C, T_m = 267 °C).

The most common backbone "flexibilizing agents" are oxygen and ester groups. Figure 7-10 and Table 2-4 show the effect of substitution of these groups.

Perhaps somewhat surprisingly, the glass transition temperatures of the

TABLE 7-2

INFLUENCE OF AROMATIC GROUPS ON MELTING POINTS OF POLYESTERS AND POLYAMIDES[a]

Repeating Unit	Melting point °C
—CO—C₆H₄—C₆H₄—COO(CH₂)₂O—	346
—CO—(CH₂)₁₂—COO(CH₂)₂O— (estd.)	87
—CO—C₆H₄—C₆H₄—COO(CH₂)₆O—	214
—CO—(CH₂)₁₂—COO(CH₂)₆O— (estd.)	76
—CO—C₆H₄—COO(CH₂)₂O—	256
—CO—(CH₂)₆—COO(CH₂)₂O— (estd.)	61–64
—CO—CH₂—C₆H₄—CH₂—COO—CH₂—C₆H₄—CH₂—O—	146
—CO—(CH₂)₈—COO—(CH₂)₈O—(estd.)	70
—CO—C₆H₄—CONH(CH₂)₆NH—	350 (decomposes)
—CO—(CH₂)₆—CONH(CH₂)₆NH—	235
—CO—CH₂—C₆H₄—CH₂—CONH(CH₂)₁₀NH—	242
—CO—(CH₂)₈—CONH(CH₂)₁₀NH—	194
—CO—(CH₂)₂—C₆H₄—(CH₂)₂—CONH(CH₂)₆NH—	280–290
—CO(CH₂)₈CONH—CH₂—C₆H₄—CH₂—NH—	268
—CO(CH₂)₈CONH—(CH₂)₈—NH—	197

[a] R. Hill and E. E. Walker, *J. Polymer Sci.*, **3**, 609 (1948).

hydrocarbon polymers are independent of tacticity. However, the atactic, isotactic, and syndiotactic forms of poly(methyl methacrylate) have glass transitions at 104 to 108 °C, 42 to 45 °C, and 105 to 120 °C, respectively. These variations arise because the extent of polar interactions between the pendant ester groups depends on the polymeric configuration. As indicated in Chapter 2, cis-trans isomerism produces substantial differences in the transitional properties of polymers. Table 7–3 compares T_g and T_m values for the cis and trans isomers of polyisoprene and polybutadiene. These differences must be a result of differences in freedom of rotation of the chain atoms on either side of the double bond.

TABLE 7–3

THE EFFECT OF CIS AND TRANS ISOMERISM ON T_g AND T_m[a]

Polymer	T_g,°C	T_m,°C
Cis-polyisoprene	−72	25
Trans-polyisoprene	−60	56, 65[b]
Cis-polybutadiene	−95	2
Trans-polybutadiene	−83	145

[a] *Polymer Handbook*, op. cit.
[b] Trans-polyisoprene is polymorphous.

If the pendant groups are stiff and the bulk is close to the chain, steric hindrance will again be raised. If, on the other hand, the substituents themselves are flexible, the effect of increasing intermolecular distance and free volume predominates and transition temperatures will be reduced. Table 2–4 shows how these competing factors operated in a study of the melting points for a series of isotactic poly(α-olefins).

A particularly interesting case of the effect of pendant groups on the glass transition of an amorphous polymer system is exhibited in the series of acrylate

$$\begin{array}{c} -CH_2-CH- \\ | \\ C=O \\ | \\ O-R \end{array}$$

and methacrylate

$$\begin{array}{c} CH_3 \\ | \\ -CH_2-C \\ | \\ C=O \\ | \\ O-R \end{array}$$

polymers. Table 7–4 shows T_g for the number of carbon atoms in n-alkyl acrylate and methacrylate polymers. In both series, T_g decreases with increasing length of the side chain until the length reaches 8 carbons for the acrylate

TABLE 7-4

THE EFFECT OF CHAIN SUBSTITUENT LENGTH ON T_g FOR
ACRYLATE AND METHACRYLATE POLYMERS[a]

n-Substituent	Acrylate (°K)	Methacrylate (°K)
Methyl	279	378
Ethyl	249	338
Propyl	225	308
Butyl	218	293
Hexyl	216[b]	268
Octyl	208[b]	253
Decyl	—	203
Dodecyl	270	263[b]
Tetradecyl	293	264
Hexadecyl	308	288[b]

[a] *Polymer Handbook*, op. cit.
[b] Brittle Point Measurements.

series and 12 carbons for the methacrylate series. The increase observed thereafter is due to crystallization of the side chains; and at low temperatures these side-chain crystallites bind the polymer into a firm, waxy structure. Points beyond the minimum T_g's actually represent the melting points of the crystalline waxes formed by these higher esters. Note that in the decreasing portion, the methacrylates have higher T_g's than the corresponding acrylates because of the stiffening effect produced in the former by the α-methyl group.

As might be expected, branching of the alcohol, as shown in Table 7-5, also raises T_g.

TABLE 7-5

THE EFFECT OF CHAIN-SUBSTITUENT
BRANCHING ON T_g FOR
ACRYLATE POLYMERS

Polymer of	Brittle Point, °C
n-butyl acrylate	−45
Isobutyl acrylate	−24
n-octyl acrylate	−65
2-ethyl hexyl acrylate	−55

Symmetry also affects transition temperatures. At first thought, one might expect polyisobutylene to have higher transition temperatures than isotactic polypropylene ($T_m = 165\ °C$ and $T_g = -10\ °C$) because the presence of two —CH_3 groups close to the backbone should tend to stiffen the chain. Such is not the case, for polyisobutylene is a rubber that crystallizes at 44 °C and that has a glass transition at −71 °C. This discrepancy resides in the fact that polypropylene has a tighter helix than polyisobutylene, which produces a stiffer chain—3 units in 1 turn compared to 8 units in 5 turns. Another example of where symmetry plays an important role is observed in comparing poly(vinyl chloride) with $T_g = 81$ °C and poly(vinylidene chloride) with $T_g = -19$ °C. As a general rule, increased symmetry decreases the transition temperatures. (The material of this paragraph illustrates the pitfalls one can encounter in trying to predict transition properties from an incomplete knowledge of molecular structure when competing mechanisms are operating.)

7.1.2 Molecular Weight

The glass-transition temperature is markedly dependent on molecular weight, although manipulation of molecular weight is not used as a technique to modify T_g. As an example, polystyrene with $\bar{M}_n = 3000$ has a T_g of 43 °C as compared to 99 °C for $\bar{M}_n = 300{,}000$. This dependence is a result of the relatively greater contribution of chain-end segments compared with internal chain segments to the amplitude of molecular motions. Since the chain-end segments are constrained only at one end, they are relatively freer to move about than those that are constrained at both ends; consequently, the chain ends assume higher degrees of motion at lower temperatures. This motion will increase the free volume available as a whole for molecular motion, so that even internal chain segments are affected. As the number of chain ends in a system increases ($\bar{M}_n$ decreases), the available free volume as a whole increases relatively faster with temperature, and the glass transition occurs at lower temperatures.

Let Θ represent the free volume contributed by a chain end. Then the total amount of free volume per cubic centimeter contributed by chain ends is given by

Chain-end free volume per cubic centimeter $= (\Theta)(2\rho N_{Av}/\bar{M}_n)$ \quad (7-2)

where ρ = polymer density

N_{Av} = Avagadro's number

This expression gives the extra amount of free volume that a polymer of finite molecular weight contains compared to a polymer of infinite molecular weight. The finite-molecular-weight polymer will have a glass-transition temperature lower than that of the infinite-molecular-weight polymer (T_g^∞), so that the additional thermal expansion will just compensate for the free-volume contribution of the chain ends. Hence

$$(\Theta) \frac{2\rho N_{Av}}{\bar{M}_n} = a(T_g^\infty - T_g) \qquad (7\text{-}3)$$

This becomes

$$T_g = T_g^\infty - \frac{K}{\bar{M}_n} \qquad (7\text{-}4)$$

where $K = 2\rho\Theta N_{Av}/a$
a = coefficient of thermal expansion

Equation (7-4) should give a straight line if T_g is plotted against $\bar{M}_n^{-1}$. This is the case for polystyrene, as shown in Figure 7-15(a). Θ can be computed from the slope of the line, since ρ, N_{Av}, and a are known. Θ is found to be 80 Å^3, which is about half the size of a styrene monomer unit, a rather reasonable result. For polystyrene and poly(methyl methacrylate), K is about 2×10^5.

As can be observed in Figure 7-15(b), however, where T_g is plotted directly against molecular weight, the effect of molecular weight on T_g is very modest in the high-polymer range.

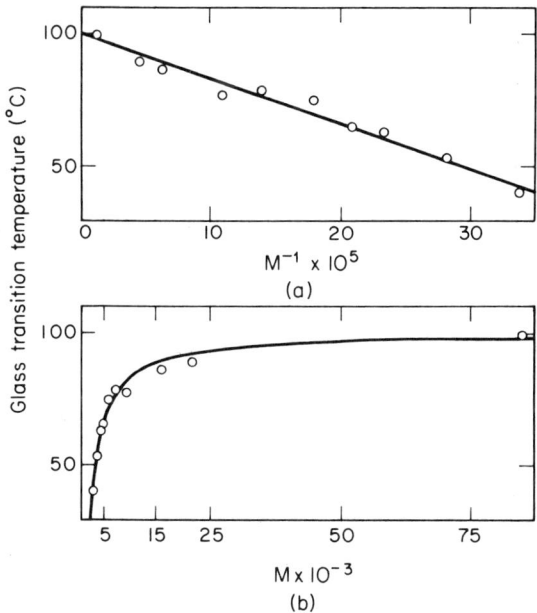

Figure 7-15. Glass transitions of polystyrene fractions vs. M^{-1} (a) and molecular weight (b). [T. G. Fox and P. J. Flory, J. Appl. Phys., **21**, 581 (1950).]

For crystalline polymers, the melting temperature does not depend on molecular weight in the high-polymer range. This insensitivity arises because the chain ends are associated with defects in the crystal structure, whereas the melting phenomenon is associated with the most perfect crystalline regions.

COPOLYMERS AND POLYBLEND SYSTEMS

One of the limiting characteristics of semicrystalline homopolymer systems, as represented in the Boyer-Beamen rule, is that T_g and T_m cannot be controlled independently. As T_g is raised or lowered in a homologous homopolymer series, such as those shown in Figure 7–10, T_m is similarly raised or lowered. To gain at least a measure of independent control, we must turn to the more complicated devices of copolymerization and polyblending. Such processes also allow us to modify the shape of the modulus-temperature curve in the transition region. This variation is exercised through control of compatability and heterogeneity.

We shall see that copolymerization and polyblending provide two of the most powerful techniques for tailor making polymer systems to meet specified engineering needs. (A third technique, which is beyond the scope of this text, is the application of composite material design.)

713 Copolymers with Structural Regularity

In these copolymers, both monomers form crystalline homopolymers. This group is restricted to condensation polymers, such as polyamides, polyesters, and polyurethanes. Figure 7–16 shows schematically the variation in T_g and T_m for isomorphous and nonisomorphous systems. For both, the glass transition varies monotonically with composition. In isomorphous systems, each component is capable of replacing the other in the crystal structure, and T_m varies smoothly over the composition range. Such behavior is rare, but it is exhibited by the reaction products of hexamethylene diamine with mixtures of adipic acid and terephthalic acid. Nonisomorphous systems, however, exhibit a behavior similar to that found in low-molecular-weight binary mixtures that do not form solid solutions. The melting point is depressed by the presence of a second constituent, but it might not show the "eutectic point" if the structure becomes sufficiently irregular. Notice how T_g and T_m vary in the same direction in the isomorphous class, whereas in the nonisomorphous class T_g and T_m vary in opposite directions from one side of the composition scale but in the same direction from the other side.

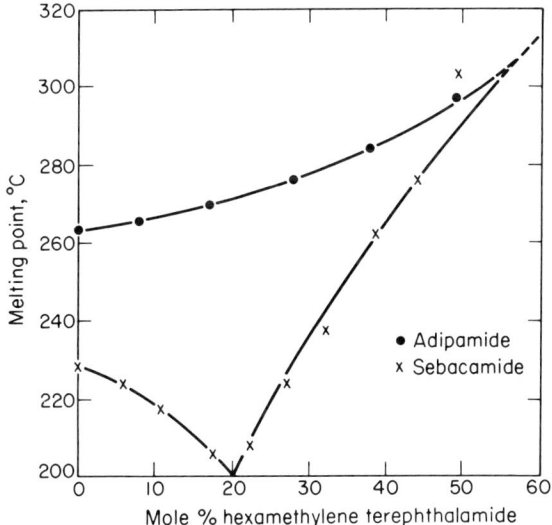

Figure 7-16. Melting points of copolymers of hexamethylene adipamide and terephthalamide, and of hexamethylene sebacamide and terephthalamide. [O. B. Edgar and R. Hill, J. Polymer Sci., **8**, 1 (1952).]

714 Homogeneous Copolymers and Compatible Polyblends

In these systems, both monomers form amorphous homopolymers. In addition, the resulting copolymers are of the random-alternating type and of homogeneous composition. For the polyblends, the homopolymers are compatible in all proportions, thus forming a solid solution.* A pair of monomers will not necessarily fulfill both of the last two conditions. For example, styrene and butadiene form copolymers that are fairly homogeneous and random, but neither the homopolymers nor the copolymers are completely compatible. If either condition is met by a pair of monomers —that is, if they form homogeneous copolymers or compatible polyblends, there will be no difference between the transition properties of homopolymer, copolymer, or polyblend as observed in the curves of modulus versus reduced temperature. These processes merely serve to shift the value of T_g but not the temperature range or modulus properties within the transition region. Figure 7-17 shows modulus-temperature curves for a series of styrene-butadiene copolymers that are obviously superposable on an appropriately

* Complete polymer-polymer compatibility over the entire composition range is rare, although compatibility may be obtained over limited ranges.

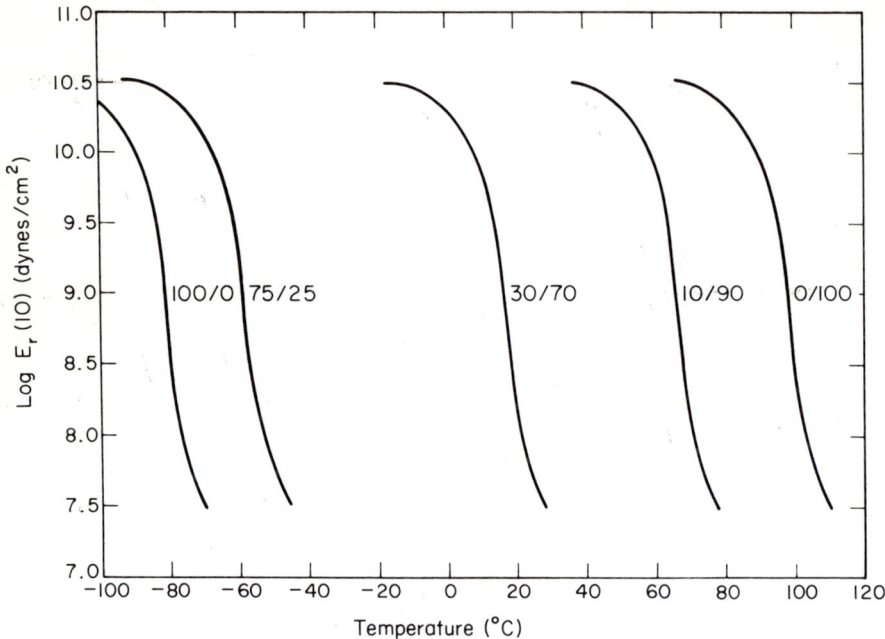

Figure 7–17. E_r (10) versus temperature for buadiene-styrene copolymers. [A. V. Tobolsky, op. cit. Figure 7–5.]

reduced temperature scale. The essential features of the polymer systems of this section are that they are homogeneous and/or compatible. For contrast, Figure 7–18 illustrates the effect of copolymer heterogeneity by comparing the transition properties of homogeneous and heterogeneous copolymers of vinyl chloride and methyl acrylate. Broadening of the transition zone results with increasing heterogeneity. This broadening arises because chains of different composition exhibit different transition temperatures. It is enhanced if the homopolymers are incompatible.

For a series of homogeneous copolymers or compatible polyblends, the values of T_g will usually fall between (but not above or below) those of the parent homopolymers in some smoothly varying fashion. Figure 7–19 shows schematically the manner in which T_g versus copolymer composition may vary. When the monomers differ considerably in their chemical nature, values of T_g may fall well below those of either of the corresponding homopolymers, and a deep well occurs in a plot of T_g versus composition. Copolymers that exhibit this behavior are formed from methyl methacrylate-acrylonitrile, styrene-methyl methacrylate, and acrylonitrile-acrylamide.

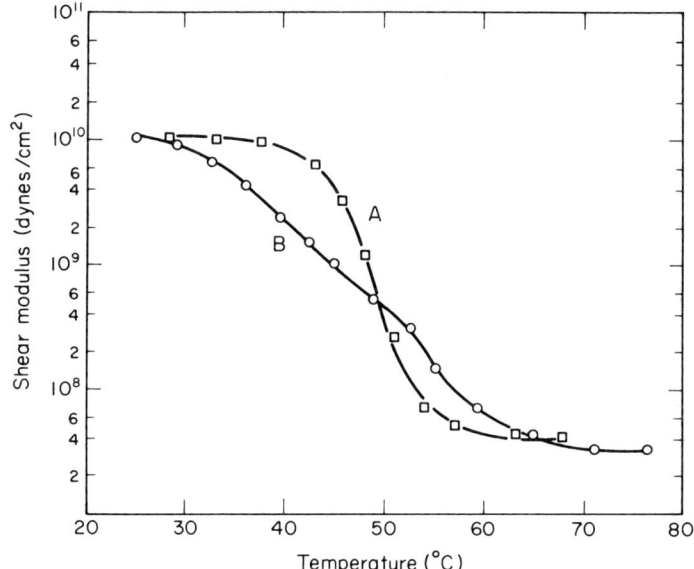

Figure 7–18. Shear modulus-temperature behavior of vinyl chloride-methyl acrylate copolymers: A. Homogeneous B. Heterogeneous composition. [L. E. Nielsen, J. Am. Chem. Soc., **75**, 1435 (1953). Reproduced with permission of the copyright owner, The American Chemical Society.]

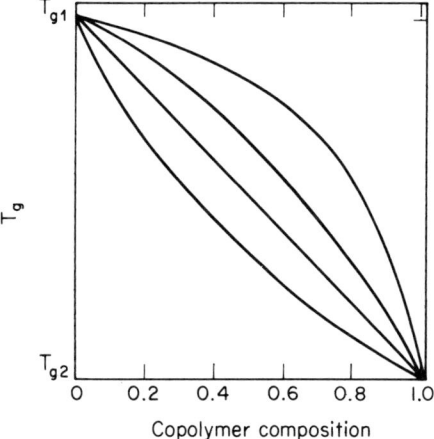

Figure 7–19. Variation in glass transition temperature with copolymer composition (schematic).

715 Semicompatible Copolymer and Polyblend Systems

Here we are interested in block and graft copolymers and polyblends in which the homopolymer portions show varying degrees of compatibility. In these semicompatible systems, one of the components will tend to distribute as droplets in a continuous matrix of the other component. The droplet size and degree of separation will depend on composition, molecular weight, the method of sample preparation, and the degree of compatibility of the homopolymer sequences.

In the extreme, where one component is completely incompatible in the other, two glass transitions will be observed. This behavior is illustrated in Figure 7–20 for an immiscible blend formed from a mixture of polystyrene

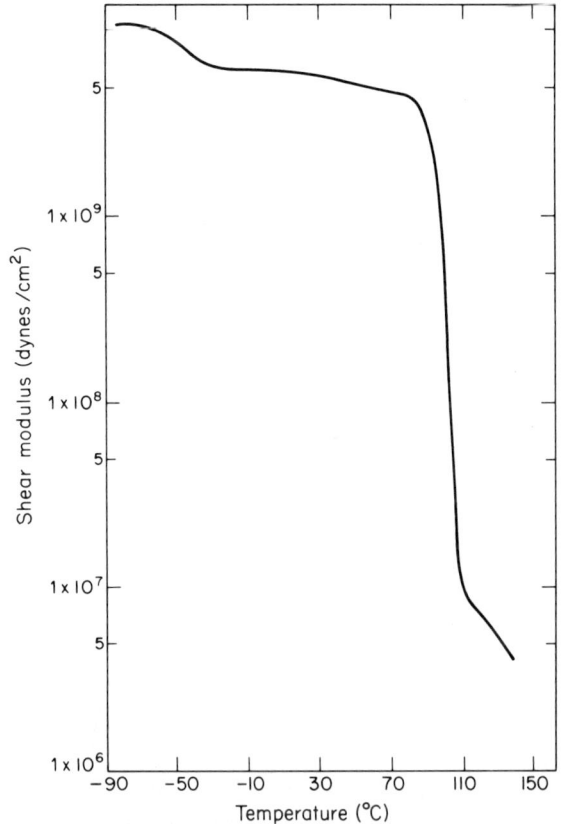

Figure 7–20. Shear modulus vs. temperature of an *immiscible polyblend*. Material is a fixture of polystyrene and a styrene-butadiene copolymer. [L. E. Nielsen, op. cit. Figure 7–14.]

and a styrene-butadiene copolymer. Another example of this type of behavior is the block copolymer formed from methyl methacrylate and isoprene. If a specimen is cast from a solvent for the rubbery isoprene ($T_g = -75\ °C$) segments but a nonsolvent for the glassy methyl methacrylate ($T_g = 105\ °C$) segments, the system will consist of hard, glasslike beads distributed in a rubbery matrix. At room temperature, it will behave as a filled or reinforced rubber. If, on the other hand, a specimen is cast from a solvent for the methyl methacrylate segments but a nonsolvent for the isoprene segments, the system will consist of rubbery beads suspended in a glassy matrix. It will exhibit highly desirable impact properties. Recall, as discussed in Chapter 5, that impact-resistant polystyrene is made by the introduction of 5 percent rubber in a graft copolymerization process.

In the other extreme, if all segments in either block or graft copolymers are fully compatible, the resulting behavior will be analogous to that of homogeneous copolymers and compatible polyblends (Section 713).

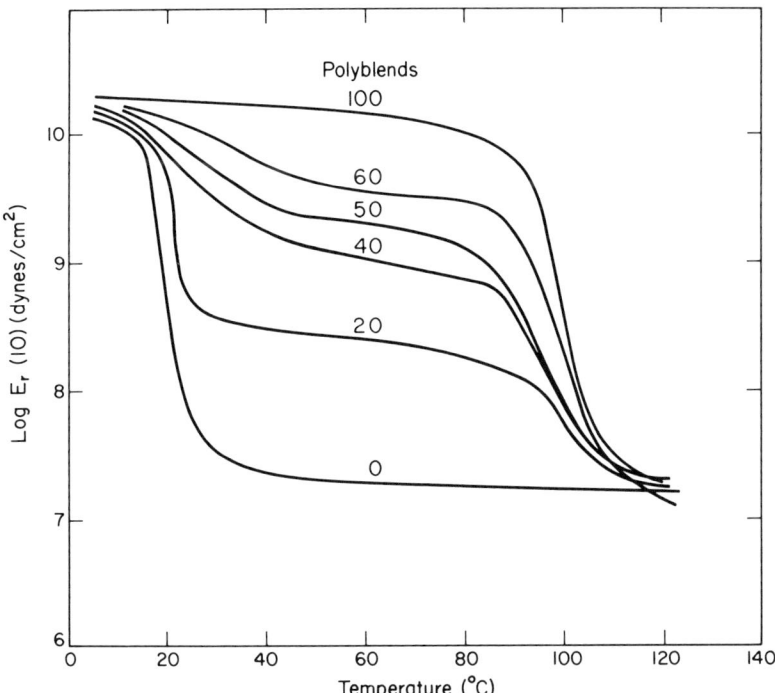

Figure 7-21. $E_r\ (10)$ versus temperature for polyblends of polystyrene and a 30/70 butadiene-styrene copolymer. Numbers on the curves to the weight per cent of polystyrene in the blend. [A. V. Tobolsky, op. cit. Figure 7-5.]

The most useful transition properties are obtained in systems that are on the borderline of compatibility and incompatibility—that is, in semicompatible systems. For example, homopolymer and copolymer systems of styrene and butadiene form a wide variety of useful polyblends, as do the systems comprised of styrene, butadiene, and acrylonitrile. Figure 7–21 shows modulus-temperature curves for several blends of polystyrene and a 30/70 butadiene-styrene copolymer. The most interesting curves are the ones in which we observe extension of the leatherlike behavior over a wider temperature range. This extension of the leatherlike properties of polymeric systems to a practical temperature range amply illustrates the importance of these materials.

8
Polymer Chain Conformation in Random Systems

What we are about to undertake is both a qualitative and quantitative description of the completely unordered state. The completely random arrangement of polymer chains and chain segments among each other represents a physical extreme that may not be fully attained in reality. Such a state is approached by real polymer chains that are sufficiently flexible and sufficiently long. The experimental exploration of the unordered state is extremely difficult and imprecise compared to the crystalline and ordered state. Analytic tools that allow for the most direct probes of structure in ordered materials fail in unordered systems, for measured responses in ordered systems to the external stimuli arise because of the order present. We have spoken or will speak qualitatively of entanglement in such contexts as network flow, viscoelastic behavior, and rubber elasticity; and yet we do not have a satisfactory means of quantifying this parameter. In other words, we cannot quantitatively say what we mean when we consider the "degree of entanglement." So we must perforce confine our discussion to an ideal state—the completely random state.

The analytic results developed here will be of direct interest to us in describing rubber elasticity theory. The concepts also find use in hydrodynamic flow problems of polymer networks, as well as polymer solution thermodynamics. The probabilistic concepts developed herein are of general interest.

801 The Randomly Coiled Chain

We first picture a single isolated polymer chain in an ideal, noninteracting environment. The only forces acting on the chain and its segments originate from random thermal fluctuations in the environment. The degree of randomness associated with the coil will be limited only by the structure of the chain itself as manifested by bond angles and the presence of bulky components. The most flexible chains will have the highest degree of randomness.* We will be conceptually interested in an infinitely flexible chain, as well as in chains ranging from the flexibility of the linear polyethylene chain to the relatively stiff cellulose chain.

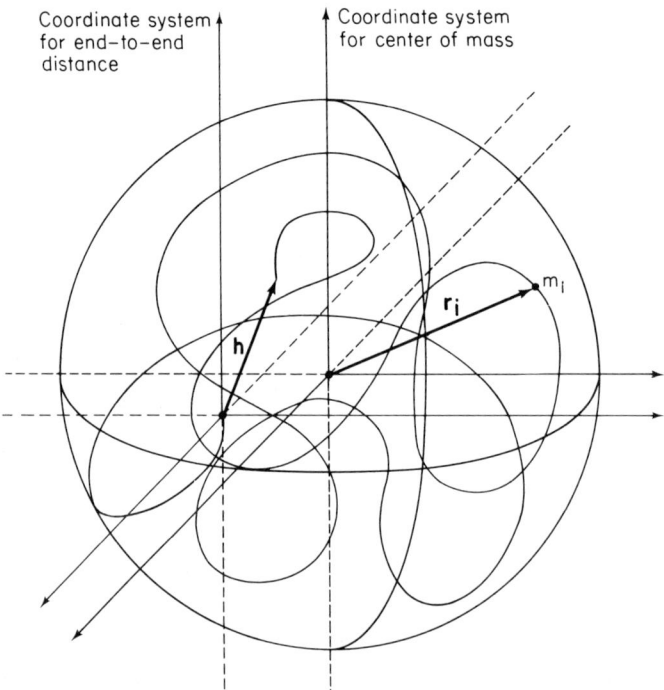

Figure 8–1. A polymer molecule represented as a random coil, contained within a spirical shell.

* The notion of the degree of randomness of a system is associated with statistical mechanics through the Planck postulate, $S = k \ln \Omega$, in which S is the entropy, k is the Boltzman constant, and Ω is the number of miscroscopic states a system may assume. The greater the number of arrangements or conformations available to a system, the greater the number of states; and the higher the entropy. High entropies are associated with systems in high states of disorder or, alternatively, with systems possessing high degrees of randomness. Thus flexible polymer chains may assume many more conformations than stiff chains; consequently, they possess higher degrees of randomness.

As illustrated in Figure 8-1, this ideal random coil is contained within a spherical envelope. Thermal fluctuations in the environment will cause segments within the envelope to shift their positions in a random fashion, and the overall shape of the envelope will remain spherical. The effect of segmental motion on the dimensions of the coil can be described in two equivalent ways: the distance of separation of the chain ends and the radius of gyration of the coil about its center of mass.

In real polymer systems, the situation is extremely complex. One must consider interaction of the chain with its own segments and with chain segments of other chains as well as with its solvent environment. The complications introduced by such interaction forces are usually treated in terms of thermodynamic arguments. The possibility of the existence of some degree of order, introduced by nonideal effects, cannot be discounted.

At this point, we are interested in polymer conformations in the solid state; and in this context one can adequately describe conformational behavior in terms of a single chain in an ideal environment. Since, in the solid state, each chain segment will be completely surrounded by identical segments, it is argued that there will be no *net* interaction effects and a high degree of randomness will prevail. Thus, in the solid state, we visualize a linear polymer system as consisting of a mass of intimately entangled chains, where each chain is still contained within a spherical envelope and where each chain assumes conformations characteristic of its ideal environment. The solid glassy or rubbery polymer state is often envisioned, somewhat graphically, as a bowl of extra-long, well-mixed, cooked spaghetti that is devoid of order.

We will begin our discussion by considering some elementary models and then develop suitable analytic representation for them. Two mathematical approaches will be used: vector averaging and application of random-walk concepts. It is necessary to use both, for each has its limitations, and the results must be combined to get a reasonably complete picture. Finally, we will see how accurately these descriptions can be used to describe real systems. In Chapter 9 these developments will be applied in the discussion of classical rubber elasticity theory.

802 Average Chain Dimensions

Two average dimensions are normally used to describe the spatial extension of a polymer chain: the average end-to-end distance and the average radius of gyration. Both are necessarily root-mean-square averages. The reason will become apparent when the problem is discussed in terms of classical, random-walk theory. The problem of determining the averages of interest can be envisioned in two ways: we can either find the average by considering a single chain and all its possible conformations, or we can consider at an instant of time a large collection of chains representing all possible conformations. Both approaches yield identical results. The approach used is determined solely by tractability.

We first locate one end of the polymer chain at the origin of a suitable coordinate system. The distance of separation of the chain ends is defined as the *end-to-end distance*. As shown in Figure 8–1, it can be represented by the vector quantity **h**. The *root-mean-square, end-to-end distance* is represented by $\bar{h}$, and it is computed as

$$\bar{h} = (\overline{\mathbf{h} \cdot \mathbf{h}})^{\frac{1}{2}} = (\overline{h^2})^{\frac{1}{2}} \tag{8-1}$$

where the averaging process is carried out over all possible conformations.

To define the radius of gyration, visualize the polymer chain as consisting of an assembly of mass elements m_i, each located at a distance r_i from the center of mass. The *radius of gyration* R for a single conformation is, from elementary physics, simply

$$R = \left[\frac{\sum_i m_i r_i^2}{\sum_i m_i} \right]^{\frac{1}{2}}$$

The root-mean-square value R_G over all conformations is given by

$$R_G = (\overline{R^2})^{\frac{1}{2}} = \left[\frac{\overline{\sum_i m_i r_i^2}}{\sum_i m_i} \right]^{\frac{1}{2}} \tag{8-2}$$

It is possible to show, using developments similar to the ones we will follow, that $\bar{h}$ and R_G are uniquely related for all polymers that are sufficiently flexible as

$$R_G^2 = \frac{\bar{h}^2}{6} \tag{8-3}$$

Our approach will center around developing relationships only for $\bar{h}$.

In the next several sections, we utilize vector-averaging techniques to compute analytic expressions for the averages of interest for increasingly complex models of polymer chains. Comparison of these results will lead to a general expression suitable for representing the end-to-end distance of any long, flexible chain.

We describe the individual covalent bonds that make up a chain by the vectors $\mathbf{l}_i$. These will be all equal in length, l, and there will be σ in number. We further concentrate all the associated atoms of the juncture of two of these bond vectors at one center of mass. For analytic purposes, we describe a chain, such as that shown in Figure 8–2, as one consisting of σ linked bonds, $\mathbf{l}_i$, where the juncture of two linked vectors, as well as the tail of the first and the tip of the last vector, locates the $\sigma+1$ centers of mass of the chain atoms. Neither the $\mathbf{l}_i$ bonds nor the $\sigma+1$ centers of mass need be equal in length or magnitude, since we can substitute average values where necessary. The analytic treatments that follow, however, do not allow for

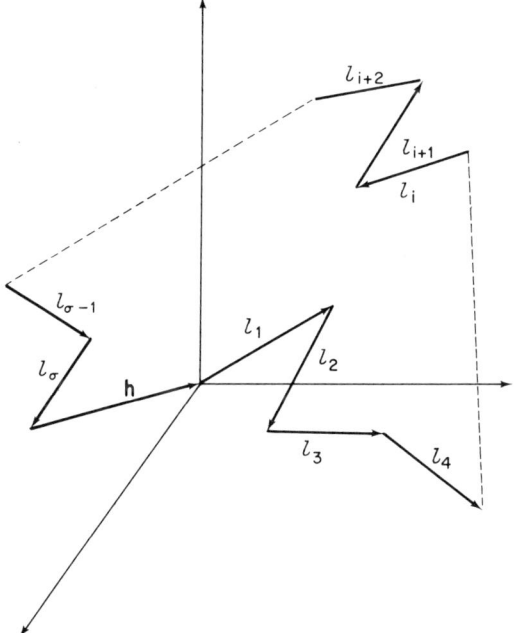

Figure 8–2. The freely orienting chain model.

the fact that the chain segments occupy a finite volume.* The level of sophistication of the models to be described now depends on how we allow the sequentially connected bond vectors to maneuver about one another.

To define the problem, we note that any particular value of $\mathbf{h}$ is merely the vector sum of all $\mathbf{l}_i$ for that conformation. Analytically, this fact is expressed as

---

* This leads to the concept of an *excluded volume*, sometimes referred to as excluded volume of the first kind. Its analytic description is an important theoretical problem.

$$\mathbf{h} = \sum_{i=1}^{\sigma} \mathbf{l}_i \qquad (8\text{-}4)$$

Then $\overline{(h^2)}$ is evaluated as a dot product.

$$\overline{(h^2)} = \overline{\left(\sum_{i=1}^{\sigma} \mathbf{l}_i\right) \cdot \left(\sum_{j=1}^{\sigma} \mathbf{l}_j\right)} = \sum_{i=1}^{\sigma} \sum_{j=1}^{\sigma} \overline{\mathbf{l}_i \cdot \mathbf{l}_j} \qquad (8\text{-}5)$$

The term $\mathbf{l}_i \cdot \mathbf{l}_j$ contained in Eq. (8–5) further expands to $|l_i||l_j| \cos \theta$, where θ is the angle between $\mathbf{l}_i$ and $\mathbf{l}_j$. Thus we see how the concept of available bonding angles enters the analytic representation of the problem. Equation (8–5) becomes

$$\overline{(h^2)} = \sum_{i=1}^{\sigma} \sum_{j=1}^{\sigma} \overline{|l_i||l_j| \cos \theta} \qquad (8\text{-}6)$$

This is a final statement of the problem at hand, which is to determine the average projection of one bond (vector) on another and sum over all the bonds.

803 The Freely Orienting Chain Model

The most rudimentary approach, as illustrated in Figure 8–2, is to allow all possible bond angles to occur with equal probability. Such a chain is referred to as the *freely orienting chain model*. Even though it is the simplest model and at first glance, the seemingly least representative of real polymer chains, we will see that it is actually the most useful, for it can be easily extended to describe the behavior of real chains—provided they are sufficiently long and flexible. To evaluate Eq. (8–6) for this case, we first note that for $j \neq i$ the average projection of any vector on any other vector will be zero, since all values of θ are equally probable. Nonzero terms will arise only when $j = i$, with θ consequently equal to zero. Equation (8–6) thus reduces to

$$\overline{h^2} = \sum_{i=1}^{\sigma} l_i^2$$

with the final result being

$$\overline{h^2} = \sigma l^2 \qquad (8\text{-}7)$$

where $l = |\mathbf{l}_i|$.

804 The Polymethylene Chain Model

A more realistic approach restricts the freedom of adjacent bonds to specified valence cones. The model shown in Figure 8-3 has a bond angle of 109°28′ $(180-\theta)$ and a bond length of 1.54 Å. This is the so-called *polymethylene chain model*.* Although it is more realistic than the freely orienting chain model, we shall see that it still represents an oversimplification of reality. As shown, the adjacent bonds l_i and l_{i+1} lie in the plane of the paper and define a plane that slices through the valence cone of l_{i+2} at right angles to its base. The extension of l_{i+1} intersects the center of the valence cone at 90 deg, and has a value of $l_{i+2} \cos \theta = l_a$. The vectors l_a and l_{i+2} also define a plane, with $l_b = l_{i+2} \sin \theta$ as one of its sides. Thus vector l_a is parallel to l_{i+1}, whereas l_b is perpendicular to l_{i+1} and defines the angle ϕ with the plane of l_i and l_{i+1}. To say that the bond l_{i+2} may be located anywhere on its valence cone with l_{i+1} with equal probability is to say that all values of ϕ are equally probable.

With this geometric representation in mind, we now begin to evaluate the individual terms of Eq. (8-6) for this model. Starting with $j = i$, we note that there will be σ such terms and that the contribution to the summation

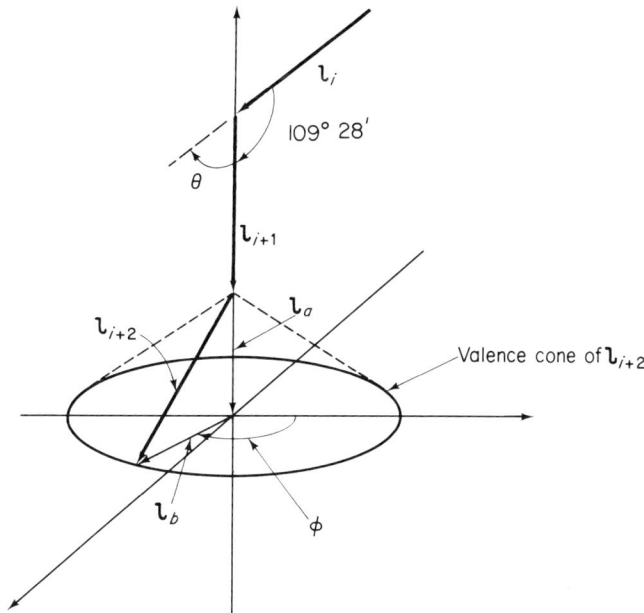

Figure 8-3. The polymethylene chain model.

* It may also be referred to as the *freely rotating chain model*.

for $j = i$ will be $\overline{\sigma l^2}$. Moving on to $j = i+1$, we note that there will be $2(\sigma-1)$ such terms: first, $(\sigma-1)$ arising as i ranges from 1 to $\sigma-1$ while j ranges from 2 to σ, and then $(\sigma-1)$ arising as j ranges from 1 to $\sigma-1$ while i ranges from 2 to σ. Clearly, $\mathbf{l}_i \cdot \mathbf{l}_{i+1} = l_i^2 \cos \Theta$, and the contribution to the summation for j will be $\overline{2(\sigma-1)l^2 \cos \theta}$.

Continuing with $j = i+2$, we note that $2(\sigma-2)$ such terms will arise. To evaluate $\mathbf{l}_i \cdot \mathbf{l}_{i+2}$, we express $\mathbf{l}_{i+2}$ in its component form to obtain

$$\overline{\mathbf{l}_i \cdot \mathbf{l}_{i+2}} = \overline{\mathbf{l}_i \cdot \mathbf{l}_a} + \overline{\mathbf{l}_i \cdot \mathbf{l}_b}$$

Since all values of ϕ are equally probable, the average projection of $\mathbf{l}_b$ on $\mathbf{l}_i$ will be zero. But $\mathbf{l}_a$ forms a constant angle θ with $\mathbf{l}_i$, so that $\mathbf{l}_i \cdot \mathbf{l}_a = l_i l_a \cos \theta = l^2 \cos_i^2 \theta$. The contribution for $j = 1+2$ is, therefore, $\overline{2(\sigma-2)l^2 \cos^2 \theta}$.

For $j = i+3$, the contribution can be shown to be $\overline{2(\sigma-3)l^2 \cos^3 \theta}$. The evaluation of $\mathbf{l}_i \cdot \mathbf{l}_{i+3}$ is performed by splitting $\mathbf{l}_{i+3}$ into two components, one parallel, the other perpendicular to $\mathbf{l}_{i+2}$. The detailed evaluation of this term is left as a student exercise. We have, at this point, evaluated enough (underlined) terms to discover that there will be $2(\sigma-k)$ terms of the type $\overline{\mathbf{l}_i \cdot \mathbf{l}_{i+k}}$ and that the contributions will be of the form $2(\sigma-k)l^2 \cos^k \theta$. For this model, evaluation of the summation of Eq. (8-6) leads to

$$\overline{h^2} = l^2[\sigma + (2\sigma-1)\cos\theta + 2(\sigma-2)\cos^2\theta \ldots \\ + (2\sigma-k)\cos^k\theta + \ldots + 2\cos^{\sigma-1}\theta] \tag{8-8}$$

For the polymethylene chain, $\cos \theta = 0.333$. Since $\cos \theta$ is significantly less than one, the $\cos^k \theta$ terms of Eq. (8-8) rapidly become negligible. Also, σ is usually very large, so that for the first several terms $(\sigma-k) \approx \sigma$. Thus Eq. (8-8) can be approximately represented by the retention of only the first several terms as

$$h^2 = \sigma l^2 (1 + 2\cos\theta + 2\cos^2\theta + 2\cos^3\theta + \ldots) \tag{8-9}$$

Again, since $\cos \theta < 1$, the cosine terms can be appropriately approximated by $(1+\cos\theta)/(1-\cos\theta)$ to yield

$$\overline{h^2} = \sigma l^2 \frac{1+\cos\theta}{1-\cos\theta} = 2\sigma l^2 \tag{8-10}$$

Inclusion of these additional restrictions to free rotation about adjacent bonds merely doubles the analogous result for the freely orienting chain model. Although the polymethylene chain model is more realistic, examination of such a chain constructed from molecular models will readily show that rotation is hardly free but can actually be quite restricted as $\phi \to \pi$. Hence steric hindrance and molecular repulsion can exert a significant effect even in so simple a chain as polymethylene, which has no bulky substituents. Note that when $\phi = 0$, $\mathbf{l}_i$, $\mathbf{l}_{i+1}$, and $\mathbf{l}_{i+2}$ all lie in the same plane, and we recognize that they are then in the planar zigzag or planar trans-conformation.

805 The Effect of Restricted Rotation

In its most elementary form, restricted rotation is encountered in the singly bonded ethane molecule. Figure 8–4(a) shows an ethane molecule as viewed along the carbon-carbon bond; the solid lines represent the forward carbon atom and hydrogen atoms leaving the plane of the paper in the direction of

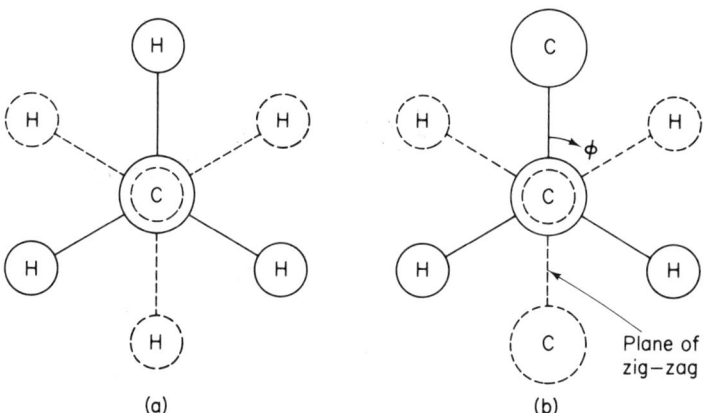

Figure 8–4. Minimum energy conformations for (a) ethane and (b) for four chain atoms of a polymethylene chain.

the reader, whereas the broken lines represent the far carbon atom and hydrogen atoms leaving the plane of the paper away from the reader. As every student of elementary chemistry knows, the most highly favored conformation is the 60 deg staggered positions shown. There is a potential barrier for free rotation through the eclipsed positions because of steric crowding. Because of symmetry, the potential energy barrier $V(\phi)$ for free rotation will contain three equal minima and maxima, and the function $V(\phi)$ will appear as described in Figure 8–5(a).

The polymethylene chain, as shown for four carbon atoms and three bonds in Figure 8–4(b), will not possess the simple threefold symmetry of the ethane molecule. But the zigzag (trans) conformation shown (the plane of the zigzag is perpendicular to the page) will be favored. A zero value for the angle ϕ, shown as measured in the plane of the paper, will thus have a high probability of occurrence. Values of $\pm 120°$ will also be quite probable. The potential energy function for rotation of the polymethylene chain is shown schematically in Figure 8–5(b). If repulsion is exceptionally high but still symmetrical, the potential function could approach that shown in Figure 8–5(c), where the slight troughs of 8–5(b) give way to inflection points. Unsymmetrical potential barriers, like those encountered in the isotactic poly(α-olefins), favor the formation of helical structures.

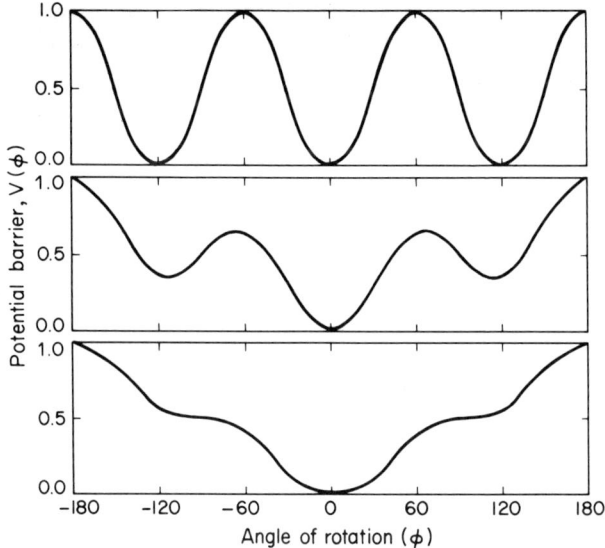

Figure 8–5. Potential energy associated with bond rotation as a function of angle: (a) symmetrical potential; (b) and (c) potential energy functions with lowest minimum at $\varphi = 0$ corresponding to the planar zigzag form of a polymethylene chain. [W. J. Taylor, J. Chem. Phys., **16**, 257 (1948).]

The effect of such restricted rotation can be treated in an elaboration of the method used in deriving Eq. (8–10). The result is

$$\overline{h^2} = \sigma l^2 \frac{1+\cos\theta}{1-\cos\theta} \frac{1+\overline{\cos\phi}}{1-\overline{\cos\phi}} \tag{8-11}$$

where $\overline{\cos\phi}$ is the average value assumed by $\cos\phi$. For free rotation $\overline{\cos\phi} = 0$, and Eq. (8–11) reduces to Eq. (8–10). Eq. (8–11) is not, however, of general utility, for the average value of $\cos\phi$ cannot be calculated theoretically.

The cellulose chain may be treated in a similar manner. As shown in Figure 8-6, the repeat unit is represented by successive bonds of length a, b, and a with fixed angles α and θ. For free rotation about the C—O bond, it can be shown that

$$\overline{h^2} = \sigma \left[(b \sin \alpha)^2 + (2a + b \cos \alpha)^2 \frac{1+\cos \theta}{1-\cos \theta} \right] \tag{8-12}$$

Similar expressions have been derived for other chains.

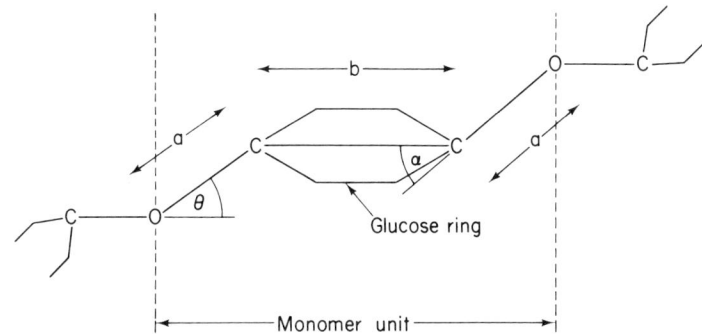

Figure 8-6. The cellulose chain, showing the parameters used in Equation 8-12.

806 General Ideal Expressions for Average Chain Dimensions

A comparison of Eqs. (8-7), (8-10,) (8-11), and (8-12) reveals that, in each, $\overline{h^2}$ depends on two terms: one term contains such parameters as l, θ, or ϕ, which depend only on the nature of the polymer; the other depends only on the chain length as represented by σ. Thus for all flexible long chains, we can write

$$\overline{h^2} = \beta^2 \sigma \tag{8-13}$$

where β^2 is a constant characteristic of the structural nature of the polymer. It can be regarded as a measure of the stiffness of a polymer chain or as a measure of its lack of rotational freedom. Values of β for typical synthetic organic polymers are about $3.5l$.

For most real polymer systems having a molecular-weight distribution, we must consider the fact that σ is not a constant and that the value of $\overline{h^2}$ determined must be an average. Let σ_i be the number of bonds in a polymer of degree of polymerization X_i. For high polymers, $\sigma_i = KX_i$, where K is a proportionality constant, and we can write three types of averages for $\overline{h^2}$: namely, those corresponding to the familiar molecular-weight averages.

$$(\overline{h^2})_n = \frac{\sum_{i=1}^{\sigma} N_i(\overline{h^2})_i}{\sum_{i=1}^{\sigma} N_i} \tag{8-14}$$

$$(\overline{h^2})_w = \frac{\sum_{i=1}^{\sigma} W_i(\overline{h^2})_i}{\sum_{i=1}^{\sigma} W_i} \tag{8-15}$$

$$(\overline{h^2})_z = \frac{\sum_{i=1}^{\sigma} N_i M_i^2(\overline{h^2})_i}{\sum_{i=1}^{\sigma} N_i M_i^2} \tag{8-16}$$

where the subscripts denote number, weight, and z averages, as before. With $(\overline{h^2})_i = \beta^2 \sigma_i = K\beta^2 X_i$, Eqs. (8–14) through (8–16) become

$$(\overline{h^2})_n = \beta^2 \bar{X}_n \tag{8-17}$$

$$(\overline{h^2})_w = \beta^2 \bar{X}_w \tag{8-18}$$

$$(\overline{h^2})_z = \beta^2 \bar{X}_z \tag{8-19}$$

Just as with molecular-weight determinations, the particular average obtained depends on the method of measurement.

Perhaps a word is in order on the effect of nonideal environments. This topic usually falls into the province of polymer solution thermodynamics. Here we will consider some rudimentary qualitative concepts. If the solvent does not provide an ideal environment in a solution of polymer molecules, interactions will exist. The polymer and solvent may be mutually attracted, in which case the solvent would be described as a *good solvent*; or the reverse may be true and such a solvent would be a *poor solvent*. If the solution is sufficiently dilute such that intermolecular polymer-polymer interactions are minimized, the polymer chain dimensions will be larger than the ideal case in a good solvent but less than the ideal case in a poor solvent. Furthermore, poor solvents tend to form unstable solutions with polymers. If the solutions are sufficiently concentrated, but not so concentrated to produce complete overlap, polymer-polymer interaction may also occur. Depending on whether the interactions are repulsive or attractive, the polymer coil may be contracted or expanded from the ideal state. Since interest in this book centers on solvent-free systems where there is complete overlap and no *net* interactions prevail, we need not consider these ramifications any further. In this case the chain is said to be unperturbed, and its dimensions, as given by Eq. 8-3 and 8-13, are referred to as *unperturbed chain dimensions*.

We now wish to make further use of the freely orienting chain model in a random-flight analysis, which will lead us to additional insights into random-chain conformational behavior. First, we must digress to develop the fundamentals of random-walk theory, as well as review some pertinent probability concepts.

807 Random-Flight Analysis

The principle of random flight can be applied to problem analysis in situations where a large number of events are occurring at random. We have already encountered a situation—a two-dimensional one—where such an analysis is applicable. With reference to Figure 8-2, we considered the conformation established by the random orientation of a large number of connected vectors. In the context of random-flight analysis, each vector could also represent an event such as a randomly directed step taken by a drunk or the path taken by a gas molecule between random collisions with other molecules. The graphical history of a series of such events would also be adequately described by Figure 8-2. The result of our analysis will be a distribution function—the Gaussian distribution. This function will describe the frequency with which the polymer chain end, drunk, or a gas molecule can be found at a specified distance from its starting point. Our initial analysis will be for a one-dimensional case, but we will extend this to the multidimensional case in applying this first result to describing polymer chain size.

Our model system will consist of a particle whose movements are restricted to unit steps along a line. The particle will start at 0; it will be able to move in the forward (f) or backward (b) direction with equal probability; and it will take a total of N steps. We now wish to compute the probability that the particle will have taken N_f steps in the forward direction after having taken a total of N steps.

To get properly oriented, first consider the trivial case of where the particle takes three steps, and compute the probability that it will take two steps in the forward direction. This could happen in any of three ways: $f, b, f; f, f, b;$ or b, f, f. If P_b and P_f represent the probabilities for a backward or forward step (both of course are equal to $\frac{1}{2}$), the probabilities for each of these series of events are equal and given respectively by $P_f \times P_b \times P_f = P_f \times P_f \times P_b = P_b \times P_f \times P_f = (\frac{1}{2})^3$. Since there are three ways in which this event can occur, the probability that the particle will take two forward steps and thus finish one step in the forward direction, if three steps are taken, is given by $3(\frac{1}{2})^3 = 0.375$. In other words, this result means that out of every 1000 sets of three steps taken, you should end up, on the average, 375 times one step forward (or backward).

Now let us generalize for a particle taking N steps, where N is a large number: The probability of any single series of steps occurring as a result of the particle taking N_f forward steps and N_b backward steps is $P_f^{N_b} \times P_b^{N_b} = \frac{1}{2}^N$. It now remains to compute the number of ways in which a particular sequence could result. To do so, we first employ the artifice that all forward steps are distinguishable and that we can plan ahead and tell exactly when the particle will take a forward step. All the backward steps will be taken when our particle is not taking a forward step. First, the forward step labeled "1" can be taken at any of N opportunities; secondly, the forward step labeled "2" at any of $N-1$ opportunities; thirdly, the forward step labeled "3" at any $N-2$ opportunities; and generally the forward step "N_f" may be taken at any of $(N_b+N_f)-(N_f-1) = (N_b+1)$ opportunities. Hence the number of ways of taking N_f distinguishable forward steps is given by $(N)(N-1)(N-2)\ldots(N_b+1) = N!/N_b!$. However, the forward steps are indistinguishable, and we account for this fact by dividing by $N_f!$ The number of ways for a particle to take N_f indistinguishable forward (or N_b backward) steps is $N!/N_b!\,N_f!$.

The intermediate desired result is the probability $P(N_f)$ that our particle will have taken N_f forward steps. This is equal to the product of the probability that the particular series of events required for the particle to do so should occur and the number of ways this event could occur. Hence

$$P(N_f) = \left(\frac{1}{2}\right)^N \frac{N!}{N_b!N_f!} \tag{8-20}$$

It is desirable, however, to cast Eq. (8–20) into a more useful form. Since N will be very large, N_b and N_f will also be large and most probably equal to $N/2$, and we can write

$$N_f = \frac{N}{2}+\xi \quad \text{and} \quad N_b = \frac{N}{2}-\xi \tag{8-21}$$

where $\xi \ll N/2$ for large N. It should be clear that this approximation excludes application to "fully extended" systems. Stirling's approximation for the factorial of large numbers can be used—that is, $\ln x! = \frac{1}{2} \ln 2\pi + (x+\frac{1}{2}) \ln x - x$—to further simplify Eq. (8–20). Also, taking cognizance of the fact that N_f is uniquely related to ξ, $P(N_f) = P(\xi)$, and Eq. (8–20) becomes

$$\ln P(\xi) = \left(N+\frac{1}{2}\right)\ln N - \left(\frac{N}{2}+\frac{1}{2}+\xi\right)\ln\left(\frac{N}{2}\right)\left(1+\frac{2\epsilon}{N}\right) \qquad (8\text{-}22)$$

$$- \left(\frac{N}{2}+\frac{1}{2}-\xi\right)\ln\left(\frac{N}{2}\right)\left(1-\frac{2\xi}{N}\right) - N\ln 2 - \frac{1}{2}\ln 2\pi$$

Since $2\xi/N \ll 1$, we can use the series approximation $\ln(1+y) = y - y^2/2$ for $(1 \pm 2\xi/N)$ to further reduce this rather formidable looking equation to

$$\ln P(\xi) = \ln\left(\frac{2}{\pi N}\right)^{\frac{1}{2}} - \frac{2\xi^2}{N}$$

and, finally, we obtain

$$P(\xi) = \left(\frac{2}{\pi N}\right)^{\frac{1}{2}} \exp\frac{-2\xi^2}{N} \qquad (8\text{-}23)$$

which is the familiar Gaussian distribution function. The physical significance of ξ will become clear in Section 809 when we apply these results to describing chain-end distributions. For large values of N, $P(\xi)$ will become a continuous distribution function such that $P(\xi)\,d\xi$ gives the probability of finding a value of ξ between ξ and $\xi + d\xi$. Equation (8-23) can be recast as

$$P(\xi)\,d\xi = \left(\frac{2}{\pi N}\right)^{\frac{1}{2}} \exp\left(-\frac{2\xi^2}{N}\right) d\xi \qquad (8\text{-}24)$$

The Gaussian distribution function has implications reaching far beyond its application to polymer systems. It is a powerful fundamental tool in statistics. The student who is not familiar with the general application of probability density functions should refer to a standard text on the subject. Here we will only briefly review some of the basic concepts.

808 Some Fundamental Probability Concepts

If $P(l)$ is a probability density function, where the value of l may range from a lower limit of $-\infty$ to an upper limit of $+\infty$, as in the preceding development, then

$$\int_{-\infty}^{\infty} P(l)\,dl \equiv 1 \qquad (8\text{-}25)$$

In other distribution functions, the limits may range from zero to infinity; or, in general, any limits may be appropriate. If the integration of a distribution function, $P'(l)$ over appropriate limits did not yield a value of unity, then it would be necessary to *normalize* the function, $P'(l)$, as

$$P(l) = \frac{P'(l)}{\int_{-\infty}^{\infty} P'(l) \, dl} \qquad (8\text{-}26)$$

to obtain $P(l)$, the associated probability distribution function. The average value of l can be obtained simply as

$$\bar{l} = \int_{-\infty}^{\infty} lP(l) \, dl \qquad (8\text{-}27)$$

It might be helpful to refer to some results of Section 3–6 to further clarify these concepts. Recall that the probability of finding a linear chain x-mer units long was given by Eq. (3–16) as

$$P(x) = n_x = (1-p)p^{x-1} \qquad (3\text{-}16)$$

First, observe that a probability density function has its analog in a number-fraction distribution function, in this case the function n_x. Recall, also, that the average value of x was determined as

$$\bar{X}_n = \sum_{x=1}^{\infty} x n_x = \frac{1}{1-p}$$

which in the limit of large x can be approximated as

$$\bar{X}_n = \int_{x=0}^{\infty} x n_x \, dx \qquad (8\text{-}28)$$

Note the exact parallel between Eqs. (8-27) and (8-28), except for the limits.

Let us now consider some additional properties of $P(l)$. Suppose this function to be the one shown in Figure 8–7 with the normalization requirement fulfilled within the limits of $\pm \infty$. Then $\int_{-\infty}^{x} P(l) \, dl$ is the probability of l assuming a value between $-\infty$ and x; $\int_{x}^{y} P(l) \, dl$ is the probability of l assuming a value between x and y: and $\int_{y}^{\infty} P(l) \, dl$ is the probability of l assuming a value between y and ∞. Note that these probabilities represent the ratio of the respective areas under the curve to the total area under the curve, which in the case of a normalized distribution is unity.

We shall make use of all these concepts in the ensuing sections as well as in Chapter 9.

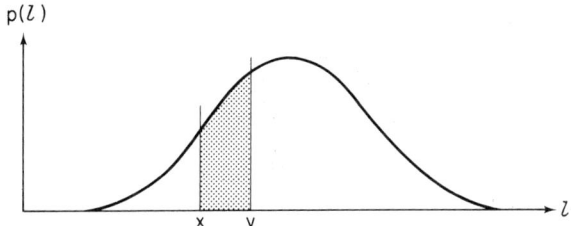

Figure 8-7. A representative probability density function.

809 Distribution Function for End-to-End Distances

We now wish to apply the concept of random walk and Eq. (8-24) to describe the distribution of the end-to-end distances of a polymer chain. For reasons of analytic tractability, we can only apply this concept to the freely orienting chain model. We will later show, however, that this represents no real

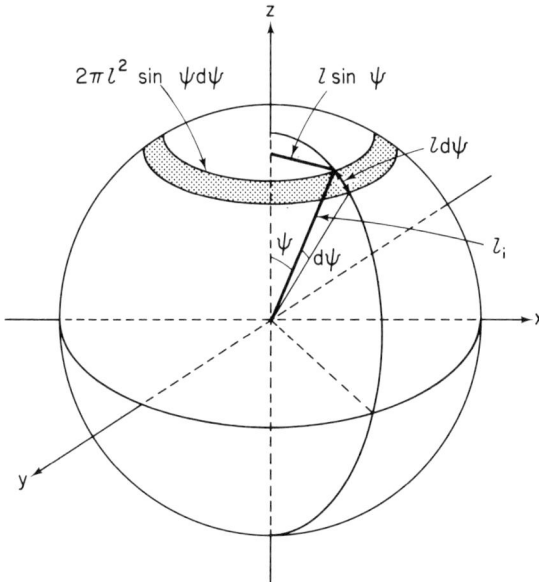

Figure 8-8. The projection of l_i on the z-axis. The bond l_i remains at angle ψ to the z axis and describes the solid angle $d\omega = 2\pi\, l^2 \sin\psi\, d\psi$.

limitation and that by appropriate modification the results can be generalized for real polymer chains.

Referring to Figure 8-8, we first consider a randomly selected bond $\mathbf{l}_i$ and its average projection on the z axis, $\mathbf{l}_{iz}$. Since all angles of $\mathbf{l}_i$ with respect to the z axis are equally probable, we can readily deduce that the average projection of $\mathbf{l}_i$ on the z axis is zero. To set the stage for subsequent discussion, we now present this result analytically. The probability that the projection of $\mathbf{l}_i$ lies between $\mathbf{l}_{iz}$ and $\mathbf{l}_{iz}+d\mathbf{l}_{iz}$ is to be represented by a distribution function $P(\mathbf{l}_{iz})\,d\mathbf{l}_{iz}$. The value of the projection is written in terms of the angle ψ that $\mathbf{l}_i$ makes the z axis—that is, $\mathbf{l}_i \cos\psi$. A specific value of $\mathbf{l}_{iz}$ can be attained in infinitely many ways as $\mathbf{l}_i$ rotates about the z axis at the angle ψ and describes the solid angle $2\pi l^2 \sin\psi\,d\psi$. The probability function, $P(\mathbf{l}_{iz})\,d\mathbf{l}_{iz}$, will be given by the ratio of this solid angle to the total solid angle $4\pi l^2$ or

$$P(\mathbf{l}_{iz})\,d\mathbf{l}_{iz} = \tfrac{1}{2}\sin\psi\,d\psi \tag{8-29}$$

This function can also be properly thought of as the ratio of the *area* that the tip of a unit vector describes as it rotates about the z axis at angle ψ, to the total area described by the tip of the unit vector. The average value of the projection $\overline{\mathbf{l}_{iz}}$, can be computed as

$$\overline{\mathbf{l}_{iz}} = \int_0^\pi \mathbf{l}_{iz} P(\mathbf{l}_{iz})\,d\mathbf{l}_{iz}$$

$$\overline{\mathbf{l}_{iz}} = \frac{1}{2}\int_0^\pi \mathbf{l}_i \cos\psi \sin\psi\,d\psi = 0 \tag{8-30}$$

This is the same result we obtained intuitively. It further illustrates the need for utilizing the root-mean-square average. The foregoing results are not very useful, so we compute l_{iz}^2

$$\overline{l_{iz}^2} = \frac{1}{2}\int_0^\pi l_i^2 \cos^2\psi \sin\psi\,d\psi = \frac{l^2}{3} \tag{8-31}$$

Note that $\mathbf{l}_i$ has lost its vector character, since $(\mathbf{l}_i)^2 = \mathbf{l}_i \cdot \mathbf{l}_i = l^2$. To utilize this result within the framework of one-dimensional, random-walk analysis, we take the square root of $\overline{l_{iz}^2}$—that is, the root-mean square of $\mathbf{l}_{iz}$—as the "step size" taken by our "particle." Thus

$$(\overline{l_{iz}^2})^{\frac{1}{2}} = \frac{l}{\sqrt{3}} \tag{8-32}$$

We now turn to formulating the results of our random-walk analysis in one dimension in terms of the problem at hand, first for the projection of the overall end-to-end distance along the z axis. We then expand this result to the three-dimensional case.

With σ large, each bond will make a contribution along the z axis, $(\overline{l_{iz}^2})^{\frac{1}{2}}$, plus or minus values occurring with equal probability. The number making a plus contribution will be N_+, while the number making a minus contribution will be N_-. The net contribution will be given by $(N_+ - N_-)$, and the projection of the end-to-end distance along the z axis, h_z, will be given by

$$h_z = (N_+ - N_-)(\overline{l_{iz}^2})^{\frac{1}{2}}$$

As per Eq. (8-21), $(N_+ - N_-) = 2\xi$, and by substitution we obtain

$$h_z = \frac{2l}{\sqrt{3}} \xi \qquad (8\text{-}33)$$

Rearrangement and differentiation yield

$$\xi = \frac{\sqrt{3} h_z}{2l} \qquad (8\text{-}34)$$

and

$$d\xi = \frac{\sqrt{3}}{2l} dh_z \qquad (8\text{-}35)$$

Substituting these results into Eq. (8-24), we obtain

$$P(h_z)\, dh_z = \left(\frac{b^2}{\pi}\right)^{\frac{1}{2}} \exp(-b^2 h_z^2)\, dh_z \qquad (8\text{-}36)$$

where

$$b^2 = \frac{3}{2\sigma l^2}$$

The function $P(h_z)\, dh_z$ represents the probability of finding the projection of the polymer chain end-to-end distance along the z axis between the values of h_z and $h_z + dh_z$.

The foregoing result is, of course, not unique to any axis, and we can write analogous expressions for the projections on the x axis and the y axis. Multiplying these three probabilities, we obtain

$$P(V)\, dV = \left(\frac{b}{\pi^{\frac{1}{2}}}\right)^3 \exp(-b^2 h^2)\, dV \qquad (8\text{-}37)$$

where $h^2 = h_x^2 + h_y^2 + h_z^2$, $dV = dh_x dh_y dh_z$, and $P(V)\,dV$ represents the probability of finding the chain end contained within the differential volume element dV at the location given by h_x, h_y, h_z. Equation (8-37), plotted as $P(V)$ versus h (see Figure 8-9), is a typical Gaussian bell-shaped curve. Another useful interpretation of Eq. (8-37) is that $P(V)\,dV$ represents the density distribution of chain-end distances for a system of many identical chains. Note from the curve how the most probable (or, alternately, the most frequently occurring) end-to-end distance is zero. Equation (8-37) does not render a valid representation of real situations for distances approaching the fully extended length σl. An expression suitable for high extensions can be derived from rubber elasticity theory.

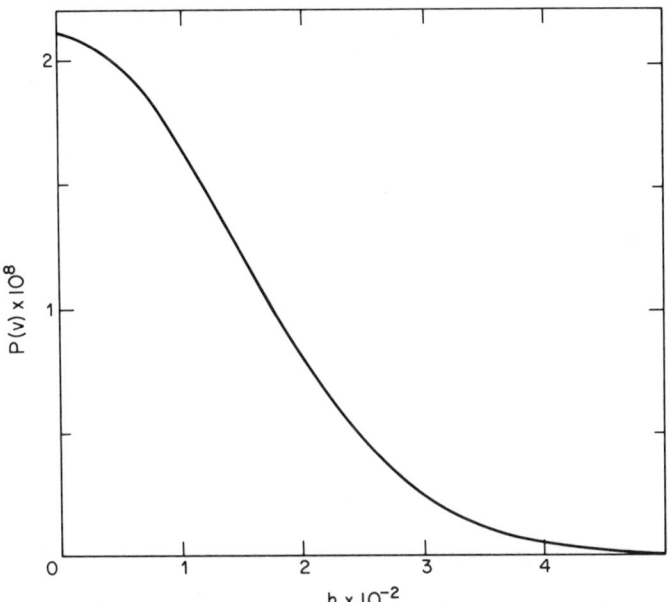

Figure 8-9. Gaussian density distribution of the chain displacement vectors for chain molecules consisting of 10^4 freely jointed segments, each of length $1 = 2.5$Å. The end-to-end length h is in Angstrom units and $P(V)$ is expressed in Å^{-3}.

An infinite set of values for h_x, h_y, h_z will lead to the same value for h^2. We are, however, interested in the magnitude of h^2 regardless of direction.

Equation (8-37) is cast into spherical coordinates by substituting $dV = h^2 \sin \theta \, dh \, d\theta \, d\phi$ and integrating over θ and ϕ as

$$W(h) \, dh = \int_{\theta=0}^{\pi} \int_{\phi=0}^{2\pi} \left(\frac{b}{\pi^{\frac{1}{2}}}\right)^3 \exp(-b^2 h^2) h^2 \sin \theta \, d\theta \, d\phi \, dh$$

$$W(h) \, dh = 4\pi \left(\frac{b}{\pi^{\frac{1}{2}}}\right)^3 \exp(-b^2 h^2) \, dh$$

(8-38)

The so-called *radial distribution function* $W(h) \, dh$ gives the probability of finding the value of the chain end-to-end distance between h and $h+dh$, regardless of direction; alternatively, it represents the fraction or density of molecules in a collection of many identical molecules with end-to-end distances between h and $h+dh$. The differential volume element is now a spherical shell rather than an infinitesimal cube. Eq. (8-38) is shown plotted in Figure 8-10. Note that in this form the most-favored distance is no longer zero.

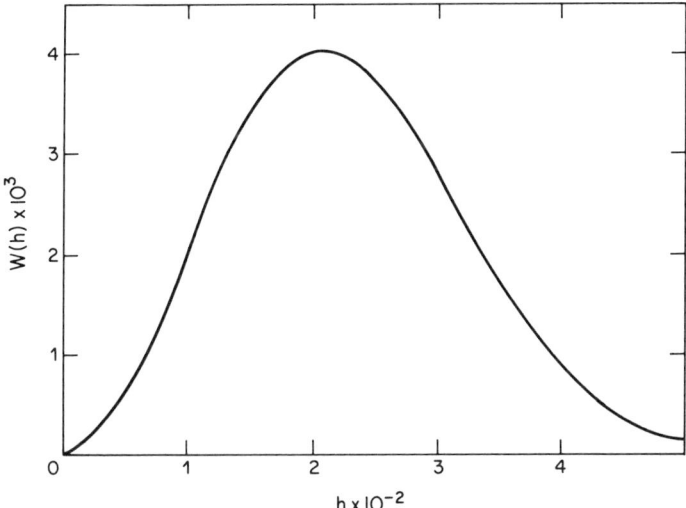

Figure 8-10. Radial distribution function $W(r)$ of the chain displacement vectors for the same polymer chains as in Figure 8-9 $W(h)$ is expressed in Å^{-1}.

To relate these results for the freely orienting chain model to actual polymer chains, refer to Figure 8-11 where a two-dimensional hindered chain is shown. The random-walk nature of this chain is restricted to steps that allow changes in direction along successive bonds ranging from $\pi/2$ to

$-\pi/2$. Every fifth bond has been connected by a dashed line on which no directional restrictions are placed. It becomes obvious that a sufficiently long restricted chain can be replaced by a less-restricted equivalent chain with fewer bonds and with an average equivalent bond length. As the length of these chains increases, the equivalent chain will approach the character of a freely orienting or completely unrestricted chain. Hence the form of Eq. (8-37) and (8-38) should be appropriate for real chains as well. This application is still restricted, of course, to polymer chains that are sufficiently long and flexible. From our discussion centering around Eq. (8-13), we need only to substitute β^2 for l^2 to relate numerically Eq. (8-37) and (8-38) to real polymer chains.

Figure 8-11. A two dimensional chain in hindered rotation. The angles between successive bonds has been restricted to the range of $-\pi/2$ to $\pi/2$.

810 The Distribution of Segments Relative to the Center of Mass

The results of the preceding section can be extended to describe the distribution of the chain segments about the center of mass. Equation (8-38) provides, in effect, a measure of the probability of finding the locus of one chain segment relative to another. In developing Eq. (8-38), we were interested in the first and last chain segments—the chain ends—but we could have chosen one very close to the center of mass as our point of reference. This point leads us to believe that the density distribution of chain segments about the center of mass can be represented by an equation identical in form to Eq. (8–38). Thus we define the segmental density distribution function $\rho(r)$ as

$$\rho(r) = A \exp(-B^2 r^2) \tag{8-39}$$

where r is the distance from the center of mass and A and B are constants to be determined. To evaluate A and B, we make use of two constants whose values we know—σ and R_G. A spherical layer of volume, $4\pi r^2\, dr$, will contain $4\pi r^2 \rho(r)\, dr$ segments such that the total number of segments will be given by

$$\sigma = \int_{r=0}^{\infty} 4\pi r^2 \rho(r)\, dr$$

Substituting Eq. (8–39) and evaluating the integral,

$$\sigma = \frac{\pi^{3/2} A}{B^3} \tag{8-40}$$

Similarly, R_G^2 is the average value of r^2 for all segments in a given conformation. Since there are $4\pi r^2 \rho(r)\, dr$ segments for which r has a given value, we obtain

$$R_G^2 = \frac{\displaystyle\int_{r=0}^{\infty} 4\pi r^4 \rho(r)\, dr}{\displaystyle\int_{r=0}^{\infty} 4\pi r^2 \rho(r)\, dr}$$

After substituting Eq. (8–38) and evaluating,

$$R_G^2 = \frac{3\pi^{3/2} A}{2\sigma B^5} \tag{8-41}$$

By combining Eqs. (8–40) and (8–41) and then solving for A and B, we obtain

$$A = \sigma \left(\frac{3}{2\pi R_G^2}\right)^{3/2} \tag{8-42}$$

$$B = \frac{3}{2} R_G^2 \tag{8-43}$$

and, finally,

$$\rho(r) = \sigma \left(\frac{3}{2\pi R_G^2}\right)^{3/2} \exp\left(\frac{-3r^2}{2R_G^2}\right) \tag{8-44}$$

9
Rubber Elasticity

Of all the materials known to man, no other class can match or even approximate the remarkable behavior of polymeric rubberlike materials. Consider the ordinary rubber band (cis-polyisoprene): It can undergo rapid deformation from five to ten times its original length without rupturing. In fact, it becomes stronger as it is stretched, achieving its highest modulus of elasticity at maximum elongations. Even at elongations as high as 400 or 500 percent it maintains the capacity to recover spontaneously its original dimensions while suffering little permanent set.

In this chapter we will first discuss the properties of polymeric rubber materials, or *elastomers* as they are frequently called, with regard to the molecular properties that are required for a material to exhibit this behavior. In the process, we shall discuss a few important technological aspects associated with rubber elasticity. We will then seek to relate the conformational properties and parameters, described in Chapter 8, through quantitative theories with such macroscopic parameters as stress, strain, and temperature. The material to be presented here is important not only because elastomers are important but also because the development of rubber-elasticity theory represents one of the most promising attempts to relate our knowledge of the behavior of a polymeric material on a molecular level with its behavior on a gross, macroscopic level.

901 Definition and Properties of Elastomers

An *elastomer* is a polymeric material that exhibits rubber elasticity. We can define rubber elasticity in two contexts: according to the mechanical properties a material must exhibit to be so classified and according to the molecular or structural properties a material must possess to exhibit these mechanical properties. In the mechanical vein, elastomers must (1) stretch rapidly under tension with little loss of energy as heat. The extent of stretching may reach as high as 500 to 1000 percent. (2) On the release of stress, they must recover their original dimensions with snap or rebound. (3) This recovery must be achieved with little permanent set or deformation.

In order for a material to exhibit rubber elasticity (1) it must be a high polymer. Indeed, the ability to exhibit rubber elasticity is a property unique to polymers. (2) In order to maintain the high-segmental mobility for rapid stretching, it must be used well above the glass transition in the unstretched state. Therefore, for a material to exhibit rubber elasticity under the usual environmental conditions, it must possess weak intermolecular bonding forces and have a flexible chain structure. (3) Also, to ensure high segmental mobility, it should possess a minimum degree of order in the stable unstressed state. (4) To prevent permanent deformation and to allow for rapid retraction, the chains must be permanently crosslinked. The extent of crosslinking should be low to maintain segmental mobility. Review of Sections 703 through 705 will provide further clarifying discussion.

Figure 9-1 shows the importance of vulcanization* with regard to an

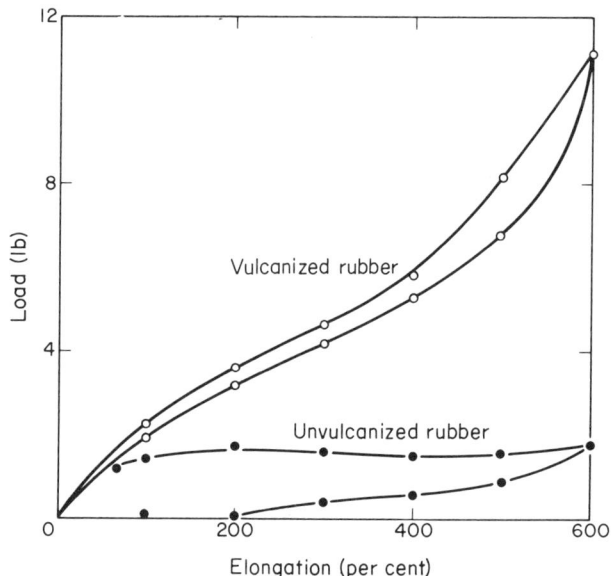

Figure 9-1. Stress-strain curves to 600% elongation and back, typical of unvulcanized and vulcanized natural rubber. [L. Hock and S. Bostrom, Kautschuk, **2**, 130 (1926).]

* Recall from Section 108 that vulcanization is the process by which rubber materials are usually crosslinked.

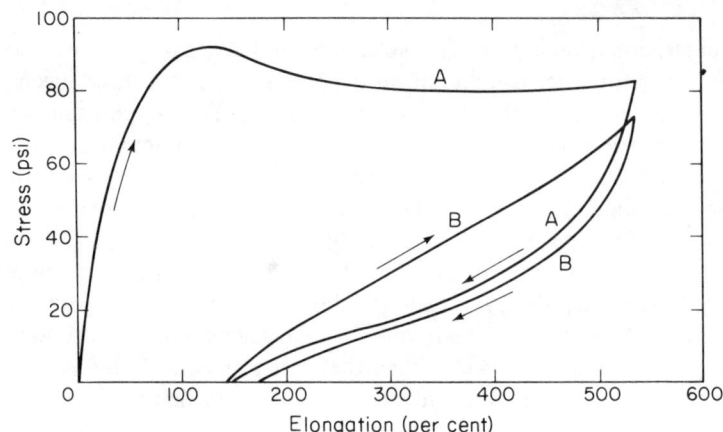

Figure 9-2. Mechanical conditioning during the first (AA) and second (BB) cycles of stress-strain, typical of unvulcanized natural rubber. [L. Hock and S. Bostrom, op. cit. Figure 9-1.]

elastomer's ability to retain its shape during a cycle of loading and unloading. The upper curve is for vulcanized rubber (cis-polyisoprene), whereas the lower curve is for the unvulcanized rubber. Note that the vulcanized form maintains a higher strength with increasing elongation and exhibits less of a hysteresis effect on cycling. As shown in Figure 9-2, mechanical conditioning occurs in an unvulcanized elastomer upon repeated cycling. Figure 9-2 shows the effect of just two cycles, but the changes in resistance to stretching, tensile strength, energy absorption, and permanent set become smaller on repeated cycling. Vulcanized rubbers also undergo a similar conditioning effect, although to a lesser extent.

Two other important aspects of rubber behavior are found in the reinforcing effect of carbon black and the crystallization that may take place upon stretching stereoregular materials. The lower curves in Figure 9-3 compare the behavior of three unreinforced but vulcanized elastomers. Of the three, SBR [styrene-butadiene-rubber or poly(butadiene-co-styrene) (75/25)] is noncrystallizable, whereas the natural and butyl rubbers are. The ability of these latter materials to sustain stress at high elongations increases due to the onset of crystallization. Crystallization is induced by the orientation of the molecules that occurs with increasing elongation, and it may reach as high a value as 80 percent. Figure 9-4 shows the x-ray diffraction patterns for isobutyl rubber in the unstretched state and at 1400 percent elongation; the induced crystallinity and orientation is clearly evidenced.

The upper curves of Figure 9-3 show the marked effect of carbon black as a reinforcing agent. Note, particularly, the close correspondence of reinforced SBR and natural rubber. The mechanism by which reinforcement operates is not clearly understood, but the carbon particles appear to introduce a network of weak "fix points" in addition to the network established

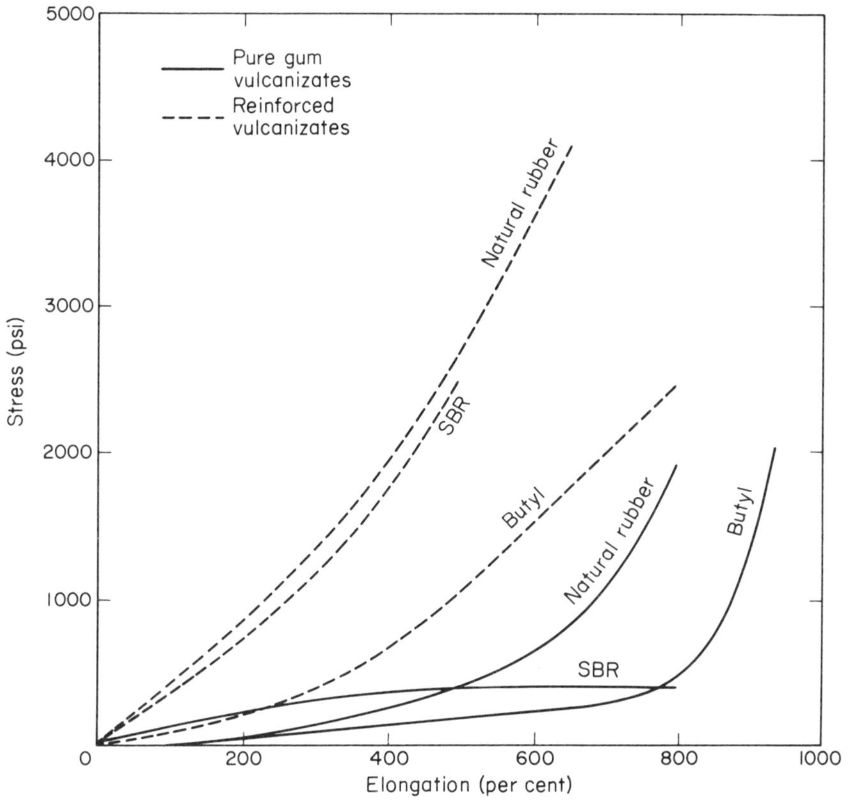

Figure 9-3. Stress–strain curves typical of several vulcanized and reinforced elastomers. [A. X. Schmidt and C. A. Marlies, *Principles of High Polymer Theory and Practice*, McGraw-Hill Book Co., Inc., New York, 1948.]

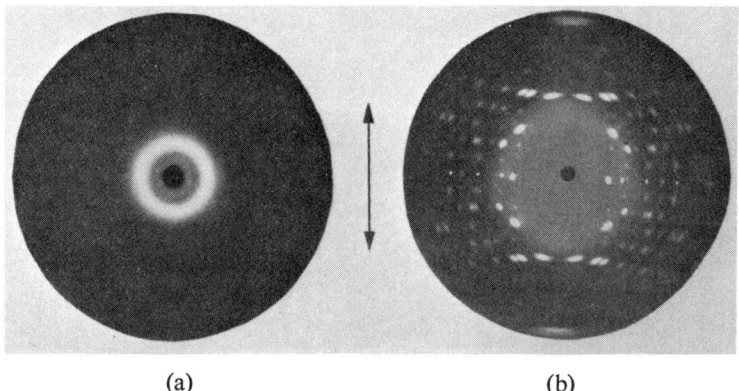

(a)   (b)

Figure 9-4. X-ray diffraction patterns for (a) unstretched polyisobutylene and (b) polyisobutylene stretched to the maximum. The fiber axis is vertical. [C. S. Fuller, G. J. Frosch, and N. R. Page, J. Am. Chem. Soc., **62**, 1905 (1940). Reproduced with permission of the copyright owner, The American Chemical Society.]

by the covalent crosslinks. Figure 9-5 shows the effect of a variety of additives on the tensile strength and elongation at break for natural rubber. Carbon black is by far the most important reinforcing agent for elastomers, and the one used in compounding vehicle tires.

As important and interesting as the preceding material may be, to discuss the various technological aspects of elastomer behavior in greater detail is beyond the scope of this text. Rather, in the remainder of this chapter, discussion will focus on the theory describing the behavior of unreinforced

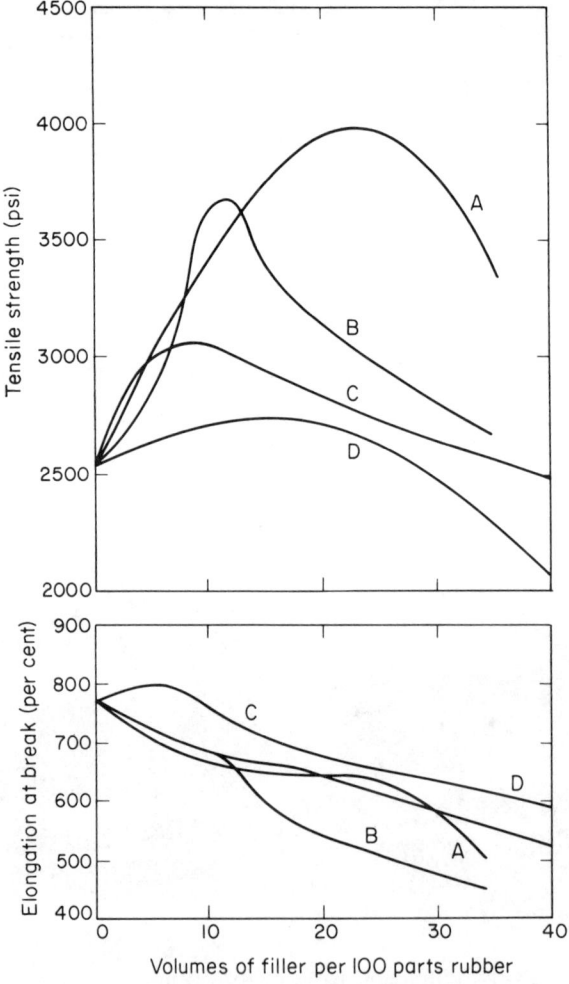

Figure 9-5. Effects of filler loading on tensile properties typical of natural rubber (2): curve A, carbon black; curve B, china clay; curve C, zinc oxide; curve D, whiting. [H. Barron and F. H. Cotton, Trans. Inst. Rubber Ind., **7**, 209 (1931-32).]

but vulcanized (or pure-gum vulcanizates) elastomers, as exemplified by natural rubber and butyl rubber [poly(isobutyl-co-isoprene) (97/3)]. In particular, we will be concerned with the *equilibrium relaxation modulus*-elongation behavior at various temperatures. In ordinary uncrosslinked, linear amorphous polymers, the relaxation modulus eventually decays to zero. Crosslinked materials, however, attain an equilibrium relaxation modulus after an initial period of decay, because the crosslinks limit the free molecular slip required for complete relaxation. In the literature on rubber elasticity, this equilibrium relaxation modulus is simply referred to as the stress, but this particular stress should not be confused with its nonequilibrium counterpart, encountered in Chapters 7 and 11. The conditions under which one might expect to achieve stress equilibrium will be discussed subsequently.

THEORY OF RUBBER ELASTICITY

A twofold approach is used in developing the theory for rubber elasticity. First a thermodynamic argument is used to show that rubber elasticity arises primarily from distortion of the random conformational state and not distortion of primary valence bond lengths or bond angles. Secondly, a statistical approach is used with these conclusions to develop expressions to describe the equilibrium relaxation state in terms of the stress, elongation, and temperature. Historically, these approaches were applied in the reverse order: the thermodynamic approach following applications of the statistical method (ca. 1940) by about ten years. The order followed here was selected for pedagogical reasons.

902 Thermodynamics of Rubber Elasticity

We set out to derive relations for the equilibrium relaxation stress as a function of temperature for a fixed elongation. This development involves application of the first law of thermodynamics, $dU = \tilde{q} - \tilde{w}$, where U is the internal energy, $\tilde{q}$ is the heat absorbed by the system, and $\tilde{w}$ is the work done by the system on its surroundings. If a specimen is subjected to a strain, γ, then the work done in further straining it $d\gamma$, while undergoing a volume change dV, is

$$\tilde{w} = P\,dV - f\,d\gamma \tag{9-1}$$

where $f\,d\gamma$ is the work done on the rubber to extend it, and $P\,dV$ is the work done by the rubber against the constant external atmosphere, P. For a reversible process, $\tilde{q} = dq = T\,dS$; and f and P represent equilibrium values. Then the first law becomes

$$dU = T\,dS - P\,dV + f\,d\gamma \qquad (9\text{-}2)$$

This expression may be further developed by introducing the Gibbs or Helmholtz free-energy expressions.

By substituting the total derivative of the Helmholtz function ($A = U - TS$) for dU in Eq. (9-2), we obtain

$$dA = -P\,dV - S\,dT + f\,d\gamma \qquad (9\text{-}3)$$

For f at constant T and V

$$f = \left(\frac{\partial A}{\partial \gamma}\right)_{T,V} \qquad (9\text{-}4)$$

The partial of the Helmholtz function with respect to γ at constants T and V is given by

$$\left(\frac{\partial A}{\partial \gamma}\right)_{T,V} = \left(\frac{\partial U}{\partial \gamma}\right)_{T,V} - T\left(\frac{\partial S}{\partial \gamma}\right)_{T,V} \qquad (9\text{-}5)$$

Substituting (9-5) into (9-4), we obtain

$$f = \left(\frac{\partial U}{\partial \gamma}\right)_{T,V} - T\left(\frac{\partial S}{\partial \gamma}\right)_{T,V} \qquad (9\text{-}6)$$

This is an equation for a straight line, which contains two terms that cannot be determined by experiment. Seeking a substitute expression for $(\partial S/\partial \gamma)_{T,V}$, we first note that, in general, entropy may be a function of P, V, T, or γ. With P, T, and γ as the independent variables, and holding T constant, we can obtain

$$dS = \left(\frac{\partial S}{\partial P}\right)_{T,\gamma} dP + \left(\frac{\partial S}{\partial \gamma}\right)_{T,P} d\gamma \qquad (9\text{-}7)$$

By rearranging and imposing the condition of constant V, we obtain

$$\left(\frac{\partial S}{\partial \gamma}\right)_{T,V} = \left(\frac{\partial S}{\partial P}\right)_{T,\gamma}\left(\frac{\partial P}{\partial \gamma}\right)_V + \left(\frac{\partial S}{\partial \gamma}\right)_{T,P} \qquad (9\text{-}8)$$

At constant volume, it is inconceivable that the external pressure could ever be affected by γ, so $(\partial P/\partial \gamma)_V = 0$ and we are left with

$$\left(\frac{\partial S}{\partial \gamma}\right)_{T,V} = \left(\frac{\partial S}{\partial \gamma}\right)_{T,P} \tag{9-9}$$

By substituting the total derivative of the Gibbs function ($F = H - TS = U + PV - TS$) for dU in Eq. (9-2), we find

$$dF = V\,dP - S\,dT + f\,d\gamma \tag{9-10}$$

We further obtain

$$\left(\frac{\partial E}{\partial \gamma}\right)_{T,P} = f \tag{9-11}$$

and

$$\left(\frac{\partial F}{\partial T}\right)_{P,\gamma} = -S \tag{9-12}$$

By comparing the second derivative of $(\partial F/\partial \gamma)_{T,P}$ with respect to T at constant P and γ and the second derivative of $(\partial F/\partial T)_{P,\gamma}$ with respect to γ at constant T and P, we find

$$-\left(\frac{\partial S}{\partial \gamma}\right)_{T,P} = \left(\frac{\partial f}{\partial T}\right)_{P,\gamma} \tag{9-13}$$

Upon substitution of Eqs. (9-13) and (9-9) into Eq. (9-6), we arrive at the desired equation of state

$$f = \left(\frac{\partial U}{\partial \gamma}\right)_{T,V} + T\left(\frac{\partial f}{\partial T}\right)_{P,\gamma} \tag{9-14}$$

Thus, by Eqs. (9–14) and (9–13), the contributions of internal energy and entropy to the equilibrium stress can be determined in terms of equilibrium stress-temperature measurements conducted at constant P. Eq. (9–14) was derived by Flory* from a slightly different starting point that resulted in a somewhat more complex derivation.

* P. J. Flory, *Principles of Polymer Chemistry*, op. cit.

903 Equilibrium Stress-Elongation Experiments

The first definitive work along the lines suggested by Eq. (9–14) was reported by R. L. Anthony, R. C. Caston, and Eugene Guth* in 1942. It was first formulated on the basis of statistical arguments. They worked with sulfur-vulcanized natural rubber. This material exhibits little hysteresis and suffers little permanent set. It was also shown by x-ray studies that appreciable crystallization occurs only at elongations greater than 350 percent. The samples were tested at the highest temperatures first so as to allow the rapid attainment of equilibrium. The remaining equilibrium-stress values were measured at successively lower temperatures. As a final check, the temperatures were increased in successive steps to see if the original temperature curve was followed.

Shown in Figures 9–6 and 9–7 are the stress-temperature curves so obtained for elongations ranging from 3 to 370 percent. The crossed readings indicate readings taken during increasing temperature. These show good agreement between decreasing and increasing temperature measurements.

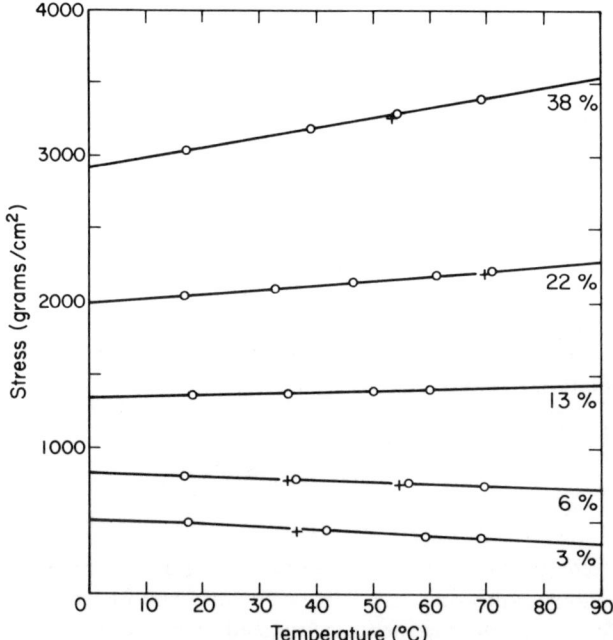

Figure 9–6. Stress–temperature curves obtained for elongations ranging from 3 percent to 370 percent. [R. L. Anthony, R. H. Caston, and E. Guth, *J. Phys. Chem.*, **46**, 826 (1942). Reproduced with permission of the copyright owner, The American Chemical Society.]

*R. L. Anthony, R. C. Caston, and E. Guth, *J. Chem. Phys.*, **48**, 826 (1942).

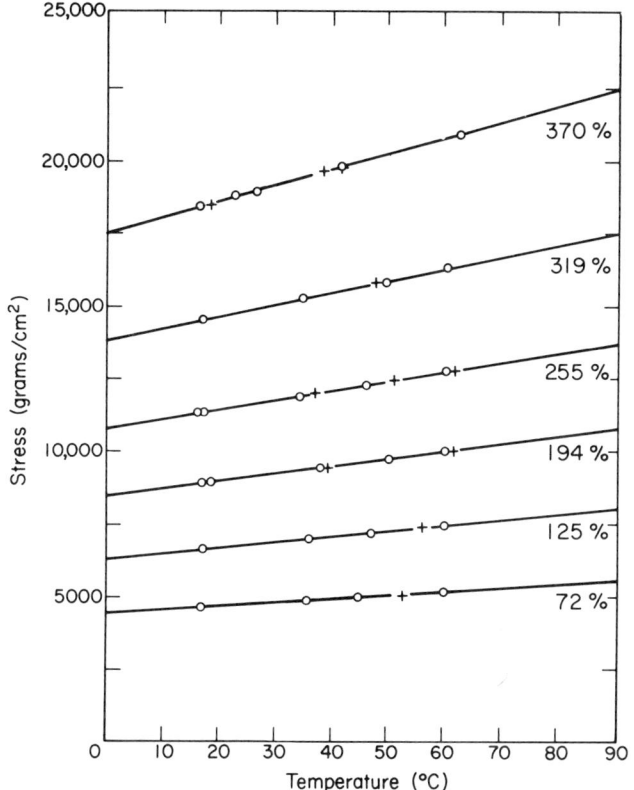

Figure 9-7. Stress-temperature curves obtained for elongations ranging from 3 percent to 370 percent. [Guth et al., op. cit. Figure 9-6.]

In agreement with Eq. (9-14), all the curves appear as straight lines. The slopes increase with increasing elongation, but for elongations less than 10 percent, the slopes are negative. If the data are cross-plotted to obtain stress-elongation curves, we find, as shown in Figure 9-8 for curves obtained at 20 °C and 70 °C, that the curves cross at 10 percent elongation. This inversion point is due to the volumetric thermal expansion present in both the stretched and unstretched state. That is, even though the samples were held at constant elongation, the unstressed length of the samples at various temperatures would be different because of ordinary volume expansion.

Corrected stress-elongation curves were obtained by applying the following correction to the elongation data in Figures 9-6 and 9-7.

$$\text{Corrected } \gamma = \frac{\text{uncorrected } \gamma - a\,\Delta T}{1 + a\,\Delta T} \tag{9-15}$$

where a is the thermal coefficient expansion and ΔT is the rise in temperature above the temperature at which the specimen was clamped—in this case 25 °C. These corrected curves are shown in Figures 9–9 and 9–10. Note the absence of an inversion point. The corrected stress-elongation curves are, in turn, obtained by cross-plotting these data to obtain the results shown in Figures 9–11 and 9–12. These curves are again straight lines in accordance with Eq. (9–14).

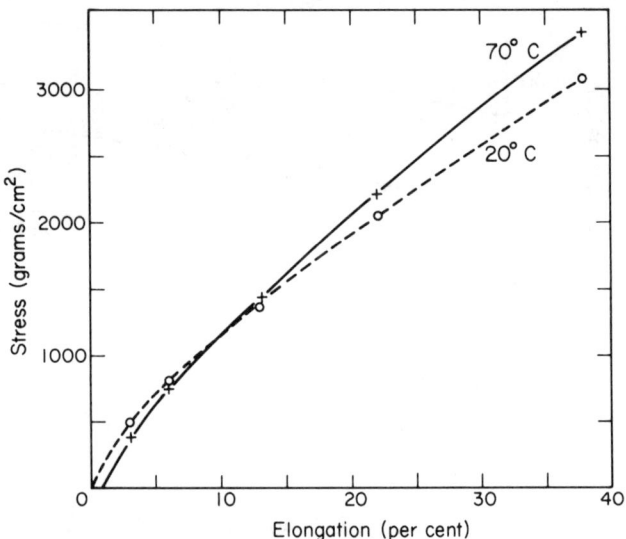

Figure 9–8. Isothermal stress-strain curves obtained at 20°C and 70°C from the stress–temperature curves of Figure 9–6. [Guth, et al., op. cit. Figure 9–6.]

The stress-temperature data can be resolved into their entropy and internal energy components by determining intercepts at absolute zero and by determining the product of the slopes and the absolute temperature. Shown in Figure 9–13 are the resolved stress elongation data for 20 °C. From these data it is clear that the change in entropy with elongation up to about 350 percent is responsible for more than 90 percent of the total stress at room temperature, whereas the contribution of the internal energy is less than 10 percent. In other words, the retractive force is due almost entirely to the tendency of the extended rubber molecules to return to their unperturbed, random conformational state, and as a first approximation the internal energy may be neglected. Above 350 percent the foregoing analysis breaks down because of the onset of crystallization, which was not allowed for in the theory.

If the rubbers were composed of ideal chains with bond lengths and angles fixed and with completely free rotation about all chain bonds—that is, with no bond deformation—then we would expect no contribution from the

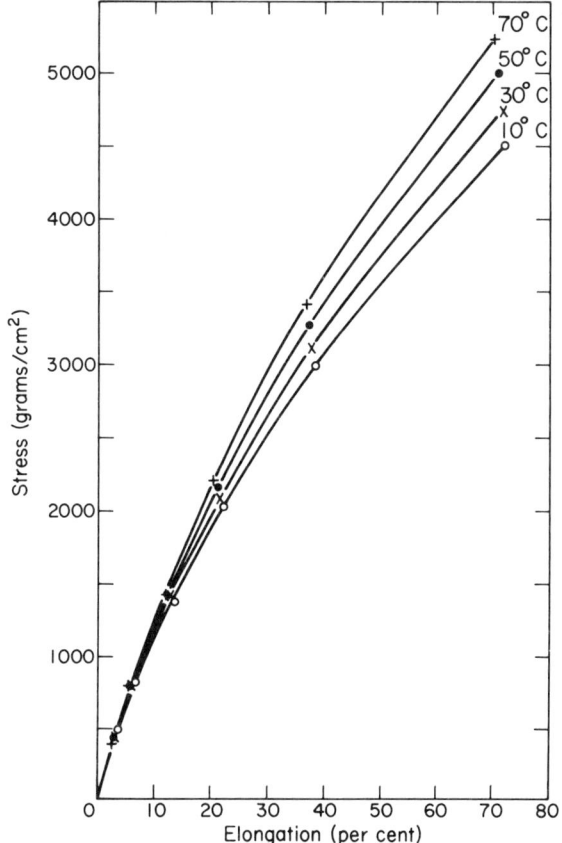

Figure 9-9. Corrected set of isothermal stress–strain curves. [Guth, et al., op. cit. Figure 9-6.]

internal energy term. Thus we define an *ideal rubber* as one in which $(\partial U/\partial \gamma)_{T,V} = 0$, and the equilibrium stress is given by

$$f = T\left(\frac{\partial f}{\partial T}\right)_{P,\gamma} = -T\left(\frac{\partial S}{\partial \gamma}\right)_{T,V} \qquad (9\text{-}16)$$

This is in analogy with an ideal gas in which $(\partial U/\partial V)_T = 0$ and $P = T(\partial P/\partial T)_V = T(\partial S/\partial V)_T$. In both an ideal gas or an ideal rubber, variations in the pressure or force of retraction arise solely due to entropy variations.

It is interesting to contrast this behavior with that of metals or ceramics. These materials are Hookean solids. They exhibit reversible extension but only up to 1 percent elongation, and the stress is directly proportional to strain. As a Hookean solid is strained, bond lengths are increased and bond angles are deformed from their equilibrium values. As a result, its internal

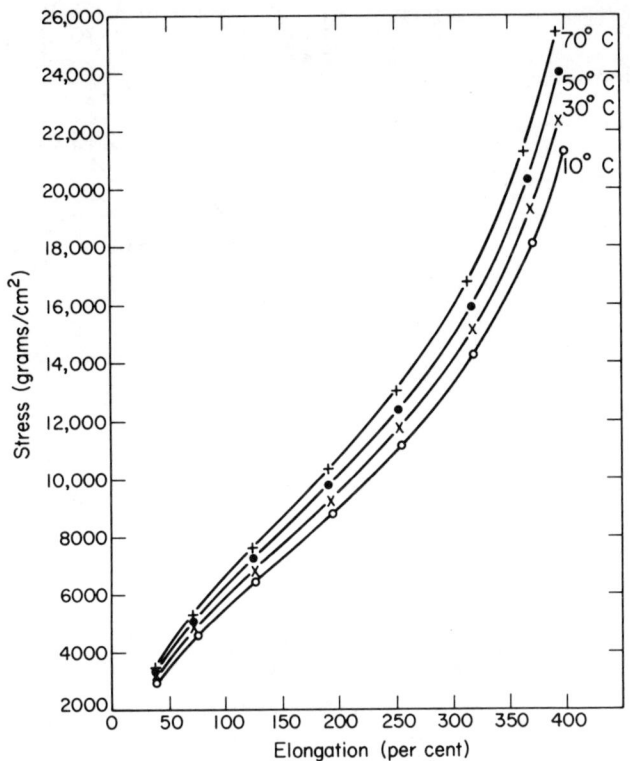

Figure 9-10. Corrected set of isothermal stress–strain curves. [Guth, et al., op. cit. Figure 9-6.]

energy increases. If the deforming force is released, within the sample's elastic region, equilibrium will be restored and the sample will return to its original dimensions. Since interatomic and intermolecular forces of attraction are effective over very short distances, an elongation of greater than 1 percent would cause the bonds of the unstrained state to break and new ones to form, thus inducing an irrecoverable change in atomic or molecular arrangement.

904 Statistical Rubber-Elasticity Theory

Consider a single chain in an ideal environment in which the preferred end-to-end distance is zero. Visualize clamps gripping the ends of this chain. For any separation of the clamps, forces will arise that, on the average, will tend to pull the clamps together. It is only when the chain ends are clamped together that the average force of restoration (to the unperturbed equilibrium conformational state) on the clamps will be zero. To compute the average force of restoration for a collection of polymer chains, not in their equilibrium conformations, we can compute the average force of restoration prevailing when the distribution of end-to-end distances is given by the Gaussian distribution. Fully equivalent, or course, is the problem of computing the average restoration force for a single chain over the range of all possible conformations. We will use the latter approach.

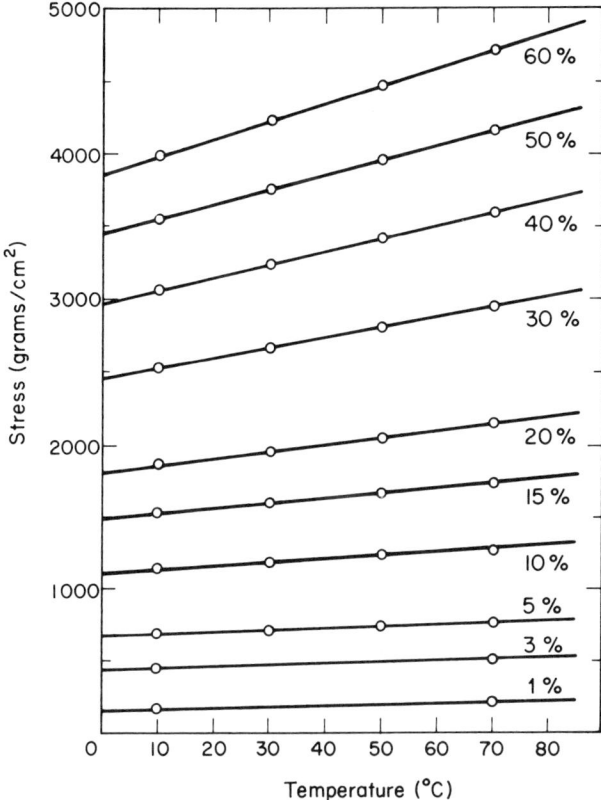

Figure 9–11. Corrected set of stress–temperature curves obtained from Figures 9–9 and 9–10. [Guth, et al., op. cit. Figure 9–6.]

Our initial efforts will be based on an *ideal network*, with the following properties:
1. The network is structurally perfect, with no defects arising from the existence of chain ends, closed loops, or permanent chain entanglements. In effect, all chain ends will be "clamped."
2. The chain-size distribution for any chain section is given by the Gaussian distribution function.

3. No energy is stored by bond distortion.
4. All chain sections have the same molecular weight.

Deviations from this idealization will be discussed subsequently. In a mechanical vein, we further assume that the sample undergoes an incompressible affine deformation.

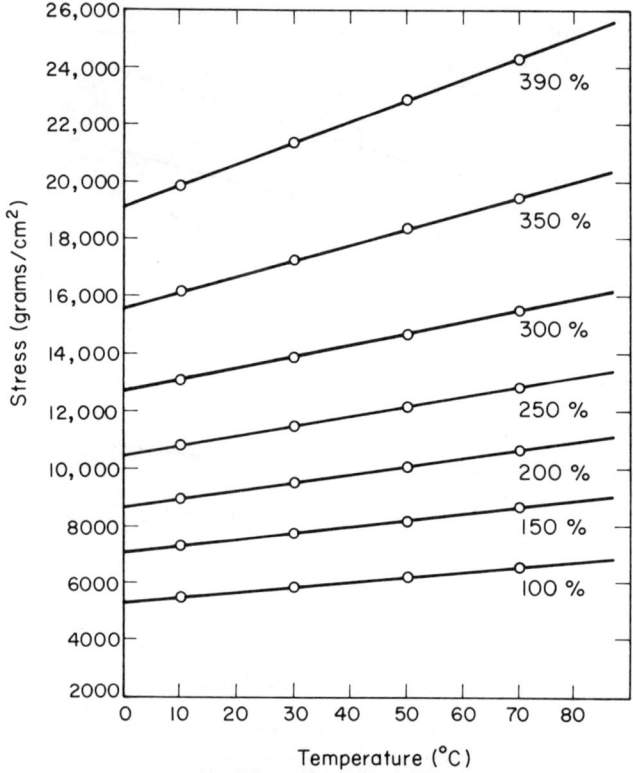

Figure 9–12. Corrected set of stress–temperature curves obtained from Figures 9–9 and 9–10. [Guth, et al., op. cit. Figure 9–6.]

Consider a single freely orienting chain section between two crosslinks and, in particular, the bond $\mathbf{l}_i$. The entire chain is under tension by forces applied at both ends in the z direction. The force on a segment is F_z. If the bond $\mathbf{l}_i$ makes an angle ψ with the z axis, the orienting energy E_i associated with this bond because of the force F_z is

$$E_i = F_z \mathbf{l}_i \cos \psi \tag{9-17}$$

According to the Boltzmann equation, the number of bonds oriented within

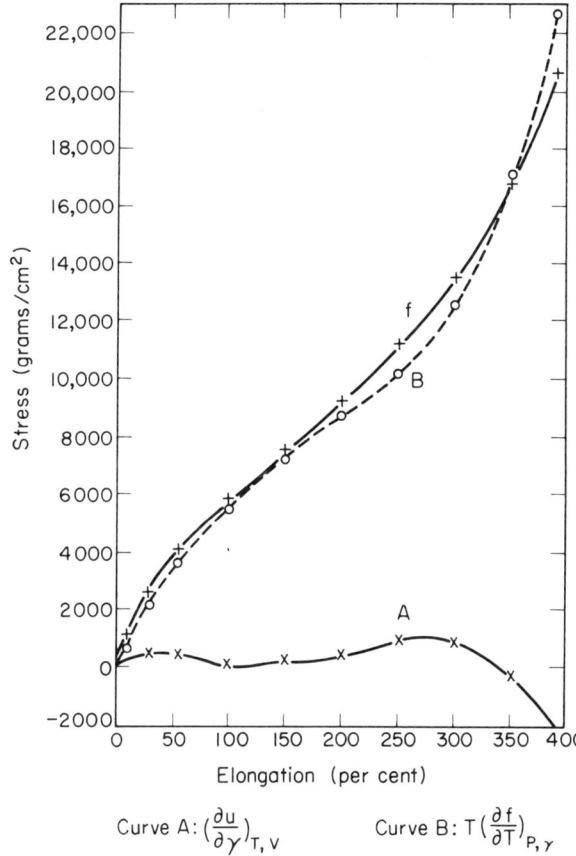

Figure 9–13. Plot of the value of each term and their sum against the corrected elongations at 20°C. [Guth, et al., op. cit. Figure 9-6.]

the solid angle $d\omega$ is then proportional to

$$P(E_i) = (\text{constant})\, e^{-E_i/kT}\, d\omega \tag{9-18}$$

The probability of finding a bond $\mathbf{l}_i$ at angle ψ with the z axis with orienting energy E_i will be given by

$$\mathbf{l}_{iz} = \frac{\int_0^{\pi} \mathbf{l}_i \cos\psi (2\pi \sin\psi\, d\psi) \exp(F_z \mathbf{l}_i \cos\psi / kT)}{\int_0^{\pi} (2\pi \sin\psi\, d\psi) \exp(F_z \mathbf{l}_i \cos\psi / kT)} \tag{9-19}$$

Integration yields

$$\mathbf{l}_{iz} = l_i \left[\coth\left(\frac{lF_z}{kT}\right) - \left(\frac{kT}{lF}\right) \right] \tag{9-20}$$

This is the well-known Langevin function, which is represented by $L(lF_z/kT)$. For the freely orienting chain, h_z will be given by

$$h_z = \sigma l L\left(\frac{lF_z}{kT}\right) \qquad (9\text{-}21)$$

To find F_z, the inverse L^* of the Langevin function is taken. Thus

$$F_z = \left(\frac{kT}{l}\right)L^*\left(\frac{h_z}{\sigma l}\right) \qquad (9\text{-}22)$$

for $h \ll \sigma l$, this can be written as

$$F_z \cong \left(\frac{kT}{l}\right)\left[3\left(\frac{h_z}{\sigma l}\right)\right] \qquad (9\text{-}23)$$

One way to look at the term $(3kT/\sigma l^2)$ is as a spring constant of zero-equilibrium length. Observe that as T increases, F_z increases, which is in line with the rubber-elasticity behavior observed in Figures 9–11 and 9–12.*

Next, consider a unit cube of material subject to an incompressible, affine deformation. Since the chain ends are all fastened to the network structure and since the deformation is affine, the changes in chain-end separation will be in the same proportion for each and every chain section and the same as that for the entire cube. If a cube is extended by the amount α† in the z direction, then, since the material is incompressible, the cube must contract by the amount $1/\sqrt{\alpha}$ in the x and y directions. If the coordinates for the end-to-end vector of a chain section are (x_0, y_0, z_0) before deformation, then after deformation they will be $(\alpha z_0, x_0/\sqrt{\alpha}, y_0/\sqrt{\alpha})$. The work for one chain undergoing this deformation, W', will be

$$W' = \int_{z_0}^{z_0 \alpha} F_z\, dz + \int_{x_0}^{x_0/\sqrt{\alpha}} F_x\, dx + \int_{y_0}^{y_0/\sqrt{\alpha}} F_y\, dy \qquad (9\text{-}24)$$

If all the forces are given by Eq. (9–23), we can find

$$W' = \left(\frac{3kT}{2\sigma l^2}\right)[(\alpha^2 - \alpha^{-1})z_0^2 + (\alpha^{-1} - 1)h_0^2] \qquad (9\text{-}25)$$

where $h_0^2 = x_0^2 + y_0^2 + z_0^2$.

* The development of this paragraph follows that for dipole-moment calculations for polar molecules in electric field.
† $\alpha = 1 + \gamma$

The average work per chain for a collection of chains will depend on the distribution of chain-end distances as given by the Gaussian distribution. Hence, to ascertain the average work per chain $\overline{W}'$, we substitute from Eq. (8-7) and (8-32) for h_0^2, σl^2 and for z_0^2, $\sigma l^2/3$ to obtain

$$\overline{W}' = \left(\frac{3kT}{2}\right)\left[\frac{1}{3}(\alpha^2 - \alpha^{-1}) + (\alpha^{-1} - 1)\right] \qquad (9\text{-}26)$$

For a collection of chains containing ν chains per unit volume, the work per unit volume, W, is

$$W = \left(\frac{\nu kT}{2}\right)[(\alpha^2 - \alpha^{-1}) + 3(\alpha^{-1} - 1)] \qquad (9\text{-}27)$$

The work unit per volume is also given by $\int_{\alpha=1}^{\alpha} f\,d\alpha$ such that $F = (\partial W/\partial \alpha)$. Thus

$$f = (\nu kT)(\alpha - \alpha^{-2}) \qquad (9\text{-}28)$$

For an ideal network

$$\nu = \frac{\rho N_{Av}}{\overline{M}_c} \qquad (9\text{-}29)$$

where ρ = polymer density, N_{Av} = Avogadro's number, and $\overline{M}_c$ = the average molecular weight of the chain segments in the network. Notice how detailed consideration of the polymer structure did not enter into the development of Eq. (9-28). Equation (9-28) has been independently derived in alternative ways by several people: W. Kuhn (1936), H. M. James and E. Gouth (1943), L. R. G. Treloar (1943), and P. J. Flory and J. Rehner (1943).* The foregoing development follows that of F. Bueche.†

The greatest difficulties in applying Eq. (9-28) or in assessing its validity arise in assigning a suitable value to ν and in establishing the conditions to achieve equilibrium relaxation. These difficulties are discussed in the ensuing sections.

905 Nonideal Networks

Real networks usually exhibit nonideal character to varying degrees for a variety of reasons. We have already treated one aspect of this problem in

---

* W. Kuhn, *kolloid-Z*, **76**, 258 (1936); H. M. James and E. Guth, *J. Chem. Phys.*, **11**, 470 (1943); L. R. G. Treloar, *Trans. Faraday Soc.*, **39**, 241 (1943); P. J. Flory and J. Rehner, *J. Chem. Phys.*, **11**, 512 (1943).
† F. Bueche, *Physical Properties of Polymers* (New York: Wiley, 1962).

establishing Eq. (9-29) by using an average molecular weight for all chain sections. In most situations, this is probably not a serious compromise. The contributions due to bond distortion have already been discussed in Section 903. Gaussian statistics may not apply for several reasons. A significant portion of the chains may be (1) too short or (2) stretched beyond the limit of the Gaussian approximation. (3) Intramolecular interaction or (4) topological restrictions due to the manner in which the chains are connected, via such structural defects as entanglements, may impair the validity of Gaussian statistics. And (5) strain-dependent orientation effects may arise. Gaussian statistics will be most applicable when (1) the primary molecular weight (before crosslinking) is sufficiently high, (2) the extent of crosslinking is sufficiently low, (3) the chains are crosslinked in the amorphous, unextended state, and (4) the extension is sufficiently low.

Three types of structural defects are commonly considered: chain ends, closed loops, and permanent chain entanglements. Of the three, the presence of chain ends is probably the most important and the only one for which a suitable analytical correction can be made. Closed loops make no contribution toward sustaining an equilibrium stress. Chain entanglements might make significant contributions if the network were sufficiently extended and if the crosslink density were sufficiently light.

906 Elastically Effective Chain Sections

A number of chain sections—namely, chain ends—will not contribute to the elasticity of the network. It is necessary, therefore, to develop an expression for the effective number of chain sections per network, ν_e. If N is the number of primary chains per unit volume, then $N-1$ is the minimum number of crosslinks per unit volume that must form to create one giant molecule. As additional crosslinks are added, the structure acquires the character of a network. Only connections in excess of N effectively fix the structure so that it responds elastically to deformation. The number of crosslinks in a perfect network will be $(\nu/2)$. The effective number of crosslinks in a real structure will be

$$\frac{\nu_e}{2} = \frac{\nu}{2} - N \qquad (9\text{-}30)$$

With Eq. (9-29) and with $N = (N_{Av}/\bar{M}_n)\rho$,

$$\nu_e = \left(\frac{\rho N_{Av}}{\bar{M}_c}\right)\left(1 - \frac{2\bar{M}_c}{\bar{M}_n}\right) \qquad (9\text{-}31)$$

This expression for v_e should be substituted for v in Eq. (9–28). Many investigations have shown that Eq. (9–31) is a suitable representation of the effective number of chain sections in a network. Considerable difficulty may arise in utilizing this relationship, however, because of the difficulty in accurately measuring or controlling $\bar{M}_c$. For example, in the vulcanization of a highly unsaturated material like polyisoprene, it is difficult to assess the crosslinking density by simple chemical analysis and stoichiometry, for the efficiency of the crosslinking process may not be known. In practice, $\bar{M}_c$ may change during experimentation or application because of oxidative or mechanical degradation. Natural rubber is particularly prone to such failings.

VERIFICATION OF RUBBER ELASTICITY THEORY

We are concerned here with the experimental verification of rubber-elasticity theory as embodied in Eqs. (9–28) and (9–31). The two principal problems associated with such verification are (1) establishing favorable conditions for the attainment of equilibrium and (2) determining and controlling the network structure such that $\bar{M}_c$ may be specified. Thus the conditions under which rubber elasticity can be applied will be dictated by the conditions limiting the attainment of equilibrium, plus the accuracy with which the network structure can be determined.

907 Testing the Stress-Strain Relation

As a rule, it is very difficult to achieve true equilibrium. The problem is compounded by the lack of a suitable test for showing when equilibrium is actually attained. Moreover, mechanical and oxidative degradation or other changes in the network structure may occur during testing or application unless special precautions are taken to guard against such complications.

According to Eq. (9–28), plots of f versus $(\alpha - \alpha^{-2})$ should yield straight lines of constant slope for any particular sample and test temperature. A more critical test is obtained by plotting the ratio $f/[T(\alpha - \alpha^{-2})]$ against α; this ratio should remain constant for all test conditions. Such plots, based on the data of Guth et al. in Figures 9–11 and 9–12, are presented in Figures 9–14 and 9–15 for 10 °C and 70 °C. Plots of f versus $(\alpha - \alpha^{-2})$ yield straight lines up to about $\alpha = 3.5$ (250 percent elongation) in accordance with theory. Very curiously, however, the ratio $f/[T(\alpha - \alpha^{-2})]$ is not constant with elongation, even though the correspondence between temperatures is rather good. Mixed confirmation of the sort just described is quite common but remains unexplained.

In general, the stress-strain behaviors of many elastomers do not seem to correlate well with Eq. (9-28), in that plots of f versus $(\alpha - \alpha^{-2})$ do not yield straight lines. Even the results of Guth et al. have been criticized on the grounds that the stresses that they report may not be *true* equilibrium stresses. In an independent study, Geoffrey Gee* followed their procedure

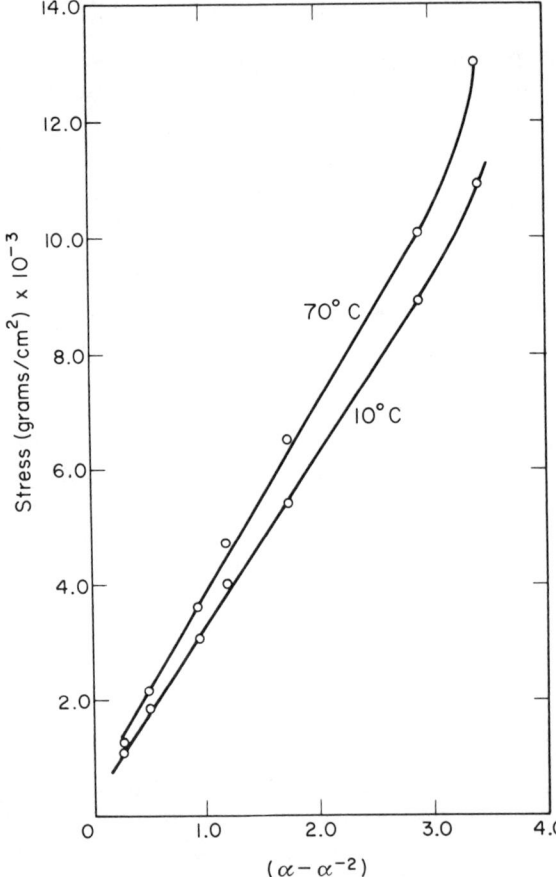

Figure 9-14. Stress vs. $(\alpha - \alpha^{-2})$ for data of Figures 9-11 and 9-12.

and found that the stress measured for any particular condition depended on the highest temperature to which the sample had been previously heated. These errors were small, and he showed that they could be eliminated by first swelling the rubber and then deswelling it while the sample was clamped in its test position. This swelling procedure apparently allowed for the rapid attainment of equilibrium due to increased chain-segment mobility.

* *Trans. Faraday Soc.*, **42**, 585 (1946).

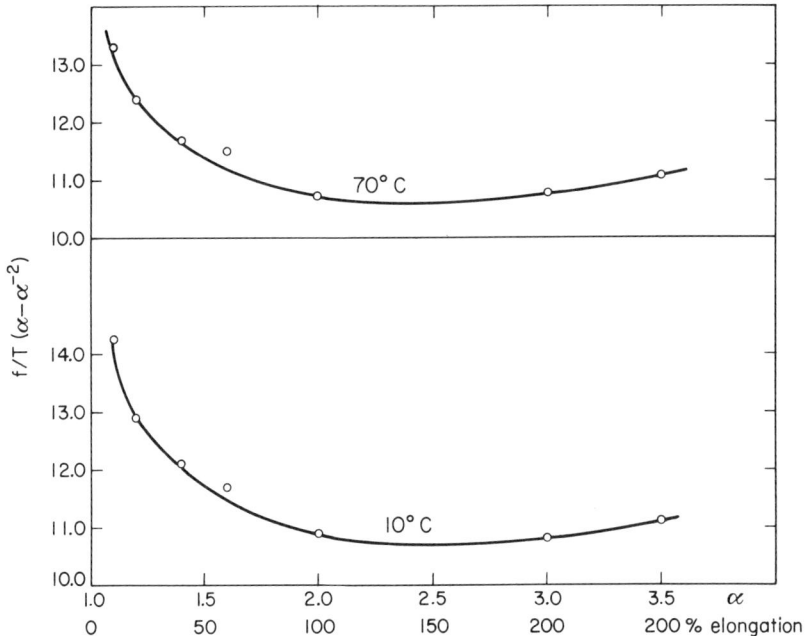

Figure 9–15. $f/T\,(\alpha - \alpha^{-2})$ vs. α for data of Figures 9–11 and 9–12.

An empirical relation that seems to represent experimental data somewhat better than Eq. (9-28) has been proposed. This equation, the so-called *Mooney Rivlin equation*, is given by

$$f = C_1(\alpha - \alpha^{-2}) + C_2(1 - \alpha^{-3}) \qquad (9\text{-}32)$$

where the constant C_1 corresponds to $(v_e kT)$ and C_2 is an empirical constant. Equation (9-32) reduces to Eq. (9-28) when $C_2 = 0$. This equation can be rearranged in terms of a *reduced stress* to yield

$$\frac{f}{(\alpha - \alpha^{-2})} = C_1 + C_2 \frac{1}{\alpha} \qquad (9\text{-}33)$$

Plots of reduced stress, $f/(\alpha - \alpha^{-2})$, versus $(1/\alpha)$ allow for the determination of both C_1 and C_2. Although it is true that in many instances Eq. (9-32) will serve as a better correlation than Eq. (9-28), the general use of such an empirical relation is hardly satisfactory, for it lacks a fundamental theoretical basis. Flory and Ciferri undertook an extensive study of several elastomer systems in an attempt to ascertain the relevance of Eq. (9-32) and the empirical constant C_2. Shown in Figure 9–16 are their reduced stress data plotted against time for several elastomers. The manifestation of stress relaxation diminishes in an order—poly(methyl methacrylate) (PMMA) at 145 °C, PMMA at 175 °C, natural rubber (NR), butyl rubber (B), and poly(dimethyl siloxane) (S)—which corresponds with the proximity of the elastomers to their glass transitions. In another aspect of their study, they elongated samples in steps up to 100 percent elongation, measuring tensions after 3

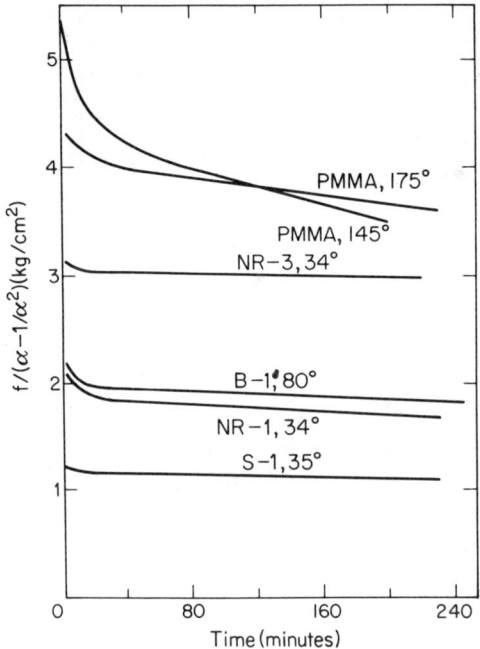

Figure 9-16. Reduced stress plotted against time at constant elongtion $\alpha = 1.50$ in all cases. NR-1: Natural rubber; NR-3: Natural rubber (higher crosslinking than NR-1); PMMA: poly(methyl methacrylate); B-1: Butyl rubber; S-1: poly(dimethyl siloxane). [A. Ciferri and P. J. Flory, J. Appl. Phys., **30**, 1498 (1959).]

minutes at each length; after reaching maximum elongation, they reversed the procedure. The divergence between the ascending and descending stages was taken as a measure of hysteresis. The results of this procedure applied to the materials of Figure 9–16 are presented in Figure 9–17. The amount of hysteresis exhibited by these data is in the same order as the diminution in the stress-relaxation data of Figure 9–16. The hysteresis data were also plotted as reduced stress versus $(1/\alpha)$. The data on the extension portion of the curve yielded straight lines in accordance with Eq. (9–33) but those in contraction did not. This indicates a fundamental weakness in the ability of Eq. (9–32) to correlate data. Nonetheless, the ascending elongation curves were used to calculate apparent values of C_1 and C_2. In line with previous observations, the C_2 values decreased with increasing departure from T_g. Similar experiments were performed on samples swollen with a suitable diluent; these showed decreasing C_2 values with increasing diluent content.

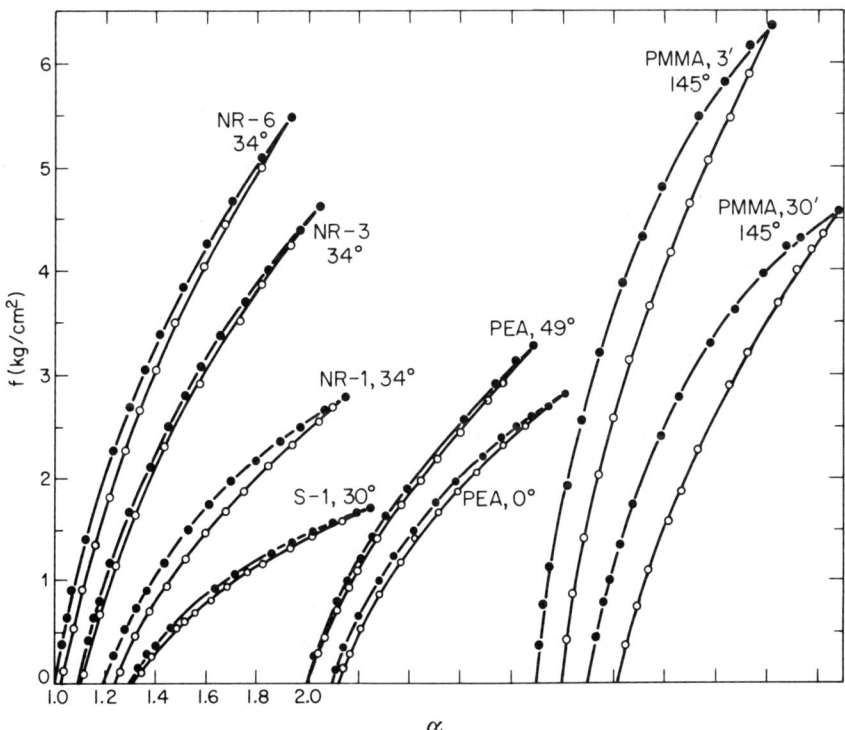

Figure 9-17. Representative stress–strain results for samples indicated. Abscissa scales shifted by integral units of 0.1. Points for elongation phase and for retraction phase of the standard procedure cycle. Time interval 3 min at elongation except for PMMA as noted on graph. [A. Ciferri and P. J. Flory, op. cit. Figure 9-16.]

To summarize, experimental evidence seems to indicate that departure from equilibrium diminishes under conditions that enhance chain mobility, (i.e., higher temperatures and diluent swelling) and thereby enhance the ability of the system to come to equilibrium. On the basis of these results, as well as others, rubber elasticity theory has a firm theoretical foundation and considerable experimental support.

908 The Effect of Network Structure

The combination of Eqs. (9-28) and (9-31) yields

$$f = (kT)(\alpha - \alpha^{-2})\left(\rho \frac{N_{Av}}{\bar{M}_c}\right)\left(1 - \frac{2\bar{M}_c}{\bar{M}_n}\right) \qquad (9\text{-}34)$$

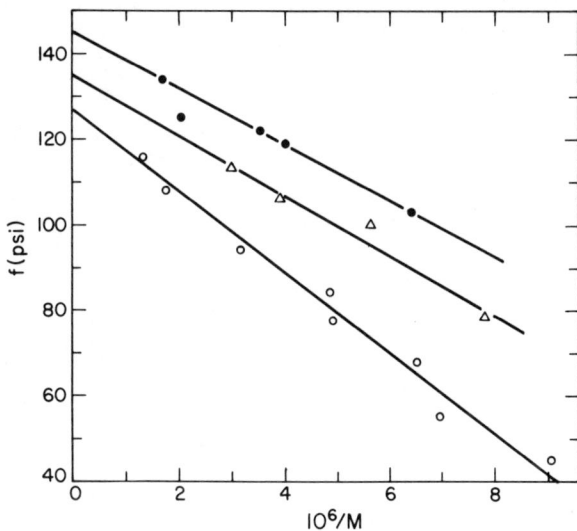

Figure 9-18. Tension versus reciprocal of original molecular weight M for three samples of butyl rubber with different $\bar{M}_c$ at constant extension ratio $\gamma = 4.0$. [P. J. Flory, Ind. Eng. Chem., **38**, 417 (1946). Reproduced with permission of the copyright owner, The American Chemical Society.]

This equation suggests that we examine the stress-elongation properties at constant elongation of a series of samples, all with identical values for $\bar{M}_c$ but different values for $\bar{M}_n$. Plots of stress at constant elongation against $(1/\bar{M}_n)$ should yield straight lines.

Flory executed the experiments suggested by Eq. (9-34) with three isobutyl rubber samples, prepared by copolymerizing isobutylene with 0.5, 1.0, and 1.2 percent isoprene. The samples so prepared were heterogeneous in molecular weight, and the isoprene units—included to provide sites for crosslinking—were presumed to be randomly distributed, since the copolymerization was ideal. Each of the three samples was fractionated according to molecular weight. The fractions were collected and crosslinked by a standard procedure so that all samples containing the same percentage of isoprene contained the same number of crosslinks. This procedure provided the requisite experimental samples: Three series of crosslinked rubbers, each series comprised of samples with identical crosslink densities (and therefor identical $\bar{M}_c$ values) but variable primary molecular weight $\bar{M}_n$. The results of Flory's experiments at 300 percent elongation ($\alpha = 4$) are presented in Figure 9-18. His results clearly support the theory. Good straight lines are evidenced, and the slope of the lines increases with increasing $\bar{M}_c$, also in accordance with theory. Finally, the values of $\bar{M}_c$ determined from these slopes agreed well with values independently predicted from the theory of rubber gelation.

PART IV

Polymer Rheology

By definition rheology is a discipline concerned with the study of the deformation and flow properties of materials. In the broadest sense it embraces all materials. A closely allied discipline is the science of mechanics in which the primary objective is to establish the governing continuity, momentum and energy relationships for any flow or deformation process. The purpose of rheology is to supply relations that describe the material properties. Both are required for the complete formulation of any given problem. As already observed, the two extremes of rheological behaviors are Newtonian (completely viscous) and Hookean (completely elastic) behavior. These lead directly to two branches of mechanics: fluid mechanics and solid mechanics. These two extremes are not of especial interest to the rheologist who is chiefly concerned with the situations in which materials exhibit intermediate character: namely that of viscoelasticity.

Although our primary objective here is to discuss the viscoelastic character of polymers, it should be understood that much of what we are to discuss is perfectly general. This is especially true of Chapter 10 in which the fundamental concepts of viscoelasticity are introduced. Further, from a rheological point of view one may consider four types of polymer systems: solids, melts, solutions, and dispersions. Here we shall only be concerned with solids and melts—Chapters 11 and 12, respectively—but much of the development will pertain to solutions and dispersions as well. The essential point is that the following discussion is really broader than the subject headings or text might indicate.

10
Introduction to Polymer Rheology

Rheology is a science devoted to the study of deformation and flow. The term is derived from the Greek and means "study of flow." In a recent brochure of the U.S. Society of Rheology, rheology is defined more precisely as a "branch of mechanics" dealing with "those properties of materials which determine their response to mechanical force."

Under a given external force, different materials will exhibit different types of deformation response. For many materials, the deformation response may be completely elastic and therefore temporary, or it may be completely viscous and hence permanent. Many materials, under ordinary stress conditions, will exhibit elastic and viscous responses simultaneously. Such materials are said to be *viscoelastic*, and polymers constitute one of the most important classes of viscoelastic materials.

Two basic approaches have been used in the rheological studies of polymers. The first is the continuum approach, in which the particular molecular structure of the material is ignored and its properties are described phenomenologically. In the second approach, the rheological behavior of the material is predicted on the basis of molecular structure.

The continuum theories have dominated the field of rheology in the past twenty years and have lead to several significant quantitative results. On the other hand, the molecular point of view of rheology is relatively unexplored, and existing molecular theories are generally applicable only to simple flow and deformation situations. Nonetheless, the importance of the molecular approach should be emphasized, particularly in the rheological studies of polymeric materials. The structural complexities of these materials and the relatively large order of magnitude of their molecular scale suggest very strongly that deep understanding of the rheology of these materials can only be gained with due account to molecular mechanism.

Both the phenomenological and molecular approaches are used in this text, although the latter is used primarily in a qualitative fashion. A notable exception to the last point was our treatment of rubber elasticity in Chapter 9, which required considerable understanding of molecular mechanisms. On the other hand, our discussion in Chapter 7 of the relaxation modulus-temperature behavior was undertaken primarily from a phenomenological standpoint, with qualitative discussion of molecular mechanisms. We will continue to use this particular dual approach in subsequent discussion.

1001 Types of Mechanical Deformation

The state of stress in a continuous body is described by means of the stress tensor, which represents the stress components acting on all the sides of a small cubic element of material. The deformation of a body is described by means of kinematic quantities, generally tensors, such as the strain and strain-rate tensors. The strain tensor, for example, is a measure of the relative change of distance between two neighboring material points, whereas the strain-rate provides a measure of the rate of change with respect to time of the distance between two neighboring material points.

The constitutive equation relates the stress in a body to its deformation. It is a mathematical relation in which the stress tensor is given as a function of the kinematic quantities on which it depends for any particular material. Together with the principles of conservation of mass, momentum, and energy, it forms a determinate system that, in principle, should be sufficient to furnish a complete solution subject to physically reasonable boundary and initial conditions. The constitutive equation does not represent a physical law; rather, it is a mathematical statement that embodies our total experience with a given material whose rheological properties we wish to describe or predict.

If the constitutive equation for a particular material is linear in all its arguments, the material is said to exhibit *linear behavior*; otherwise we speak of *nonlinear behavior*. The linear case has wide applicability for viscoelastic materials under modest strain conditions, and its use will be considered in Chapter 11 on polymeric solids. The nonlinear case must be used to describe large deformations, such as those found in the flow of polymeric melts. Actually, all viscoelastic materials exhibit nonlinear behavior, and it is only in the limit of small strains that linear behavior is observed. The advanced treatment of nonlinear behavior is founded in rather complicated mathe-

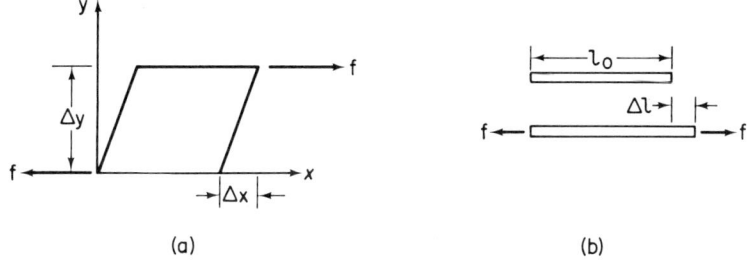

Figure 10-1. (a) Simple shear; (b) simple extension.

matical theory; consequently, situations requiring such treatment cannot be covered in this book. The application of tensor notation also involves mathematics beyond the scope of this text. Thus discussion will be restricted to linear, one-dimensional problems and to some elementary, nonlinear problems. In a sense, these limitations will enhance the presentation rather than hinder it, for we can concentrate on the properties of interest rather than on the intricacies of the mathematics that would otherwise be required.

Among the many types of deformations one can imagine, two of the most important, and the ones with which we will be concerned, are simple shear and simple extension. *Simple shear*, as illustrated in Figure 10-1(a), involves the application of lateral force f to a unit cube of material, parallel to its top surface area A. The shear stress is $S = f/A$, and the resulting shearing strain is $\gamma = \Delta x/\Delta y$, which is simply the tangent of the angle of deformation. As we shall see, this definition of strain follows from Newton's law for viscous flow. For the study of viscoelastic materials, simple shear is by far the most important type of deformation. It can be performed on liquids and solids alike and on materials with all degrees of intermediate character. For example, steady-state laminar flow of a liquid between two parallel plates set in relative motion at constant speed is a simple shear flow with a constant rate of strain.

Simple extension, as illustrated in Figure 10-1(b), occurs when a rod is subjected to a normal tensile force f. If its cross-sectional area is A, the tensile stress is $\epsilon = f/A$, and the resulting strain is $\epsilon = \Delta l/l_0$.

1002 Simple Rheological Responses

The two simplest types of material responses are the ideally elastic or purely viscous. The type of response does not depend on whether the deformation is simple shear or simple extension. To avoid unnecessary duplications, without sacrificing generality, we will adopt the practice of developing most relations in terms of shear deformations. The conversion to tensile representation will be obvious where appropriate.

The ideal elastic response

This response takes place instantaneously upon the application of stress and disappears instantaneously upon the release of stress. In an ideally elastic body, the stress is a function of the strain only. In 1676 Hooke showed that for small strains, certain solids exhibit elasticity, and the stress is directly proportion to the strain. Hooke's law for shear is

$$S = G\gamma \quad (10\text{-}1)$$

where the proportionality constant G is the elastic *shear modulus*. Its reciprocal is referred to as the elastic *shear compliance J*. For representation in tensile deformation, G is replaced by E and J by D. The elastic constants are related by the following expression:

$$E = 2(1+\mu)G \quad (10\text{-}2)$$

where μ is the Poisson ratio. For elastic materials μ is about 0.5, so that the elastic tensile modulus is three times greater than the shear modulus.

Pure viscous flow

In 1685 Newton postulated that the shearing force necessary to maintain the motion of a liquid plane at a constant velocity in laminar flow is proportional to the velocity gradient. Figure 10–2 shows two parallel flat planes of

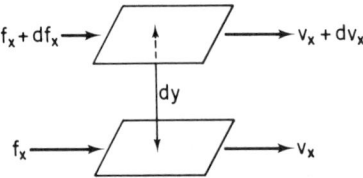

Figure 10–2. Two parallel planes of liquid in pure viscous flow.

liquid separated by a distance dy with a velocity difference of dv. Newton's observation is expressed by

$$S = \frac{dv}{dy}\eta$$

where η is a proportionality constant referred to as the *viscosity*. By the chain rule, if x is the forward direction of shear, then

$$\frac{dv}{dy} = \frac{d}{dy}\frac{dx}{dt} = \frac{d}{dt}\frac{dx}{dy} = \frac{d\gamma}{dt} = \text{the } \textit{rate of strain}$$

Newton's law for viscous flow can be written simply as

$$S = \eta \dot{\gamma} \qquad (10\text{-}3)$$

where the raised dot will indicate differentiation with respect to time.

1003 Introduction to the Viscoelastic Properties of Polymers

All polymer systems of engineering interest, as well as a wide variety of other industrially important materials, exhibit rheological behaviors far more complex than those outlined in the preceding section. As already indicated, materials exhibiting ideal elastic deformation and pure viscous flow represent the behavioral extremes of a wide spectrum of rheological behaviors. That is, we must consider a continuous spectrum of viscoelastic behavior in which ideal elastic and pure viscous responses can be thought of as special cases. Where a material fits into this viscoelastic spectrum depends on two factors in addition to its intrinsic nature—namely, temperature and the material's stress-strain history. For example, molten polymer flowing under the influence of shear rates comparable to those encountered in processing equipment will exhibit elastic retraction upon release of the applied stress. If the shear rate is reduced to sufficiently low levels, this elastic behavior will be reduced to zero and the molten polymer will flow as a highly viscous Newtonian fluid. On the other hand, water, which under most flow conditions behaves as a Newtonian fluid, could exhibit elastic properties if the shear rate were sufficiently high.

In studies of the rheological properties of polymers, it is convenient to consider four types of systems:

1. Polymer solutions
2. Polymer dispersions
3. Polymer melts
4. Polymer solids

All of these types may exhibit viscoelastic behavior, and in one form or another they are all important to the engineer. We will restrict our efforts to polymeric solids and polymeric melts, which are examples of viscoelastic solids and fluids, respectively. Much of our discussion would, however, be germane to a discussion of polymeric solutions and dispersions as well.

One of the most elementary manifestations of viscoelastic behavior exhibited by a linear amorphorus polymer is a process known as *relaxation* in which the molecules undergo a rearrangement to relieve an applied stress. If a polymeric solid is strained by the application of an external stress, the material exerts a retractive (elastic) stress response. As time passes, however, the force required to hold the initial strain decays from the initial value to zero. This stress decay is a manifestation of its viscous nature. On a molecular level, we can picture the polymeric mass as existing in the unstrained state as an intimate entanglement of chains. In their equilibrium conformation of the unstrained state, the individual chains are pictured as being enclosed within spherical envelopes. When the entangled network structure is strained, these conformations are elongated. In seeking to return to their unperturbed state, the molecules give rise to the measured stress response. Under the continued influence of the applied stress, the molecules begin to slide past one another as they seek to reestablish their equilibrium conformations. As this molecular flow occurs, the stress is relieved.

If in the relaxation experiment, the clamp maintaining the constant strain is released, the specimen will tend to return to its original dimensions. The extent to which the specimen will recover its original dimensions depends on the magnitude of elastic stress remaining in the specimen. The earlier the strain is relieved, the higher will be the stress, and the greater the extent of strain recovery. The later the strain is relieved, the lower will be the stress, and the lower the extent of strain recovery. In a picturesque vein, we like to associate a polymer's ability to recover its original dimensions with a "memory" and the decreasing ability of a polymer to recover its dimensions with the passage of time with a *"fading memory."* In the extremes, a purely elastic body has a perfect memory and a purely viscous body has no memory. Hence a viscoelastic body has a fading memory.

The manner in which the viscoelastic properties of a material are measured and the form of the data in which they are reported depend on the form of the material. A wide variety of rheological measuring devices are employed. Where each type is used depends on where the material fits into the viscoelastic spectrum. As one can readily imagine, the devices used to measure the properties of a viscoelastic solid would be quite different from those used to measure the properties of a viscoelastic liquid. We will not undertake an extensive study of instruments or experimental methods; for this, the reader is referred to the reference works of van Wazer *et al.*, Ferry, or McCrum et al.

For a further introduction of the concepts of viscoelasticity in this chapter, we will rely on two of the simplest types of experiments usually performed on polymeric solids. Relaxation and the effect of time can be studied quite easily in experiments that hold either the stress or strain constant. In the *relaxation experiment*, the sample is strained promptly to a given deformation and maintained, while the decreasing stress is measured as a function of time. In the *creep experiment*, a given stress is applied to the specimen and maintained constant while the increasing strain is measured as a function of time.

SOME SIMPLE LINEAR VISCOELASTIC MODELS

The behavior of linear viscoelastic materials is most conveniently introduced in terms of mechanical models or analogs. We use the spring as the analog for ideal elastic or Hookean behavior and the dashpot as the analog for pure viscous or Newtonian behavior. These elements are then simply combined in parallel and series arrangements to give one-dimensional, linear models of viscoelastic behavior. Three-dimensional systems are treated in advanced treatises on linear viscoelasticity.

The rudimentary models to be discussed initially are useful solely as pedagogical tools, for they do not accurately represent the behavior of real materials. The physical insight that they lend to our understanding of viscoelastic materials, however, makes their use worthwhile on an introductory level. Presently we will develop more general linear models of viscoelasticity, models that do have practical value in that they represent the actual behavior of linear, amorphous polymer systems. In fact, they represent the behavior of any viscoelastic material in the linear region.

1004 The Maxwell Model

The series combination of an elastic element and a viscous element is termed a *Maxwell model*. This model is illustrated in Figure 10–3 with the rheological equations governing the behavior of each component element. It is most helpful in representing viscoelastic relaxation. Since the elements are in series, the stress across each element will be common; the total strain and strain rate will be the sum of the elemental strains and strain rates. To obtain the rheological equation for the Maxwell model, we simply add the strain rates

$$\dot{\gamma} = \frac{1}{G}\dot{S} + \frac{1}{\eta}S \qquad (10\text{–}4)$$

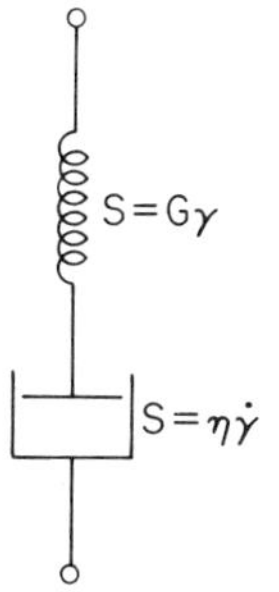

Figure 10–3. The Maxwell model.

This equation is commonly rearranged as follows:

$$\dot{S} = G\dot{\gamma} - \frac{1}{\lambda}S \tag{10-5}$$

where $\lambda = \eta/G$.

λ is an extremely important viscoelastic parameter, which will be the subject of much discussion in later sections. Note that λ has the units of time and that it characterizes the viscoelastic nature of the element very concisely, as the ratio of the viscous portion of the response to the elastic portion. This naturally occurring parameter is taken to be the *response time* of the model. We shall see that this parameter arises in other situations as well and that the concept of a material response time is quite natural and universal.

The time response of the Maxwell model for relaxation conditions is obtained by observing that when $\dot{\gamma} = 0$. Eq. (10-5) integrates to

$$\frac{S}{S(0)} = e^{-t/\lambda} \tag{10-6}$$

where $S(0)$ is the original stress and λ is the time required for $S/S(0)$ to decay from unity to $1/e$ or 0.37. This stress decay behavior is shown in Figure 10-4.

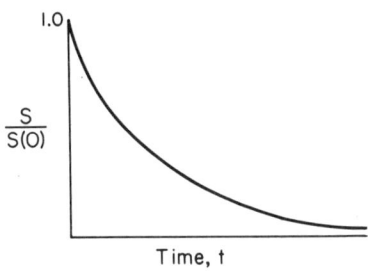

Figure 10-4. Stress decay curve for a Maxwell model in relaxation.

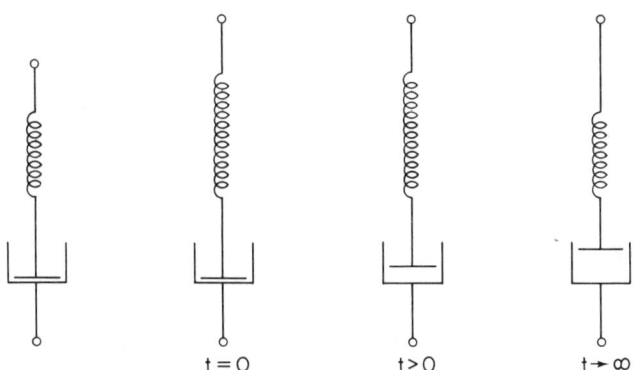

Figure 10-5. Stress decay in a Maxwell model.

The mechanical response of the model to an applied stress is illustrated in Figure 10–5 for comparison to the response predicted by Eq. (10–6). At $t = 0$, the spring deforms instantaneously to absorb the total strain, since the dashpot has not had time to move. As time passes, the dashpot flows under the influence of the applied stress, and in doing so it gradually relieves the strain and hence the stress on the spring. The relief of stress causes the rate of decay to decrease, since the rate of flow to the dashpot is proportional to the stress, and an infinite amount of time is required for the stress to decay to zero.

The creep response of the Maxwell element can be determined by setting $\dot{S} = 0$ and integrating Eq. (10–5) to yield

$$\gamma = \gamma(0)(1 + t/\lambda) \tag{10-7}$$

where $\gamma(0) = S(0)/G$, the strain at $t = 0$. This behavior is illustrated in Figure 10–6.

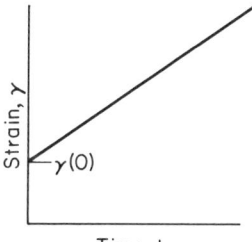

Time, t

Figure 10–6. Creep curve for a Maxwell model.

1005 The Voigt Model

The parallel combination of the elements under consideration is termed a *Voigt model*, which is illustrated in Figure 10–7. It is useful for representing viscoelastic creep and creep recovery. However, it does not possess a mechanism for stress relief. With the elements in parallel, the strains and strain

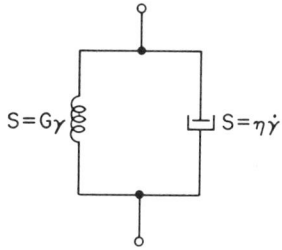

Figure 10–7. The Voigt model.

rates will be common; and the total stress is obtained by adding stresses in the elements. This procedure yields

$$S = \dot{\gamma}\eta + G\gamma \tag{10-8}$$

For constant stress conditions, Eq. (10–8) integrates to

$$\gamma = \frac{S(0)}{G}(1-e^{-t/\lambda}) \tag{10-9}$$

where $S(0)/G$ is the strain at infinite time. This behavior is illustrated in Figure 10–8. The Voigt model is said to exhibit *retarded elastic deformation* in creep experiments and *retarded elastic recovery* in creep-recovery experiments. Note that λ plays a role here analogous to the one it played in the Maxwell model.

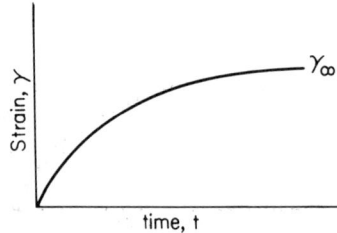

Figure 10–8. Retarded elastic deformation curve for a Voigt model.

In the Voigt model, the total stress is borne initially by the dashpot, for the spring would elongate instantaneously if it could. Under the continued influence of the constant stress, the dashpot begins to flow, thereby transferring part of the load to the spring. As more of the load is transferred to the spring, the rate of creep decreases, for the rate of strain is proportional to the stress on the dashpot. Eventually, as the element reaches its equilibrium strain, the load is borne entirely by the spring.

1006 Series Combination of Maxwell and Voigt Models

A crude, qualitative representation of the molecular mechanisms responsible for the viscoelastic behavior of linear amorphous polymers under creep conditions is provided by the mechanical analog constructed by placing Maxwell and Voigt models in series. This representation is illustrated in Figure 10–9. The four parameters of this model are indicated as G_V and η_V for the Voigt element and as G_M and η_M for the Maxwell element. The total strain for creep conditions will be due to an instantaneous elastic deformation

Sec. 1006 Series Combination of Maxwell and Voigt Models

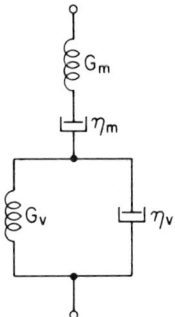

Figure 10-9. Maxwell and Voigt models in series.

(Maxwell element spring), an irrecoverable viscous flow (Maxwell element dashpot), and a recoverable retarded elastic deformation (Voigt element). Molecular mechanisms may be assigned as follows:

1. Instantaneous elastic deformation is due to bending and stretching of primary valence bonds.
2. Irrecoverable viscous flow is due to slippage of the polymer chain or chain segments past one another.
3. Retarded elastic deformation arises from the transformation of a given equilibrium conformation into a biased conformation in which elongated and oriented structures are favored.

The flow equation for constant loading conditions is given by the sum of the analogous terms for the constituent elements. From Eqs. (10-7) and (10-9), the strain is

$$\gamma = \frac{S}{G_m} + \frac{S}{\eta_m} t + \frac{S}{G_v}(1 - e^{-t/\lambda_v}) \quad (10\text{-}10)$$

The creep curve for this model, as described by Eq. (10-10), is shown in Figure 10-10 along with the strain-recovery curve. Strain recovery begins at time t_1 when the load is released. A portion of the strain is immediately

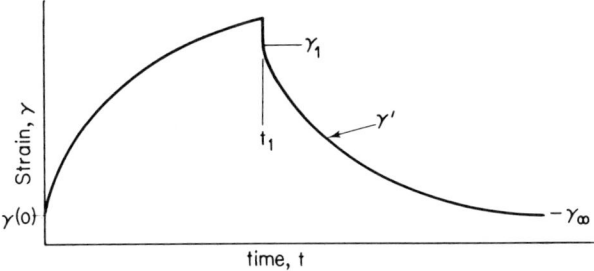

Figure 10-10. Creep and creep recovery curve for Maxwell and Voigt models in series.

recovered, part of the strain recovery is retarded, and another portion of the strain is irrecoverable. If

γ_1 = strain when the stress is first removed
γ_∞ = strain after complete possible recovery
γ' = strain remaining in the model after time t_1

then the fractional approach to recovery, f, is given by

$$f = \frac{\gamma_1 - \gamma'}{\gamma_1 - \gamma_\infty} = \left[\frac{1 - \exp(t_1 - t')}{\lambda_v}\right] \qquad (10\text{-}11)$$

where t' is the time greater than t_1.

The rheological equation for this model can be obtained by summing strains, since the stresses are constant, and by combining them with the rheological equations of the elements. The result is given by

$$\ddot{S} + \left[\frac{G_m}{\eta_v} + \frac{G_m}{\eta_m} + \frac{G_v}{\eta_v}\right]\dot{S} + \left[\frac{G_m G_v}{\eta_m \eta_v}\right]S = G_m \ddot{\gamma} + \left[\frac{G_m G_v}{\eta_v}\right]\dot{\gamma} \qquad (10\text{-}12)$$

This equation could have been used with proper boundary conditions to obtain the results of Eqs. (10–10) and (10–11).

Another interesting feature of the model is that it exhibits the phenomenon known as *hysteresis*. The amount of work required to cause the irrecoverable deformation is dissipated as heat, and the path followed for retraction is not the same as that for extension. Such an effect is noted in all real materials, the magnitude of which depends on the specific material involved. In polymers, the dissipated energy per cycle can be quite substantial.

1007 The Material Response Time

At this point, we should consider how λ serves to specify the viscoelastic character of materials. To do so, consider the response curves of a Maxwell element under relaxation conditions. Equation (10–6) gives rise to a family of curves, as shown in Figure 10–11, depending on the value of λ. The curves with low values of λ show a relatively rapid decay and are indicative of a liquidlike behavior, whereas those with higher values of λ maintain relatively high values of stress for longer periods of time and are indicative of solidlike behavior. In the limit, as $\lambda \to 0$ or as $\lambda \to \infty$, completely viscous or elastic behavior is expected.

Since λ depends on η and η is temperature sensitive, we can note the effect of temperature on the viscoelastic properties of materials. As tem-

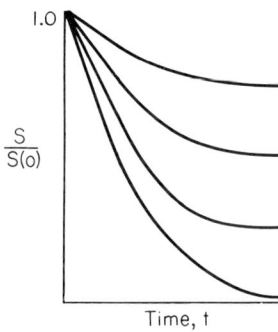

Figure 10-11. Relaxation curves for Maxwell models with different values of λ.

perature increases, the viscosity decreases and λ will similarly decrease. This behavior indicates that materials become more fluidlike with increasing temperature and more solidlike with decreasing temperature.

We must realize, of course, that even though the foregoing discussion is qualitatively in line with observation, the discussion is necessarily oversimplified. For one thing, the model used is too simple; and in order to specify completely the viscoelastic nature of a material, one must employ a continuous spectrum or distribution of relaxation times. More will be said shortly on this subject. The concepts generally apply, however, for we can characterize many real systems to some extent with an average or characteristic response time.

1008 The Deborah Number

Marcus Reiner first proposed that the relation between the characteristic material response time and the time scale of experiment or observation should be expressed as a ratio to yield a dimensionless quantity, the *Deborah number D*,

$$D = \frac{\text{material response time}}{\text{time scale of experiment or observation}} \qquad (10\text{-}13)$$

If the time scale of experiment or observation is short compared to the material response time, the viscoelastic material would appear to be more solidlike than if the denominator were larger. That is, a high Deborah number indicates viscoelastic solid behavior, whereas a low Deborah number for the same material indicates viscoelastic liquid behavior. In practice, the denominator contains the time factor dictated by the situation. It might be, for example, the residence time of a polymer melt in an extruder die of the service time of a structural element.

The foregoing concepts can be readily demonstrated with a sample of unvulcanized silicone gum rubber (silly putty). First, the material should be fashioned into a short rod by rolling it between the hands. If it is then extended under light tension, one observes marked creep, a characteristic liquidlike behavior. On the other hand, if the material is extended as rapidly as possible, one observes fracture behavior characteristic of elastic solids. The material response time is the same in both cases, but the time scale of experiment is different. The proper term for the denominator in computing the Deborah number is obviously the reciprocal of the strain rate, $\dot{\gamma}^{-1}$. With slow extension, the denominator is relatively large and the Deborah number is low, indicating liquidlike behavior. With rapid extension, the Deborah number is large and a solidlike character is indicated. At intermediate strain rates and Deborah numbers, the silicone rubber would exhibit viscoelasticity. This latter behavior can be observed by straining a rod of the material at an intermediate rate and then releasing it sharply to observe the elastic retraction. Another high-speed test involves rolling the material into a ball and then hurling it against the floor or wall; the material will bounce elastically, and when retrieved, it will exhibit little or no deformation.

Even though we will not employ the Deborah number quantitatively, the notion that the viscoelastic nature of a material depends on the experimental time parameter, as well as its inherent response time, is of paramount importance. The Deborah number was proposed as a fundamental number of rheology by Reiner in an afterdinner talk presented at the Fourth International Congress on Rheology, August 1963, Providence, R.I. The appropriate remarks appear in *Physics Today*, January 1964.

GENERALIZED LINEAR VISCOELASTICITY

In these next few sections, we define linear viscoelasticity in terms of the Boltzmann superposition principle, and we show how the concept of a relaxation or retardation time is employed in describing the behavior of real materials.

1009 The Boltzmann Principle

Consider a body that is subjected to a series of equal incremental shear strains

$$(\gamma_0 - 0), (\gamma_1 - \gamma_0), (\gamma_2 - \gamma_1), (\gamma_3 - \gamma_2) \ldots (\gamma_n - \gamma_{n-1})$$

at times $t_0, t_1, t_2, t_3, \ldots, t_{n-1}, t_n$. For a Hookean solid, the application of an incremental strain will change the stress by the same amount, no matter what load the material is already bearing. Hence the stress at the end of this loading history is

$$\begin{aligned} S = G[\gamma_0] + G[\gamma_1 - \gamma_0] + G[\gamma_2 - \gamma_1] + \cdots \\ + G[\gamma_n - \gamma_{n-1}] = G\gamma_n \end{aligned} \quad (10\text{-}14)$$

Similarly, a Newtonian dashpot will show the same increase in stress under a given strain-rate increment, independent of the load it already bears. Therefore spring and dashpot models of the type considered here should respond to an applied strain in a fixed fashion, independent of their history.

In a viscoelastic material, the modulus is not a constant but may, in general, be a function of time and the strain history. In a material exhibiting *linear viscoelastic behavior*, the modulus varies *only* with time. Again, we may imagine a viscoelastic material subjected to a series of incremental strains $(\gamma_n - \gamma_{n-1})$, each applied at some instant of time t_n of the past. Each strain increment causes a corresponding change in stress $(S_n - S_{n-1})$. In 1876 Boltzmann postulated that in a linear material the superposition principle is valid, so that the total stress $S(t)$ at the present time is simply the sum of all the incremental stresses

$$S(t) = S(t-t_0) + S(t-t_1) + S(t-t_2) + \ldots + S(t-t_n) \qquad (10\text{-}15)$$

For a relaxation experiment, the superposition principle becomes

$$S(t) = G_r(t-t_0)[\gamma_0] + G_r(t-t_1)[\gamma_1-\gamma_0] + G_r(t-t_2)[\gamma_2-\gamma_1] + \ldots \\ + G_r(t-t_n)[\gamma_n-\gamma_{n-1}] \qquad (10\text{-}16)$$

where the parentheses indicate that the modulus is a function of time and the brackets indicate multiplication. This relationship states that the total stress induced in a viscoelastic material, by a series of incremental strains, is the sum of the incremental stresses resulting from the incremental strains. The modulus used to compute each incremental stress is a function of the time during which the particular incremental strain has been applied, but *not* of the previous strain. Introducing such a possibility would, of course, destroy the linear character of the relationship expressed in Eq. (10-16). For a continuous strain history, the Boltzmann principle can be expressed as an integral equation

$$S(t) = \int_0^t G_r(t-\theta)\dot{\gamma}(\theta)\,d\theta \qquad (10\text{-}17)$$

where the function $\dot{\gamma}(\theta)$ describes the strain history.

Equations analogous to Eqs. (10-16) and (10-17) can be written for a linear viscoelastic body subject to successive incremental stresses (creep conditions) $S_0, S_1, S_2, \ldots, S_{n-1}, S_n$ at times $t = t_0, t_1, t_2, \ldots, t_{n-1}, t_n$. For these, we use the time-dependent elastic compliance defined as $J_c(t) = 1/G_c(t)$.

$$\gamma(t) = J(t-t_0)[S_0] + J(t-t_1)[S_1-S_0] + \ldots + J(t-t_n)[S_n-S_{n-1}] \qquad (10\text{-}18)$$

For continuous loading, the Boltzmann principle for creep conditions is expressed by another integral equation

$$\gamma(t) = \int_0^t J(t-\theta)\dot{S}(\theta)\,d\theta \tag{10-19}$$

where $\dot{S}(\theta)$ describes the stress history.

Equation (10–17) essentially states that the present stress is a function of the entire history of the strain rate, whereas Eq. (10–19) expresses the fact that for a linear viscoelastic material the strain may also be regarded as a function of the entire stress history. We would expect linear viscoelastic behavior to be exhibited by a material whose structure is not changing during the loading period to the extent that the modulus would depend on the strain. Amorphous polymers under sufficiently low strain conditions fall into this category. We would expect materials in which molecular structure is changing with strain to exhibit nonlinear viscoelasticity. Materials such as nonoriented crystalline polymers, strain-crystallizable elastomers, and polymeric melts in processing operations would be expected to exhibit nonlinear behavior; and in Eq. (10–15) the incremental stresses would be functions of the strain history as well as time.

RETARDATION AND RELAXATION SPECTRA

We now set out to describe two model systems that are generally applicable to real systems. The models show not a single relaxation time but a distribution or spectrum of relaxation times. The behavior of a polymer system is so complicated that we cannot represent it with a single response time —a large number of such parameters are required.

1010 The Maxwell-Weichert (Relaxation) Model

Weichert in 1893 showed that stress-relaxation experiments could be represented as a generalization of Maxwell's equation. The mechanical model corresponding to Weichert's formulation is shown in Figure 10–12; it consists of a very large (or infinite) number of Maxwell elements coupled in parallel. The rheological equation for the ith element can be written as

$$\dot{\gamma} = \dot{S}_i/G_i + S_i/G_i\lambda_i \tag{10-20}$$

As can be seen, this model necessarily possesses many relaxation times. For

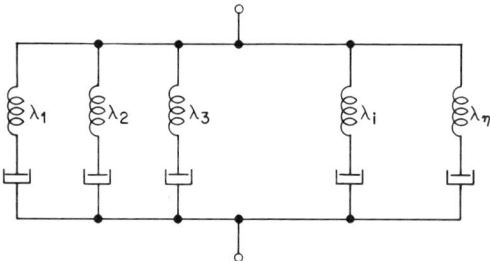

Figure 10-12. The Maxwell-Weichert model.

real materials we postulate the existence of a continuous spectrum of relaxation times. A spectrum skewed toward lower times would be characteristic of a viscoelastic fluid, whereas a spectrum skewed toward longer times would be characteristic of a viscoelastic solid. If the spectrum were skewed heavily toward very long or infinite relaxation times we would expect the real system to contain crosslinks. In generalizing, we will allow λ to range from zero to infinity. The concept that a continuous distribution of relaxation (or retardation) times should be required to represent the behavior of real systems would seem to follow quite naturally since we already know real polymeric systems also exhibit distributions in molecular weight, distance between crosslinks, conformational size, copolymer composition, etc.

Since the strain in each element is common, we sum stresses to obtain the total stress as a function of time.

$$S(t) = \sum_{i=1}^{n} S_i \tag{10-21}$$

For relaxation with constant strain, $\gamma(0)$ we combine Eq. (10–20) and (10–21) to obtain

$$S(t) = \gamma(0) \sum_{i=1}^{n} G_i \exp(-t/\lambda_i) \tag{10-22}$$

If the number of units is allowed to become infinite with the characteristic parameters becoming continuous functions of relaxation time, the summation over the differential units of the model can be replaced by an integration over all relaxation times. Thus, as $i \to \infty$, the range of allowable relaxation times becomes zero to infinity. From the notion that the stresses in the individual elements, S_i, are functions of time and relaxation times, $S_i = S_i(t, \lambda_i)$, we define a continuous function $S(t, \lambda)$ such that the total stress, $S(t)$, is given by

$$S(t) = \int_0^\infty S(t, \lambda) d\lambda \tag{10-23}$$

The continuous function $G(\lambda)$ is defined by

$$S(t, \lambda) = \gamma(0)G(\lambda)e^{-t/\lambda} \qquad (10\text{-}24)$$

such that

$$S(t) = \gamma(0) \int_0^\infty G(\lambda)e^{-t/\lambda}d\lambda \qquad (10\text{-}25)$$

Since $G_r(t) = S(t)/\gamma(0)$, we find that we have developed an expression suitable for representing the time dependence of the relaxation modulus, i.e.,

$$G_r(t) = \int_0^\infty G(\lambda)e^{-t/\lambda}d\lambda \qquad (10\text{-}26)$$

The function $G(\lambda)$ is referred to as the *distribution of relaxation times*. In principle once $G(\lambda)$ is known, the result of any other type of mechanical experiment can be predicted. In practice $G(\lambda)$ is determined from experimental data on $G_r(t)$. If the relaxation modulus data could be represented analytically, an analytic function for $G(\lambda)$ could be obtained since Eq. (10-26) is of the Laplace transform type. In practice this cannot be really accomplished and various approximation techniques must be used. A more complete discussion of these aspects of linear vicsoelasticity will be undertaken in Chapter 11 after we have discussed the physical nature of polymeric solids embodied in $G_r(t)$ data.

1011 The Generalized Voigt (Creep) Model

Equations appropriate for creep conditions can be derived which are similar in form to those of Section 1010. The generalized model for creep consists of a large number of Voigt elements in series. Analysis leads to the result that a continuous *distribution of retardation times*, $J(\lambda)$, can be defined such that

$$J(t) = \int_0^\infty J(\lambda)[1 - \exp(-t/\lambda)]d\lambda. \qquad (10\text{-}27)$$

11
The Linear Viscoelastic Behavior of Polymeric Solids

In general studies of viscoelastic behavior, one seeks to establish the relations between stress, strain, and time or temperature for a particular strain history. With materials that exhibit linear viscoelastic behavior, this objective reduces to the experimental determination of the time and temperature dependence of the viscoelastic parameters corresponding to a particular deformation. One does not usually vary time and temperature simultaneously but holds one constant while measuring the effect of the other. It is important to remember, as outlined in Sections 1009 to 1011, that the concept of linear viscoelasticity is a mathematical one. When we say that a material exhibits linear viscoelastic behavior, we mean that its rheological response to an applied stress or strain can be predicted and described in terms of the linear equations developed there. For a material to exhibit linear viscoelastic behavior, the strains must be held at very modest levels.

In this chapter we consider only simple shear and tensile deformation experiments. With regard to loading patterns, we consider step-function and oscillatory experiments. In *step-function experiments*, such as relaxation, creep, and stress-strain experiments, the applied stress, strain, or strain rate is applied in a stepwise fashion, and the responses vary monotonically with time. In *oscillatory experiments*, the applied stress or strain is caused to vary periodically with time. Another type of experiment (not to be described here) involves, first, the measurement of creep and then creep recovery after removal of the load. To summarize: the subject matter of this chapter treats the rheological, time-temperature response of polymeric solids exposed to conditions of modest strains in tension or shear for various loading patterns.

In the next four sections, we consider the general character of step-function and oscillatory experiments. We cannot discuss apparatus or techniques in any detail here; the interested reader is referred to the excellent reference works of Ferry, van Wazer et al. and McCrum et al. for this information. We will first note that the elastic modulus in tension or shear is a parameter that can be obtained in both step-function and oscillatory experiments, and, within the limits of linear viscoelastic behavior, the values obtained are independent of the experimental loading pattern and depend only on the experimental time scale and temperature. We should also note that more detailed information regarding polymeric composition and properties can be obtained from stress-strain and oscillatory experiments than either creep or relaxation experiments.

1101 Creep Experiments

In creep experiments, a stress is suddenly applied and maintained at a constant value while strain is measured as a function of time. Creep involves a change in shape and dimension at a constant volume. It is one of the most important and widely used engineering tests of viscoelastic behavior. The reason? Most useful polymeric objects must be designed to withstand modest loads for long periods of time without changing shape or dimension. For example, a plastic raincoat, which must be soft and flexible, must also be able to hang for weeks without flowing so that it reaches the floor. A rubber tire must not develop flat spots on its surface even though the car may be parked for some time. Fibers must not deform under prolonged stress, or our clothing would become undesirably misshapen. A plastic pipe, such as is used in industry to convey corrosive chemicals, must be capable of withstanding the constant stress (hoop stress) imposed by the pressures required to convey the liquid.

The results are reported as the creep modulus in tension $E_c(t, T)$ or in shear $G_c(t, T)$. The data are obtained as percent elongations versus time and then converted to modulus values. If a change in dimension does take place, it must be accounted for in making the conversion. Figure 11–1 shows creep data typical for polymeric materials. When the load is first applied, instantaneous elongation that is directly proportional to the creep compliance occurs. This initial period of deformation is followed by a period of rapid creep, which, in turn, is followed by a period of constant creep rate. Note from Figure 11–1 that the greater the load, the greater the creep elongation and creep rate. The moduli are functions of time and independent

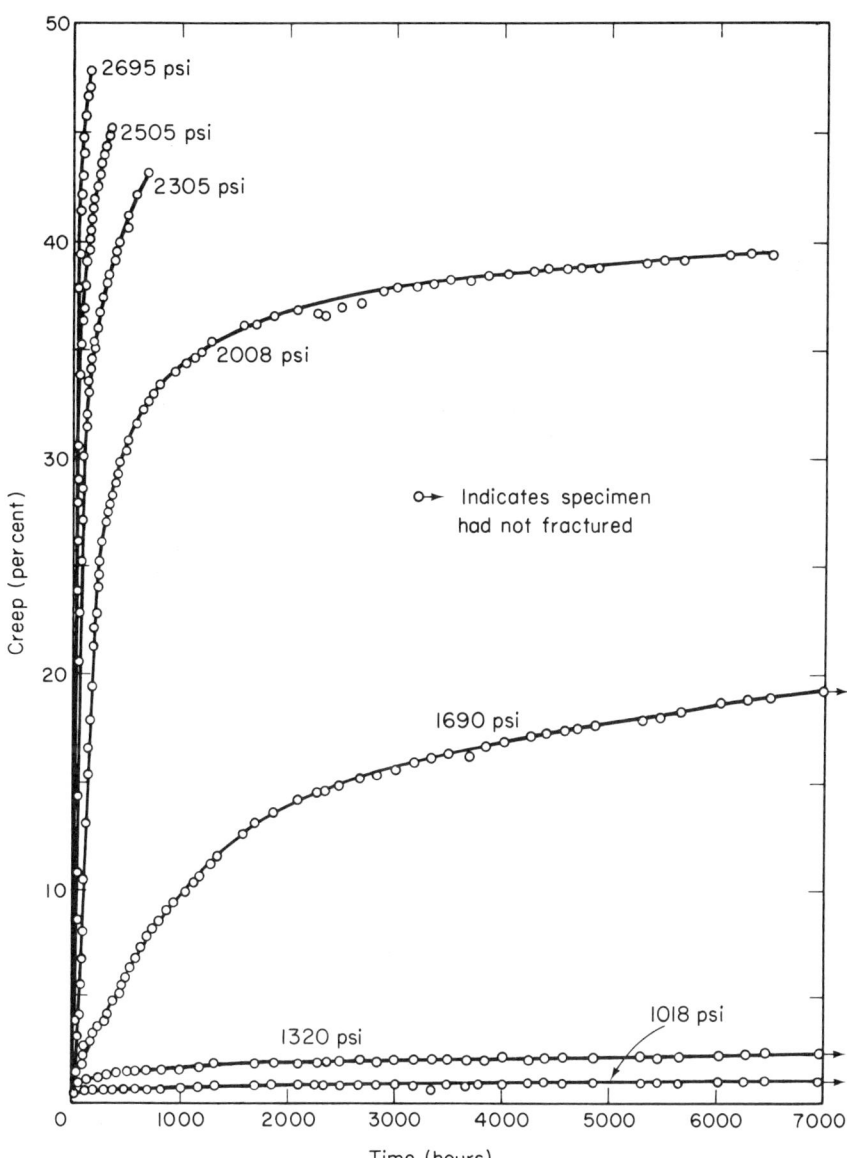

Figure 11–1. Creep of cellulose acetate at 25°C. [W. N. Findley, Modern Plastics, **19**, 71 (August, 1942). Reprinted by permission, Modern Plastics Magazine, McGraw-Hill, Inc.]

of stress only at low values of stress. Initial values of strain are experimentally difficult to measure because of their very low values. Higher values—beyond the range of linear viscoelasticity—of strain are easily measured, and such data are of considerable interest in engineering design.

1102 Stress-Relaxation Experiments

In stress-relaxation experiments, the specimen is suddenly brought to a given strain, and the stress required to maintain that strain is measured as a function of time. The stress is a maximum when the stress is first applied; then it gradually decays with time to a minimum constant value for crosslinked materials or zero for linear materials. Stress-relaxation curves at several temperatures are shown for a linear amorphous polymer—poly(methyl methacrylate)—in Figure 11-2. This experiment is somewhat easier to

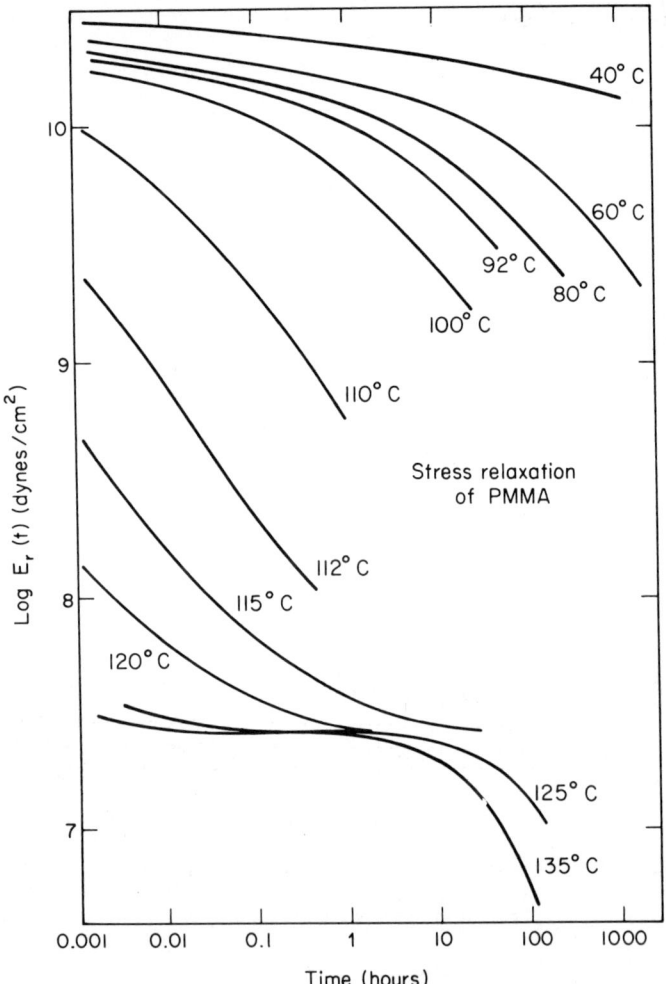

Figure 11-2. Log $E_r(t)$ vs. log t for unfractionated polymethyl methacrylate ($\bar{M}_v = 3{,}600{.}000$) between 40 and 135°C. [After J. R. McLoughlin and A. V. Tobolsky, J. Colloid Sci., **7**, 555 (1952).]

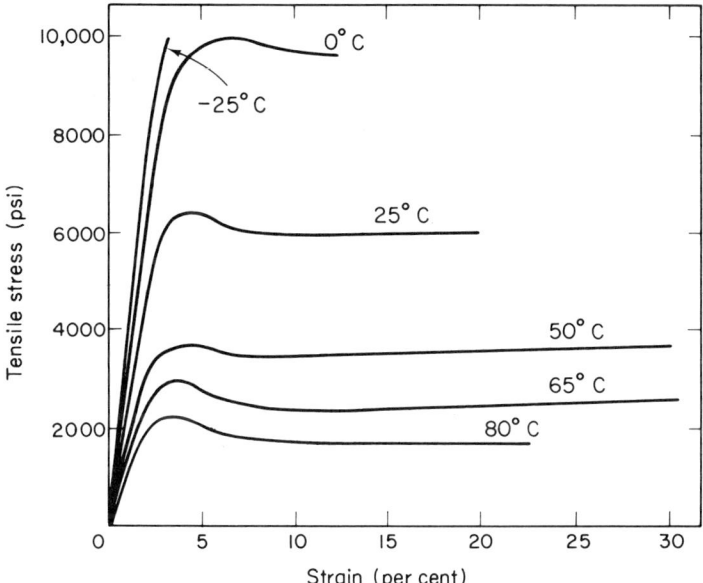

Figure 11-3. Effect of temperature on the stress-strain properties of cellulose acetate. [T. S. Carswell, and H. K. Nason, Modern Plastics, **21**, 121 (June, 1944). Reprinted by permission, Modern Plastics Magazine, McGraw-Hill, Inc.]

perform than creep measurements since the specimen does not change dimensions during the test and initial values of stress are readily measured. Here again, the data are reported as the relaxation modulus $E_r(t, T)$ in tension and $G_r(t, T)$ in creep.

1103 Stress-Strain Experiments

In these experiments, the strain is increased linearly with time, and the stress is measured as a function of strain. This is probably the most widely used engineering test; however, the rheological responses are among the least understood from a theoretical point of view. Stress-strain measurements made over a wide range of temperature and testing speeds are very important for the engineering design of polymeric materials. Such measurements are among the few that indicate something about the ultimate strength and toughness of a material. They are especially useful for studying the process of drawing crystalline filaments, and the process of elongating elastomers. The stress-strain behavior typical of a fibrous material is shown in Figure 11-3. The elastic modulus (usually reported as Young's modulus) is obtained from the initial linear portion of the curve. The flat portion represents the period during which necking is taking place (Section 610).

1104 Oscillatory Experiments

In these tests, the applied force and the resulting deformation both vary sinusoidally with time. The frequency of the test may vary from 10^{-5} to 10^8 cycles/sec., thereby providing the widest range of experimental time scales. In such tests, it is possible to separate the elastic and viscous responses so that an elastic modulus and a mechanical damping are reported. Dynamic data are extremely useful in studying variations of properties with molecular architecture. Such data are essential in applications where mechanical or acoustical vibrations play an important role or where toughness is a consideration. Dynamic mechanical testing is the most suitable experimental technique for studying such materials as tire rubbers.

The data are reported either as a function of the frequency or of temperature. The elastic modulus is represented by $E'(\omega, T)$ for tension and $G'(\omega, T)$ for shear. Other parameters derived from oscillatory experiments will be discussed in detail beginning with Section 1112.

THE ELASTIC MODULUS

It is important enough to restate the following experimental observation: for linear amorphous polymers, within the limits of linear viscoelastic behavior, the measured values of the elastic modulus are nearly the same in each of the techniques described above. The values for tension or shear will, of course, differ by an amount given by Eq. (10–2) and Poisson's ratio, μ. Since μ ranges from zero to 0.5, the tensile and shear moduli can differ only by a factor of three or less. (μ is 0.5 for an ideal rubber and about 0.25 or 0.33 for most crystalline and glassy materials.) This is an important point. From our discussion in Chapter 7, we know that for linear amorphous polymers the modulus changes by at least three orders of magnitude in undergoing a glass transition. This means that all the examples of modulus versus temperature behavior for linear amorphous polymers cited there would be of the same character, including orders of magnitude modulus values, regardless of whether the modulus was measured in tension or shear.

What we are now about to do is complete the discussion begun in Chapter 7, where we used the elastic modulus to illustrate transitional phenomena.* Next, we will briefly discuss additional properties derived from stress-strain experiments. Oscillatory tests will be considered in some detail; and, finally, we will discuss the application of the concept of relaxation or retardation-time spectra to some specific situations.

* Before continuing, the reader should review Chapter 7, especially Sections 703 to 706, 709, 710 and 713 to 715.

1105 Time-Temperature Equivalence

To restate the case for linear viscoelasticity, the elastic modulus is a function of time and temperature only. During an experiment, either the time scale or temperature is held constant; and the moduli are reported for either a constant time scale as a function of temperature or a constant temperature as a function of time. As a matter of feasibility, relaxation and creep experiments are run at constant temperature; the temperature dependence is found by first running a series of tests at various temperatures, then selecting a time, and finally cross-plotting the moduli for this selected time scale as a function of temperature. In oscillatory tests, it is not always convenient to study the modulus as a function of time, for it is experimentally difficult to vary frequency smoothly. Instead, the frequency may be held constant and the modulus studied as a function of temperature.

One of the most important observations centering around the modulus characteristics of linear amorphous polymers is that the modulus-time and modulus-temperature curves are of identical shape—that is, they all show five regions of viscoelastic behavior, and all the modulus values are within an order of magnitude of each other in each of the regions. Figure 11-2 shows modulus-time curves for poly(methyl methacrylate) at several temperatures for comparison with the modulus-temperature curves of Figure 7-5 for polystyrene. The five regions of viscoelastic behavior are clearly discernible, and in each region the modulus value is of the same order as shown in Figure 7-5. In fact, since the glass-transition temperature for both polymers is the same, a cross plot of the data from Figure 11-2 at $t = 10$ seconds should yield a curve superposable on the curve of Figure 7-5. This comparison of modulus-time and modulus-temperature behavior must be made at several temperatures in Figure 11-2, since the time required to show the entire spectrum of viscoelastic behavior at one temperature would extend to several centuries. Such remarkable parallels in behavior can only mean that time and temperature have an equivalent effect on the elastic modulus. As it turns out, and as we shall discuss subsequently, this time-temperature equivalence also applies to other linear viscoelastic properties of polymers as well. This time-temperature equivalence principle cannot, however, be extended *a priori* to crystalline polymers because of the changes in morphology that such a material may undergo with increasing temperature, time, and strain.

1106 The Time-Temperature Superposition Principle

In engineering practice, it is frequently necessary to design for the use of a material over an extended period of time—many years, for example. A common parameter to use in design work is the elastic modulus. We know, however, that the modulus decreases with increasing time. Laboratory tests can only be conveniently conducted for about a week or so. Consequently, one must use a method of extrapolating short-time tests over several decades of time so that a lower limit of the modulus can be determined for use in design. On the other hand, it is also sometimes difficult to obtain very short time-scale data, and a method is needed to extrapolate data obtained under practicable experimental conditions to these short time scales. An empirical extrapolation method is available for amorphous polymer systems and, in general, for systems where structure does not change during the period of testing.

We have already noted that time and temperature have essentially equivalent effects on the modulus values of amorphous polymers in that the shapes of the curves are similar and the modulus values in each region are about equal. The left-hand side of Figure 11–4 shows modulus data taken at several temperatures for polyisobutylene. Because of the equivalent effect

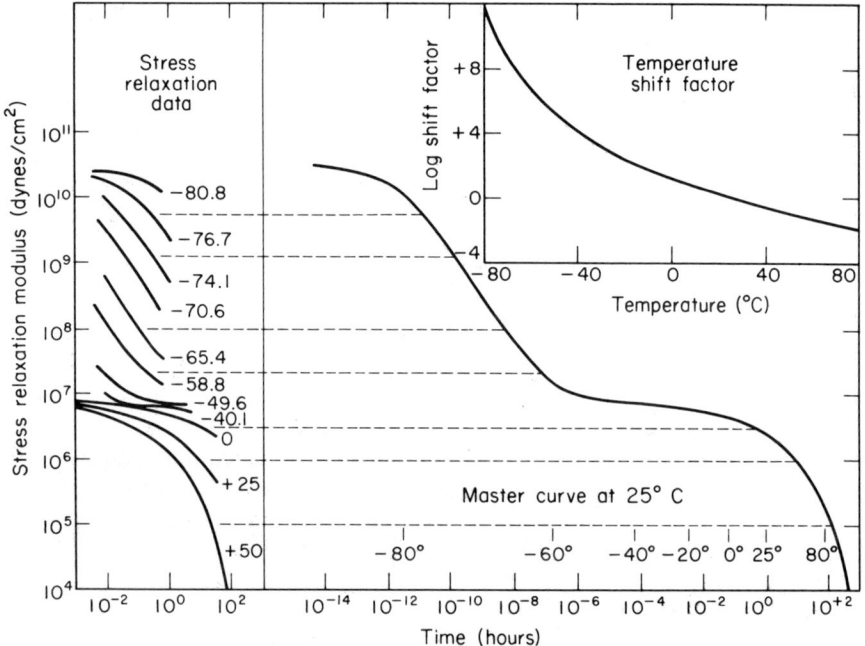

Figure 11–4. Time-temperature superposition principle illustrated with polyisobutylene data. The reference temperature of the master curve is 25°C. The inset graph gives the amount of curve shifting required at the different temperatures. [From *Mechanical Properties of Polymers* by L. E. Nielsen. Copyright 1962 by Reinhold Publishing Corporation by permission of Van Nostrand Reinhold Co.]

of time and temperature, data at different temperatures can be superposed on data taken at a specified reference temperature merely by shifting individual curves one at a time and consecutively along the log t axis about the reference temperature T_0. This *time-temperature superposition* procedure has the effect of producing a single continuous curve of modulus values extending over many decades of log t for the reference temperature. The constructed curve on the right-hand side of Figure 11-4 is known as the *master curve*. It has a reference temperature of 25°C.

Before the curves can be shifted to make the master curve, the modulus values should be corrected for density and temperature to obtain *reduced modulus* values, $E(t)_\text{reduced}$

$$E(t)_\text{reduced} = \left(\frac{T_0}{T}\right)\left(\frac{\rho_0}{\rho}\right)E_r(t) \tag{11-1}$$

where T_0 = reference temperature, °K
ρ_0 = density at T_0
ρ = density at T

The density correction (ρ_0/ρ) is not usually very large. The temperature correction (T_0/T) is suggested by rubber-elasticity theory (Chapter 9). Reduced data are shown in Figure 11-4.

This procedure asserts that the effect of temperature on viscoelastic properties is equivalent to multiplying the time scale by a constant factor at each temperature. In mathematical terms, it is expressed as

$$E_T[a(T)] = E_{T_0}(t) \tag{11-2}$$

where the parentheses indicate functional dependence and the brackets indicate multiplication. The quantity $a(T)$ is called the *shift factor*, and it must be obtained directly from the experimental curve by measuring the amount of shift along the log time scale necessary to match the curve. The parameter $a(T)$ is chosen as unity at the reference temperature and is a function of temperature alone, decreasing with increasing temperature. The shift factor is shown in the upper right-hand corner of Figure 11-4.

An important empirical correlation has been developed by which the shift factor can be computed for temperatures between T_g and $T_g + 100$ °C. This correlation was developed by Williams, Landel, and Ferry, and the so-called WLF equation is expressed as

$$\log a(T_0) = \frac{-C_1(T-T_0)}{C_2+T-T_0} \tag{11-3}$$

For $T_0 = T_g$, $C_1 = 17.44$ and $C_2 = 51.6$, while for $T_0 = T_g + 45$ °C, $C_1 = 8.86$ and $C_2 = 101.6$.

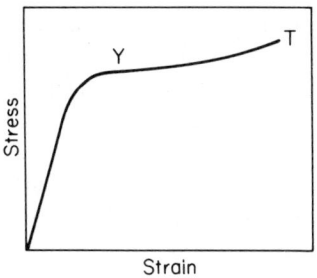

Figure 11-5. Schematic stress-strain curve.

1107 Five Basic Properties

In addition to the elastic modulus, other properties (which are beyond the limits of linear viscoelasticity) may be derived from stress-strain experiments. Figure 11-5 shows schematically a stress-strain curve from which the following properties may be evaluated:

1. *Stiffness.* This property indicates the ability to carry stress without changing dimension. It is measured as the modulus of elasticity over the linear portion of the curve *OP*. This is the region in which Hooke's law is obeyed.
2. *Elasticity.* This parameter stipulates the ability to carry stress without suffering permanent deformation. It is indicated by the yield point or *elastic limit* at *Y*. Beyond *Y*, the material behaves as a plastic solid, the degree of ductility being indicated by the length of the horizontal portion of the curve after *Y*.
3. *Resilience.* This characteristic indicates the ability to absorb energy without suffering permanent deformation. It is reported as resilient energy and is measured as the area under the elastic portion of the stress-strain curve.
4. *Strength.* This factor measures the ability to carry dead load. It is measured as stress and is reported as the *ultimate strength* or *tensile strength* and is measured at point *T*.
5. *Toughness.* This factor indicates the ability to absorb energy and undergo large permanent deformation without rupturing. It is defined as the ultimate energy resistance and is measured as the total area under the stress-strain curve.

OSCILLATING EXPERIMENTS*

Experiments in which an oscillating stress or strain is applied to the specimen constitute an important class of experiments for studying the visco-

* Also referred to as dynamic mechanical experiments.

elastic behavior of polymeric solids. In addition to the elastic modulus, it is possible to measure the viscous response directly. In this context, the viscous behavior of a polymer is usually reported in terms of characteristic damping parameters. Damping is an engineering material property with which the student should already be somewhat familiar. Furthermore, the observed responses, especially damping, are much more sensitive to the polymer constitution than in step-function experiments, and these methods offer a powerful technique to study molecular structure and morphology. A final feature significance is the breadth of the time-scale spectrum available with these methods—10^{-5} to 10^8 cycles/second.

To introduce the basic concepts of dynamic mechanical testing, we again turn to model analysis. In Section 1113 we will return to our discussion of the phenomenological aspects of polymeric behavior.

1108 Rheologically Simple Bodies

Consider the responses of elastic and viscous bodies subjected to the sinusoidal strain

$$\gamma = \gamma_0 \sin(\omega t) \tag{11-4}$$

where γ_0 is the maximum value of the strain, and ω is the angular frequency; ω is related to f, the frequency in cycles per second by $\omega = 2\pi f$. For the elastic body, the stress will vary as

$$S = G\gamma_0 \sin(\omega t) \tag{11-5}$$

Comparison with Eq. (11-4) shows that the strain and stress response will be in phase. For the viscous body, the stress will vary as

$$S = \eta\omega\gamma_0 \cos(\omega t) \tag{11-6}$$

and we see that the stress response will be 90 deg out of phase with the applied strain. With a viscoelastic body, our intuition leads us to expect a phase angle, δ, which is between zero and 90 deg. In our treatment of viscoelastic responses, we will include such an angle in our analytical relationships. The stress responses for these three situations are illustrated schematically in Figure 11-6.

The work per unit cube W is given by

$$W = \int_0^{1/f} S \, d\gamma$$

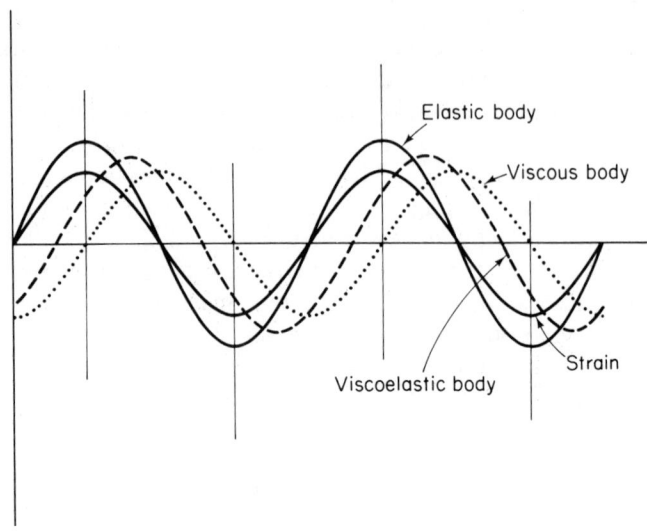

Figure 11-6. Three schematic stress responses to the sinusoidal stress, $\gamma = \gamma_0 \sin(\omega t)$: elastic, viscous, and viscoelastic.

which by the chain rule becomes

$$W = \int_0^{1/f} S\dot{\gamma}\, dt \tag{11-7}$$

For an elastic body, as we must anticipate, energy is neither dissipated nor stored over a complete cycle. The amount of energy stored during the first and third quarter cycles, $G\gamma_0^2/2$, is recovered during the second and fourth quarter cycles. For a viscous body, the imparted energy is completely dissipated over a full cycle; it is given by

$$W = \int_0^{1/f} \eta\omega^2\gamma_0^2 \cos^2(\omega t)\, dt = \pi\eta\omega\gamma_0^2 \tag{11-8}$$

This latter analysis leads us to expect that, in a viscoelastic body, part of the energy will be dissipated as viscous losses.

1109 Introduction to Viscoelastic Responses

We find it most convenient to develop the elemental concepts for the response of viscoelastic bodies to oscillating stresses and strains within the framework of complex variable theory. The initial development is the most general possible. Here we will be seeking to define parameters useful for describing the responses of viscoelastic bodies to oscillating loading patterns. We will then turn to analysis of the response of models based on Voigt and Maxwell elements. Models comprised of single elements are, of course, useful only as pedagogical tools. As in Sections 1010 and 1011, they can be combined in series or parallel to yield useful models. Discussion of this aspect of model analysis will be undertaken in Section 1121.

The *complex strain* γ^* and the *complex stress* S^* are defined in polar and exponential forms as

$$\gamma^* = \gamma_0[\cos(\omega t) + i \sin(\omega t)] = \gamma_0 e^{i\omega t} \qquad (11\text{-}9)$$

$$S^* = S_0[\cos(\omega t - \delta) + i \sin(\omega t - \delta)] = S_0 e^{i(\omega t - \delta)} \qquad (11\text{-}10)$$

where $\gamma_0 = |\gamma^*|$ and $S_0 = |S^*|$. Notice, in these definitions, how we have accounted for the fact that we do not expect the stress and strain to be in phase by including the phase angle δ in Eq. (11-10). The quantities γ^* and S^* are shown in Figure 11-7 as rotating vectors. Also shown are the real (γ', S') and the imaginary (γ'', S'') components of γ^* and S^*. In component form, γ^* and S^* are defined as

$$\gamma^* = \gamma' + i\gamma'' \qquad (11\text{-}11)$$

$$S^* = S' + iS'' \qquad (11\text{-}12)$$

The *complex strain rate* $\dot{\gamma}^*$ is obtained from Eq. (11-9) as

$$\dot{\gamma}^* = \omega \gamma_0 \left[\cos\left(\omega t + \frac{\pi}{2}\right) + i \sin\left(\omega t + \frac{\pi}{2}\right) \right] \qquad (11\text{-}13)$$

Notice that γ^* and $\dot{\gamma}^*$ are 90 deg out of phase.

The *complex modulus* G^* is defined as

$$G^* = \frac{S^*}{\gamma^*} = \frac{S_0}{\gamma_0} \cos \delta + i \frac{S_0}{\gamma_0} \sin \delta \qquad (11\text{-}14)$$

or in component form as

$$G^* = G' + iG'' \qquad (11\text{-}15)$$

where $G' = (S_0/\gamma_0) \cos \delta$ and $G'' = (S_0/\gamma_0) \sin \delta$. Since G' is in phase with the real components of the γ^* and S^*, it is (analogous to an elastic body) associated with elasticity and energy storage. G' is therefore referred to as

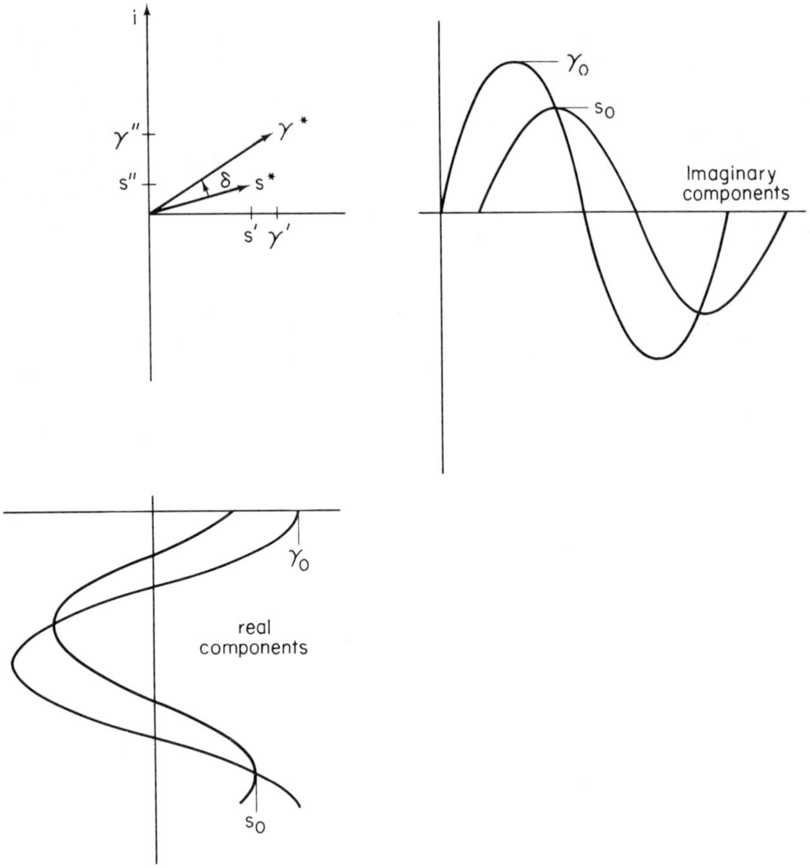

Figure 11-7. Complex stress and strain.

the *dynamic modulus*. It is approximately equivalent to the elastic modulus determined in creep and relaxation experiments for linear viscoelastic behavior. The quantity G'', on the other hand, is 90 deg out of phase with the real components of γ^* and S^*, and (analogous to a viscous body) it is associated with viscous energy dissipation. Hence G'' is called the *loss modulus*.

Another important parameter is the so-called *dissipation factor* or *loss tangent*, defined as

$$\tan \delta = \frac{G''}{G'} \tag{11-16}$$

It can be shown to be proportional to the ratio of energy dissipated per cycle to the maximum potential energy stored during a cycle. The significance of the parameters G', G'' and tan δ is that they can be determined experimentally. For a discussion of the manner in which this can be accomplished the reader is again referred to the general references cited. Note, too, that although G' has its counterpart in relaxation and creep experiments, G'' and tan δ are uniquely determined in oscillatory experiments.

In addition to the complex quantities already introduced, it is sometimes convenient to discuss dynamic mechanical properties in terms of a *complex viscosity* η^* and a *complex compliance J**. The equations and parameters associated with these quantities are developed below for the reader's reference.

$$\eta^* = \frac{S^*}{\dot{\gamma}^*} = \left(\frac{S_0}{\omega \gamma_0}\right)\left[\cos\left(S - \frac{\pi}{2}\right) + i \sin\left(\delta - \frac{\pi}{2}\right)\right]$$

$$\eta^* = \left(\frac{S_0}{\omega \gamma_0}\right)[\sin \delta + i \cos \delta] \tag{11-17}$$

Finally,

$$\eta^* = \eta' - i\eta'' \tag{11-18}$$

where $\eta' = (S_0/\omega\gamma_0) \sin \delta$ and $\eta'' = (S_0/\omega\gamma_0) \cos \delta$. The relationships between G', G'', and η', η'' are

$$G' = \omega \eta'' \tag{11-19}$$

$$G'' = \omega \eta' \tag{11-20}$$

and

$$\frac{G''}{G'} = \frac{\eta'}{\eta''} \tag{11-21}$$

$$J^* = \frac{1}{G^*} = J' - iJ'' \tag{11-22}$$

where J' = storage compliance = $\dfrac{G'}{G'^2 + G''^2}$

J'' = loss compliance = $\dfrac{G''}{G'^2 + G''^2}$

1110 The Voigt Element

Consider the response of a Voigt element subjected to the sinusoidal stress $S^* = S_0 \exp(i\omega t)$. This stress, combined with the rheological equation for the Voigt element, yields

$$S_0 \exp(i\omega t) = G\gamma^* + \eta\dot{\gamma}^* \tag{11-23}$$

where G and η represent the usual model parameters† and γ^* is the unknown strain response. The solution to this equation is of the form $\gamma^* = \gamma_0 \exp(i\omega t)$. Substituting into Eq. (11-23) and solving for γ_0 yields $\gamma_0 = (S_0/G)(1+i\omega\lambda)^{-1}$. The solution to Eq. (11-23) is expressed as

$$\frac{\gamma^*}{S_0 \exp(i\omega t)} = \frac{\gamma^*}{S^*} = \frac{1}{G}(1+i\omega\lambda)^{-1} \qquad (11\text{-}24)$$

where (γ^*/S^*) corresponds to the definition of J^*. Thus, for the Voigt model, the complex compliance, $J^*(\omega\lambda)$, is

$$J^*(\omega\lambda) = \frac{1}{G}(1+i\omega\lambda)^{-1} \qquad (11\text{-}25)$$

On separating $J^*(\omega\lambda)$ into real and imaginary components, we obtain

$$J^*(\omega\lambda) = \frac{1}{G}(1+\omega^2\lambda^2)^{-1} - i\left(\frac{\omega\lambda}{G}\right)(1+\omega^2\lambda^2)^{-1} \qquad (11\text{-}26)$$

$$J^* = J' - iJ''$$

$$J' = \left(\frac{1}{G}\right)(1+\omega^2\lambda^2)^{-1}$$

$$J'' = \left(\frac{\omega\lambda}{G}\right)(1+\omega^2\lambda^2)^{-1}$$

In complex variable notation, the work per cycle is expressed as

$$W = \int_0^{1/f} S^* \dot{\gamma}^* \, dt$$

where $\dot{\gamma}^* = J^*\dot{S}^* = J^*S_0 i\omega \exp(i\omega t)$, and, in its most general form, W is given by

$$W = (J^*S_0 i\omega)\int_\delta^{1/f} S^* \exp(i\omega t) \, dt \qquad (11\text{-}27)$$

With the application of the real stress $S^* = S_0 (\cos \omega t)$, Eq. (11-27) becomes, upon retaining only the real terms,

$$W = S_0^2 \omega \int_\delta^{1/f} [-J' \sin(\omega t)\cos(\omega t) + J''\cos^2(\omega t)] \, dt \qquad (11\text{-}28)$$

† A word of caution is in order at this point. The reader should not confuse the symbol G for the Hookean model constant with the measured experimental parameters, G_r, G_c, G', or G''.

Integrating over a full cycle, we obtain

$$W = 0 + \pi S_0^2 J''$$

We observe that the term containing J'' gives rise to an energy loss whereas that containing J' does not.

The integral of the first term of Eq. (12-28) will change from plus to minus and vice versa for $t = n/4f$, where n is a positive integer. Integrating the first term over the interval $t = 1/4f$, we find the maximum potential energy stored during a cycle to be $\pi S_0^2 J'/2$. These latter results, coupled with Eq. (11-16), show that for the Voigt element the dissipation factor is indeed proportional to the ratio of energy dissipated to the maximum energy stored per cycle. The dissipation factor, given by

$$\tan \delta = (\omega \lambda) \quad (11\text{-}29)$$

is seen to increase monotonically with $(\omega\lambda)$. The behavior of $J'(\omega\lambda)$ and $J''(\omega\lambda)$ is shown in Figure 11-8. As we shall see in Section 1117 when we

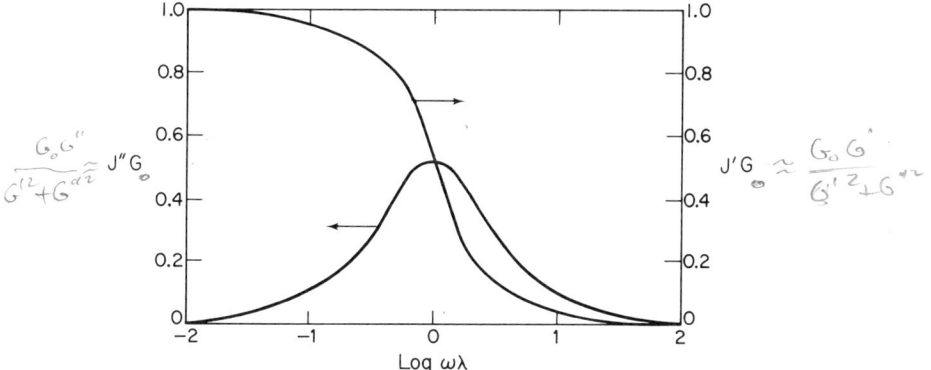

Figure 11-8. Frequency dependence of the dynamic mechanical properties of the Voigt model.

view the phenomenological aspects of dynamic mechanical experiments, the behavior of $J'(\omega\lambda)$ and $J''(\omega\lambda)$ is qualitatively in line with those determined for linear amorphous polymers. This is not true for $\tan \delta$, which exhibits a maximum and then a minimum. The variation of J', J'' and the strain amplitude with increasing frequency can be explained qualitatively as follows:

1. *Low frequency.* Since the rate of strain is low, the viscous drag of the dashpot will be low. Hence the loss compliance and energy loss should be low. The major portion of the applied stress must be borne by the spring; hence the elastic compliance will be high, as will the amplitude of vibration (strain).

2. *Higher intermediate frequencies.* The dashpot moves more swiftly to increase the viscous forces and J''. Since the applied stress must stretch the spring and counteract a large viscous force, the elastic response J' and the strain amplitude must decrease.
3. *Very high frequencies.* The dashpot will offer great viscous resistance and consume the major part of the applied stress so that J' will approach zero. The strain amplitude will be very low, so that the energy loss per cycle and J'' will also decay to zero. As already discussed, the overall power loss increases continuously.

We find that the loss tangent (tan δ), is zero for low frequencies and very large for high frequencies. Therefore the phase angle δ ranges from zero to $\pi/2$ in going from low to high frequencies.

The effect of temperature on the loss and elastic compliances can also be noted. This effect is manifested in $\lambda = \eta/G$. Increasing the temperature decreases the viscosity. Hence an increase in temperature at a constant frequency is equivalent to a decrease in frequency at constant λ. All previous discussion for low frequency would apply to high temperatures and vice versa. The application of increasing frequency would be equivalent to decreasing the temperature.

1111 The Maxwell Element

Here we outline the results for a Maxwell element and allow the reader to fill in the details as an exercise. By assuming the sinusoidal strain $\gamma^* = \gamma_0 \exp(i\omega t)$, one can obtain for a Maxwell element a complex compliance

$$G^* = (G\omega^2\lambda^2)(1+\omega^2\lambda^2)^{-1} + i(G\omega\lambda)(1+\omega^2\lambda^2)^{-1} \tag{11-30}$$

$$G^* = G' + iG''$$

$$G' = (G\omega^2\lambda^2)(1+\omega^2\lambda^2)^{-1}$$

$$G'' = (G\omega\lambda)(1+\omega^2\lambda^2)^{-1}$$

The work per cycle becomes

$$W = \int_{\delta}^{1/f} \gamma_0 \omega [G' \sin(\omega t) \cos(\omega t) + G'' \cos^2(\omega t)] \, dt \tag{11-31}$$

$$W = 0 + \gamma_0^2 \pi G''$$

Again we observe that the G'' term gives rise to an energy loss, whereas the G' term does not. The dissipation factor is given by

$$\tan \delta = \frac{1}{\omega \lambda} \qquad (11\text{-}32)$$

which, in contrast to the Voigt element result, decreases continuously with increasing frequency.

Curves showing the behavior of $G'(\omega\lambda)$ and $G''(\omega\lambda)$ are shown in Figure 11-9. These curves are also in qualitative agreement with experimental

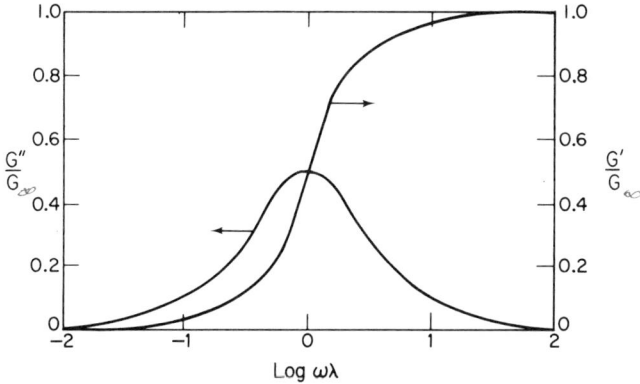

Figure 11-9. Frequency dependence of the dynamic mechanical properties of the Maxwell model.

results for linear amorphous polymers. A qualitative explanation for the behavior observed can be given, but it is left as an exercise. The effect of increasing temperature is again equivalent to decreasing frequency.

1112 The Maxwell Element with Mass

It is instructive to consider as a final model the Maxwell element with mass. This element is to be subjected to an initial strain and then allowed to vibrate freely. A force balance yields the governing linear differential equation

$$m\ddot{\gamma} + \eta\dot{\gamma} + G\gamma = 0 \qquad (11\text{-}33)$$

This equation has several solutions, depending on the values of the parameters in the characteristic equation

$$\gamma = -\frac{\eta}{2m} \pm \frac{1}{2m}(\eta^2 - 4Gm)^{\frac{1}{2}}$$

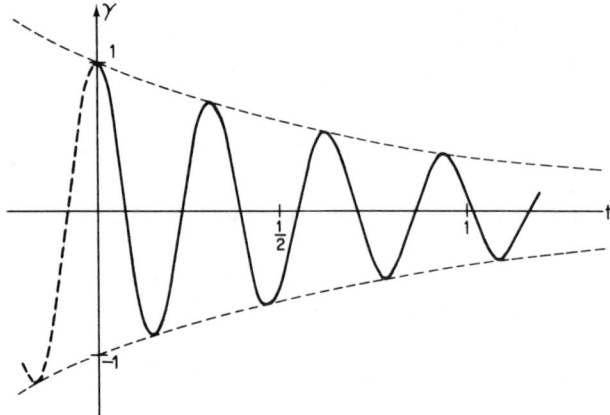

Figure 11-10. Damped oscillation.

A situation analogous to the actual testing of a viscoelastic solid would have the elastic response predominating, so the square root term would be negative. The solution for this case is given by

$$\gamma = e^{-at}(A \cos bt + B \sin bt) \qquad (11\text{-}34)$$

where $a = \eta/2m$, $b = 1/2m(4Gm - \eta^2)^{\frac{1}{2}}$, and A and B are arbitrary constants. The motion described by Eq. (11-34) is known as *damped motion*, and its general appearance is shown in Figure 11-10. This curve is also characteristic of those obtained in many types of free vibration experiments with polymers. Of particular interest is the ratio of successive extreme displacements on the same side of the equilibrium position. The extremes in γ occur when $\dot{\gamma} = 0$, and from Eq. (11-34) we have

$$\dot{\gamma} = -ae^{-at}[A \cos bt + B \sin bt]$$
$$+ e^{-at}[-bA \sin bt + Bb \cos bt] = 0$$

That is, the extremes occur when

$$\tan bt = \frac{bB - aA}{aB + bA}$$

or, finally, when

$$t = \frac{1}{b} \tan^{-1}\left(\frac{bB - aA}{aB + bA}\right) + \frac{n\pi}{b}$$

$$t = \frac{T}{b} + \frac{n\pi}{b}$$

$$= \tan^{-1}\left(\frac{bB - aA}{aB + bA}\right)$$

For the ratio of interest, we obtain from Eq. (11-34)

$$\frac{\gamma_n}{\gamma_{n+2}} = \frac{e^{-a(T+n\pi/b)}[A\cos(T+n\pi)+B\sin(T+n\pi)]}{e^{-a(T+(n+2)\pi/b)}[A\cos(T+(n+2)\pi)+B\sin(T+(n+2)\pi)]}$$

which simply reduces to

$$\frac{\gamma_n}{\gamma_{n+2}} = e^{2\pi a/b}$$

The important feature of this result is that the ratios of successive maximum displacements remain constant throughout the entire free motion of the system (for elastic effects predominating). Taking the natural logarithm of the last expression, we obtain

$$\ln\left(\frac{\gamma_n}{\gamma_{n+2}}\right) = \frac{2\pi a}{b} \tag{11-35}$$

This quantity is known as the *logarithmic decrement*, Δ. In actual free vibration experiments, the ratios of successive peak heights are approximately constant and these are measured to obtain this quantity.

PHENOMENOLOGICAL ASPECTS OF DYNAMIC MECHANICAL TESTING

Indirectly, we have already discussed certain phenomenological aspects of polymeric materials subjected to dynamic mechanical testing. That is, the elastic or dynamic modulus values measured in oscillatory experiments are essentially equivalent to those measured in transient experiments (for equivalent time scales and within the limits of linear viscoelasticity). However, because oscillatory test methods furnish a much wider spectrum of experimental time scales than transient methods, we are provided with an extended view of this property. As the preceding sections have shown, dynamic mechanical testing also provides additional information in terms of the loss modulus and the damping parameters, tan δ and log decrement Δ. These parameters are especially sensitive to molecular architecture, and particular attention will be paid to the damping characteristics of polymers in subsequent sections.

1113 Time-Temperature Relations

As discussed in conjunction with the elastic modulus in Section 1105, the dynamic mechanical properties of polymers are also time and temperature dependent in an equivalent fashion. As a consequence, the time-temperature superposition principle can be applied to the measurement of dynamic mechanical parameters as well. The limits of application of these methods are, of course, the same.

The time-dependent dynamic properties of a typical linear amorphous polymer, in this case poly(vinyl acetate), are shown in Figure 11-11. The shear moduli G' and G'' and the dynamic viscosity η' are shown as a function of the logarithm of the reduced frequency log ωa_T, where a_T is the shift factor. Since the frequency range covered in Figure 11-11 was too great to be studied at a single temperature, experiments were run at several temperatures and

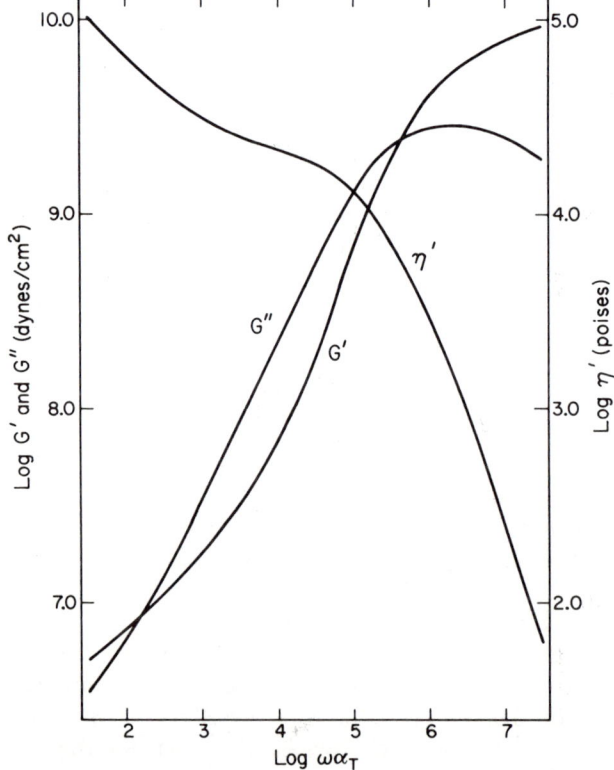

Figure 11-11. Master curves for the real and imaginary parts of the complex shear modulus and the real part of the viscosity of polyvinyl acetate, reduced to 75°C, plotted logarithmically against reduced frequency. [M. L. Williams and J. D. Ferry, J. Colloid Sci., **99**, 479 (1954).]

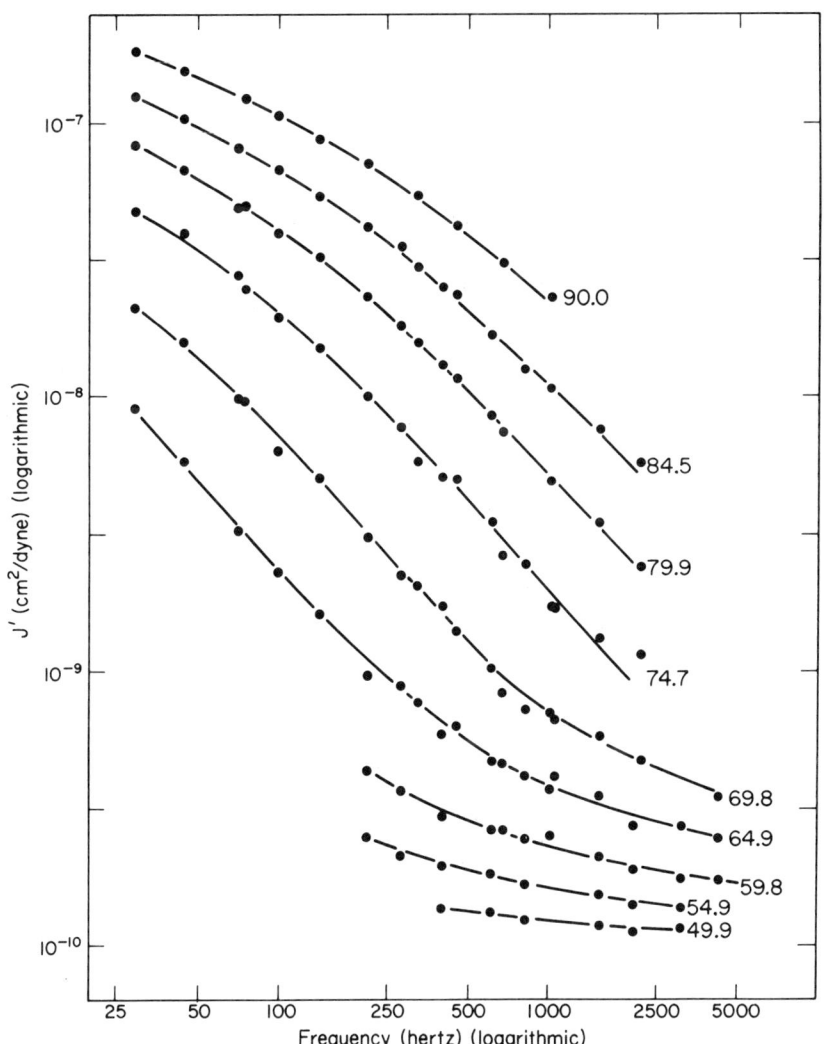

Figure 11-12. Variation of the real part of the dynamic shear compliance J' with frequency for polyvinyl acetate at the nine temperatures indicated on the curves. [Op. cit. Figure 11-11.]

the master curves were obtained by the superposition of the data shown in Figures 11-12 and 11-13, using 75 °C as a reference temperature. The glass transition of poly(vinyl acetate) is 29 °C, and for a reference temperature of 75 °C, we would ordinarily expect rubberlike behavior with an elastic modulus on the order of $10^{6.5}$ to $10^{7.5}$ dynes/cm^2. Both the storage and loss moduli are of this order at low frequencies (Figure 11-11), since the frequencies are

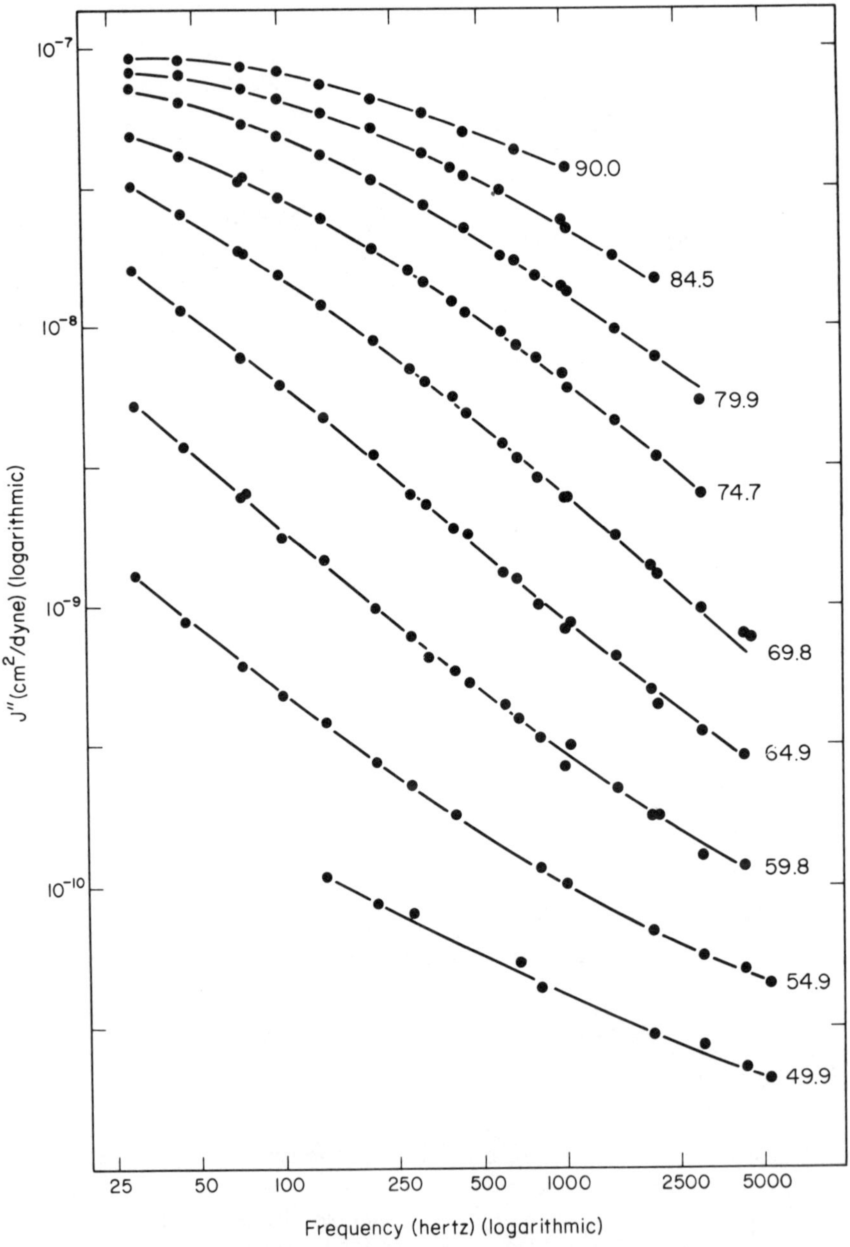

Figure 11-13. Variation of the dynamic loss compliance J'' with frequency for polyvinyl acetate at nine temperatures as indicated. [Op. cit. Figure 11-11.]

of the same order as those utilized in the transient experiments discussed in Chapter 7. However, they increase rapidly as the frequency increases by several decades. At very high frequencies (> 10^7 cycles/sec) the storage modulus eventually reaches a plateau on the order of 10^{10} dynes/cm^2 and the material assumes glasslike character. In other words, the material becomes "stiffer" as the frequency increases. Notice, too, that the shape of the storage modulus is identical (only inverted on the horizontal scale) to those shown for the modulus-temperature behavior of linear amorphous behavior discussed in Chapter 7. These results show the dramatic effect of the experimental time scale on material properties in a manner impossible with transient experiments. In the region where the storage modulus is changing most rapidly, the loss modulus, and hence the damping factor, is going through a maximum.

The temperature dependence of the dynamic mechanical properties of a typical linear amorphous polymer (in this case, a styrene-butadiene copolymer) is shown in Figure 11-14. These data were obtained in a torsion pendu-

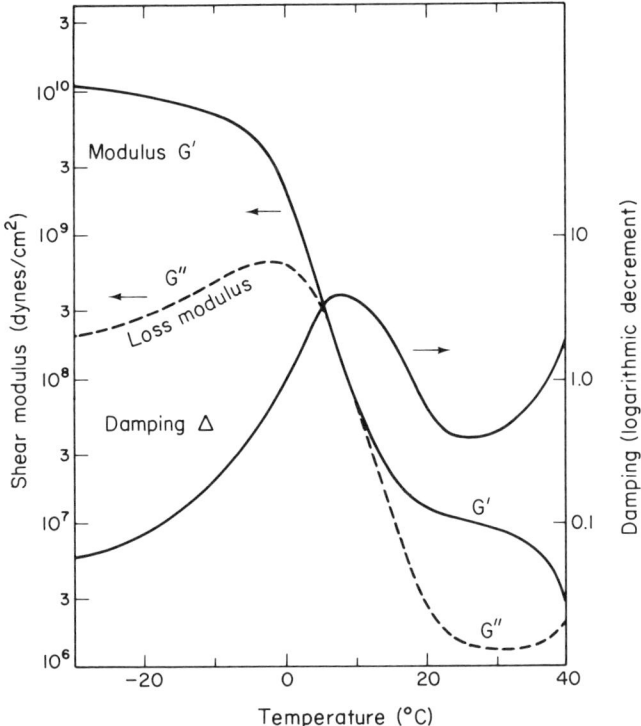

Figure 11-14. Typical dynamic mechanical behavior of linear amorphous polymers. The material is a copolymer of styrene and butadiene. [L. E. Nielsen, op. cit. Figure 11-4.]

lum experiment (a free vibration experiment) at a frequency of approximately one cycle per second. The storage modulus shows the five regions of viscoelastic behavior as well as their corresponding numerical values. The excellent correspondence with the date of Figure 7–5 results from the approximate correspondence of the experimental time scales.

Of particular interest in Figure 11–14 is the damping behavior shown—in this case—as the logarithmic decrement. (It may also be characterized by tan δ or J'' and G''.) As the temperature increases, the damping goes through a maximum, a few degrees above T_g, in the transition (leathery) region and then a minimum in the rubbery region. Such behavior is typical of all linear amorphous polymers. It can be partially explained on a molecular basis. The damping is low below T_g, for the chain segments are "frozen in." Below T_g, the deformations are thus primarily elastic and molecular slip resulting in viscous flow is low. Above T_g in the rubbery region, the damping is also low because molecular segments are very free to move about; consequently, there is little resistance to their flow. Hence when the segments are either completely frozen in or free to move, damping is low. In the transition region, on the other hand, a portion of the segments are free to move about whereas the remainder are not so free. This may be visualized as a dynamic situation with segments alternately becoming free to move and then frozen in. As a frozen segment stores energy through deformation, it ultimately releases it as viscous energy when it becomes free to move. The interplay

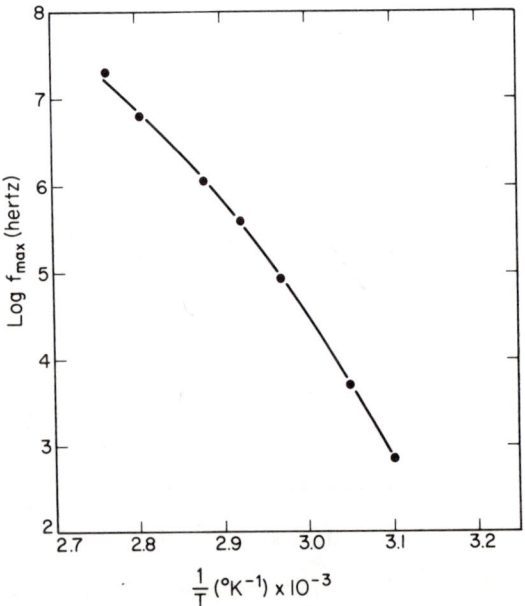

Figure 11–15. Log f_{max} vs. (1/T) for poly(vinyl acetate).

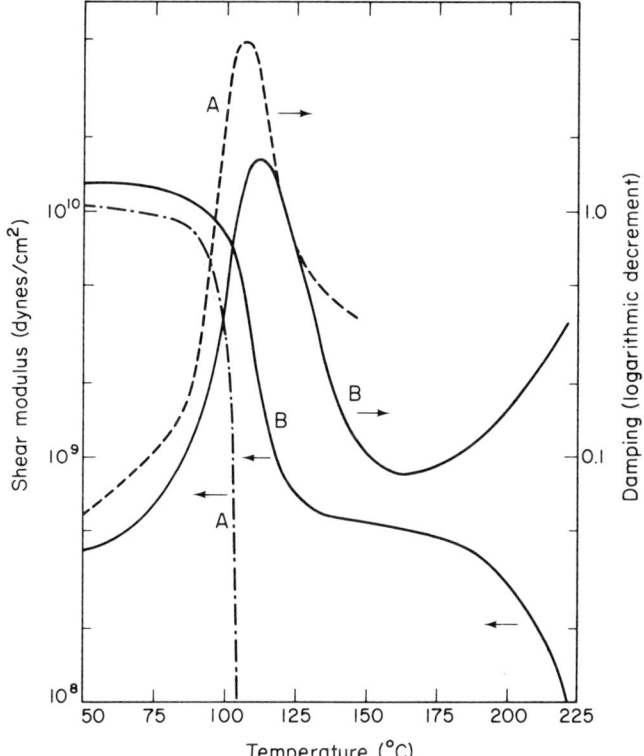

Figure 11-16. Dynamic mechanical properties of isotactic polystyrene. [S. Newman and W. P. Cox, J. Polymer Sci., **46**, 29 (1960).]

between the time scales of mechanical and molecular motion (vibrational oscillation) is readily apparent.

The loss modulus exhibits a peak a few degrees below the damping peak. At one cycle per second, this peak temperature corresponds rather closely to T_g determined in volume-temperature measurements.

Also of fundamental importance is the variation in the frequency at maximum peak height for the damping curves (log decrement Δ, tan δ, or G'' and J'') with temperature. An analogous aspect of this problem was previously introduced in Section 706, where the variation in relaxation modulus-temperature curves and glass-transition temperature with time scale of experiment was discussed. For dynamic mechanical experiments, such data are presented as $\log f(c/s)$ at maximum peak height versus $1/T$; when plotted in this form, the data are nearly linear. Data for poly(vinyl acetate) are presented in Figure 11-15.

The dynamic mechanical properties of a typical semicrystalline polymer (in this case, isotactic polystyrene) are compared with those of the same polymer in the amorphous state in Figure 11-16. As expected, the crystalline form maintains its mechanical integrity far above its glass transition, exhibiting tough, semirigid behavior. The damping is less than for the amorphous form, and it exhibits amorphouslike behavior, thereby showing the predominant damping mechanism to be associated with the amorphous regions of the crystalline polymer.

1114 Damping Peaks

The characteristics of the damping peak observed in the transition region of amorphous polymer systems is of paramount importance in engineering applications. For example, for sound-absorption qualities, we would look for a polymer with a damping peak in the range of audible sonic frequencies, 10^3 sec^{-1}, at the use temperature. As another example, the characteristics of vehicular tires will depend markedly on the character of the damping peak in the frequency and temperature-operating range of the tire: a high peak would mean superior riding qualities but excessive heat buildup and hence wear; whereas a low peak would result in poorer riding qualities with less wear. We are generally interested in four features of these peaks: temperature or frequency at peak height, plus peak height and width.

In addition to the damping peak observed in the glass-transition region, additional peaks are observed in conjunction with short-range structural motion, which occurs in both crystalline and amorphous regions. These so-called *secondary transition peaks* occur at temperatures above T_g within crystalline regions and below T_g within amorphous regions. Most polymers exhibit multiple damping peaks. The nature of these secondary transitions are particularly dependent on polymer structure. Certain peaks are especially important determinants of toughness.

In the next series of sections, we will first discuss the character of the glass-transition region damping peak and then the character of the secondary transition peaks. The latter will be accomplished by considering some representative polymer systems with important multiple transition behavior.

1115 The Effects of Crosslinking

The general effect of crosslinking on the mechanical properties of a linear amorphous polymer is amply illustrated in Figure 11-17 for a phenol formaldehyde resin crosslinked with hexamethylene tetramine at concentrations of 2, 4, and 10 percent. As previously noted, crosslinking raises T_g and the elastic modulus beyond T_g. There is a corresponding temperature shift in the damping peak, accompanied by an increase in the breadth but a decrease in the height of the peak. In highly crosslinked systems, there is no glass transition and no damping up to the polymer-decomposition temperature.

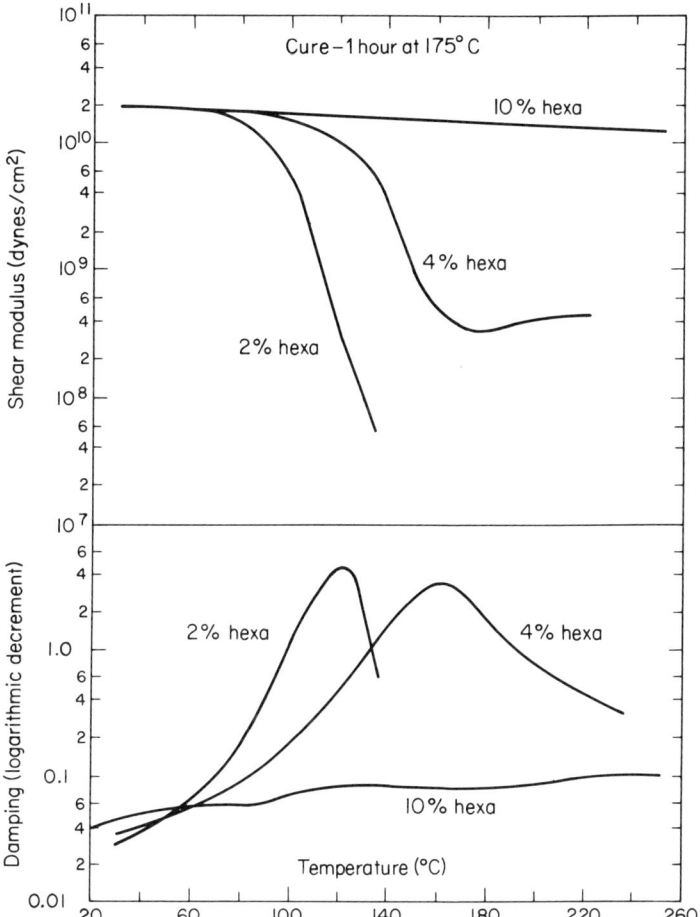

Figure 11-17. Dynamic mechanical properties of a phenolformaldehyde resin crosslinked with hexamethylene tetramine at concentrations of 2%, 4% and 10%. Resin was cured. [M. F. Drumm, C. W. H. Dodge, and L. E. Nielsen, Ind. Eng. Chem., **48**, 76 (1956). Reproduced with permission of the copyright owner, The American Chemical Society.]

1116 The Effects of Plasticization

The effect of plasticization on the glass-transition peak of a linear amorphous polymer is illustrated in Figure 11-18, which shows poly(vinyl chloride) plasticized with various amounts of diethylhexyl succinate. The corresponding modulus-temperature data are shown in Figure 7-13. Plasticization lowers the temperature at maximum peak height, just as it lowers T_g, and it broadens the transition peak, according to the polymer-plasticizer compati-

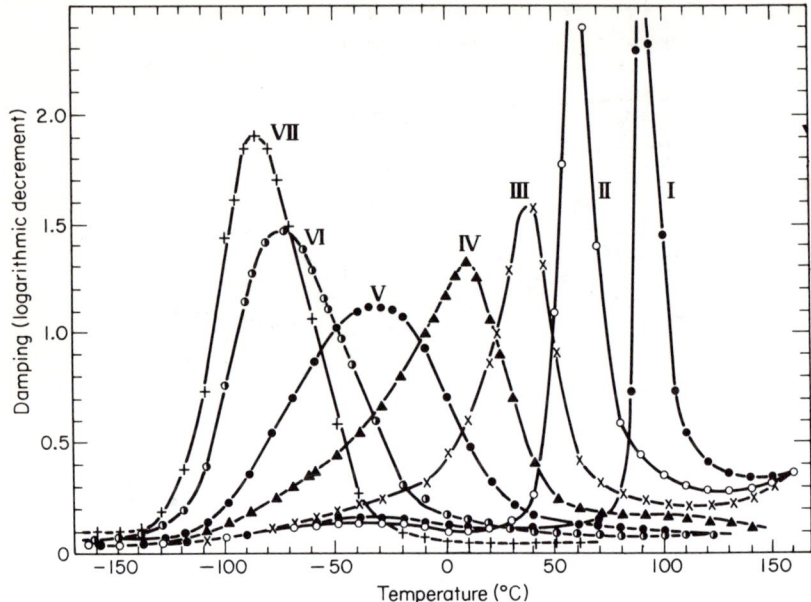

Figure 11–18. Log decrement plotted against temperature for measurements at approximately 1 cycle/sec. on polyvinyl chloride plasticized with diethylhexyl succinate, with compositions as indicated: I, 100% polymer; II, 91%; III, 79%; IV, 70.5%; VI, 51.8%; VII, 40.8%. [K. Schneider and K. Wolf, Kolloid-Z., **127**, 65 (1952).]

bility. The effect of plasticizer compatibility is reflected in Figure 11–19, which shows poly(vinyl chloride) plasticized with various plasticizers. The peak width increases with decreasing compatibility.

1117 The Effect of Copolymerization

As previously explained in Chapter 5 on copolymerization, a major role is played by both the compatibility of the homopolymers and the homogeneity of the copolymers. For example, the glass-transition peaks of a homogeneous and heterogeneous vinyl chloride-methyl acrylate copolymer are shown in Figure 11–20. The corresponding modulus-temperature data are shown in Figure 7–18. The increase in peak width induced by heterogeneity is enhanced if the homopolymers are incompatible. Increasing width results from the fact that chains of varying composition have different glass-transition temperatures. Consequently, the more heterogeneous the composition, the wider the transition region. Figure 11–21 shows the transition character of a miscible polyblend and a homogeneous copolymer compared to a semicom-

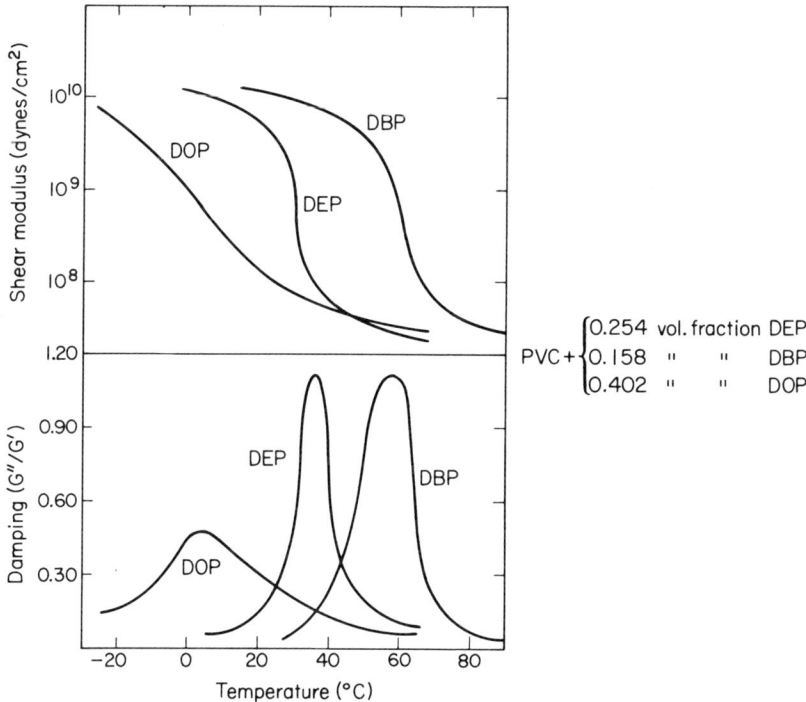

Figure 11-19. Dynamic mechanical properties of plasticized polyvinyl chloride. Diethyl phthalate (DEP, 0.254 volume fraction), dibutyl phthalate (DBP, 0.158 volume fraction), n-dioctyl phthalate (DOP, 0.402 volume fraction). [L. E. Nielsen, R. Buchdahl, and R. J. Levreault, J. Appl. Phys., **21**, 607 (1950).]

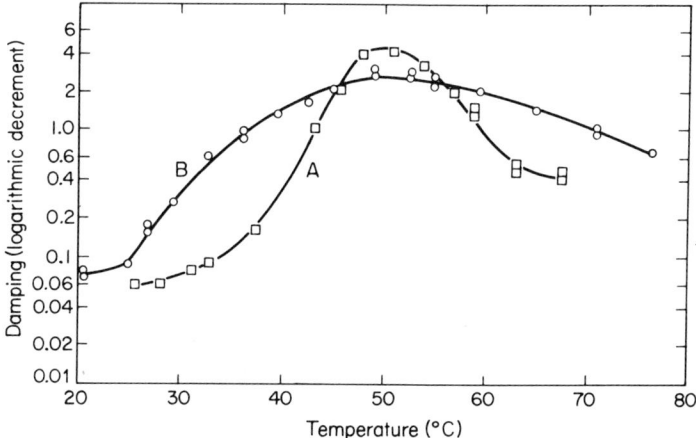

Figure 11-20. Dynamic mechanical properties of vinyl chloride-methyl acrylate copolymers: A. Homogeneous; B. Heterogeneous. [L. E. Nielsen, J. Am. Chem. Soc., **75**, 1435 (1953). Reproduced with permission of the copyright owner, the American Chemical Society.]

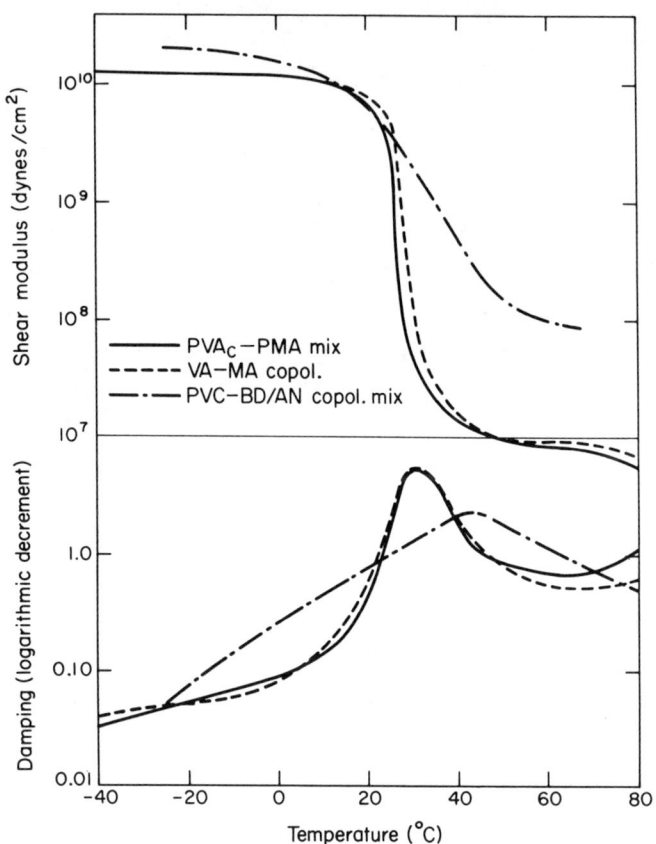

Figure 11-21. Dynamic mechanical properties of miscible polyblends and copolymers. A. 50/50 molar mixture of polyvinyl acetate and polymethyl acrylate. B. Vinyl acetate–methyl acrylate copolymer. C. Mixture of polyvinyl vinyl chloride and a copolymer of butadiene and acrylonitrile. [L. E. Nielsen, op. cit. Figure 11-4.]

patible polyblend. A miscible polyblend is, in fact, a solid solution and differs negligibly in physical properties from a linear amorphous polymer. Note that there is no difference between the properties of the homogeneous copolymer and compatible polyblend but that the peak width is considerably increased in the semicompatible blend. A completely incompatible blend system, on the other hand, exhibits two transition peaks, each nearly corresponding to those of the component homopolymers. This behavior is illustrated in Figure 11-22. The properties of graft and block copolymers are analogous to those of polyblends, with their particular characteristics depending on the degree of compatibility of the various chain segments as well as on composition.

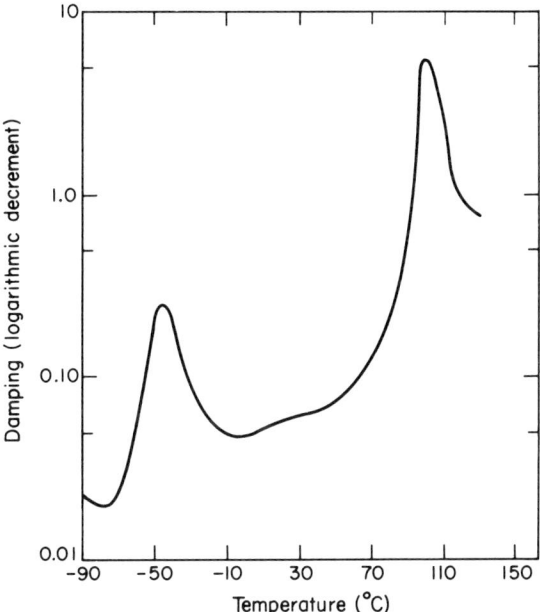

Figure 11-22. Damping properties of an immiscible polyblend. Material is a mixture of polystyrene and a styrene-butadiene copolymer. [L. E. Nielsen, op. cit. Figure 11-4.]

1118 Multiple Relaxation (Transition) Peaks in Polymers

As previously mentioned, most polymers display more than one relaxation or damping peak. By way of illustration, consider the modulus-log decrement behavior for poly(methyl methacrylate) in Figure 11-23 and for high and low density (designated HD and LD, respectively) polyethylene in Figure 11-24. Amorphous polymers exhibit secondary peaks below the glass transition—the primary transition—and crystalline polymers exhibit them above and sometimes below the glass transition. The temperature peaks below T_g are normally attributable to motion of short segments of the chain backbone or to motion of substituent groups. Those above T_g in crystalline polymers are thought to be associated either with the motion of the folded chains or with the motion of lattice defects. The presence of nonpolymeric components, such as plasticizers or absorbed water, may also induce secondary transitions.

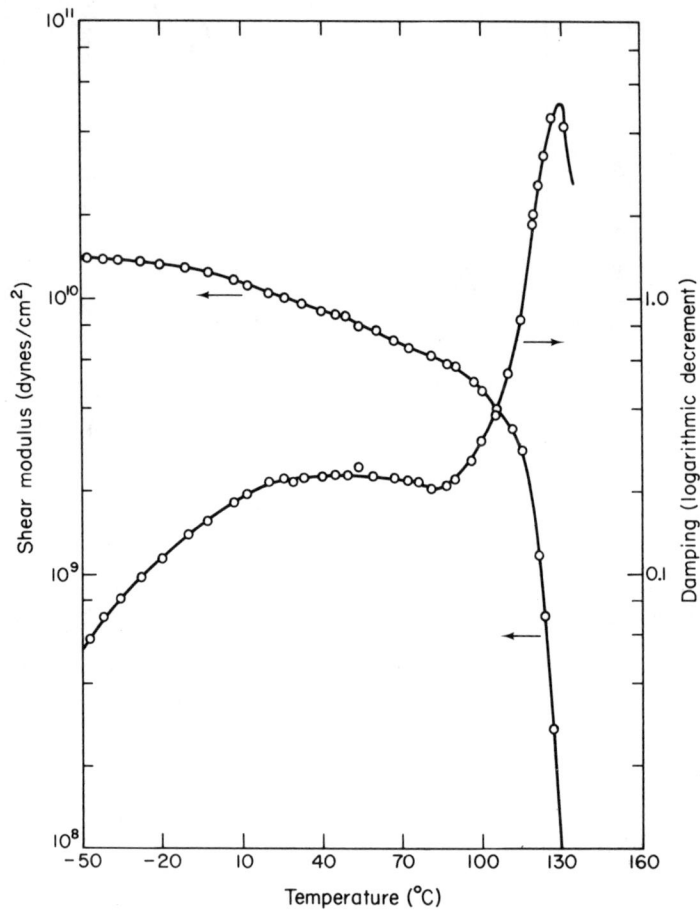

Figure 11-23. Dynamic mechanical properties of poly(methyl methacrylate). [L. E. Nielsen, Plastics Eng. J., **16**, 525 (1960).]

The importance of the secondary damping peaks in determining the mechanical properties of a polymer varies according to their individual character (peak height and width plus temperature and frequency at peak height) and the source of the motion inducing the peak. To amplify the latter point: secondary transitions involving the motion of side chain substituents are not as effective as secondary transitions involving the motion of backbone groups in improving impact resistance. It is well known that polymers that have a pronounced secondary transition involving backbone motion well below their use temperature have superior impact resistance. Such a transition seems to provide a highly effective mechanism for energy dissipation.

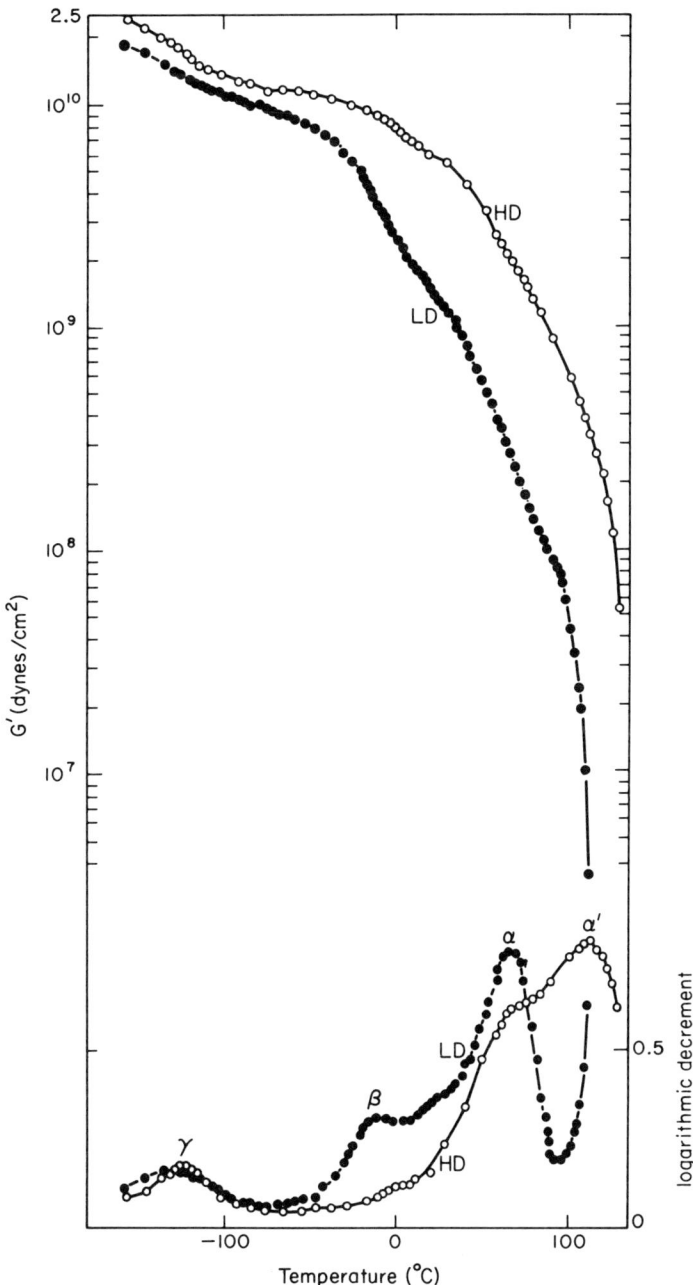

Figure 11-24. Temperature dependence of shear modulus and logarithmic decrement of high density (HD) and low density (LD) polyethylene at ca.1 c/s. [H. A. Flocke, Kolloid-Z., **180**, 118 (1962).]

The peaks are usually identified, as in Figures 11-23 and 11-24, by labeling them with the Greek letters α, β, γ, etc. According to this system of nomenclature, the α-transition is the highest temperature transition observed at a given frequency or the lowest frequency transition observed at a given temperature. The β- and γ-transitions apply to other relaxation regions in order of decreasing temperature or increasing frequency. For amorphous polymers, the α-transition is associated with the glass or primary transition. In crystalline polymers, the α-transition is mostly (but not always) associated with motions in the crystalline regions. Hence there is no correspondence of identification between crystalline and amorphous polymers or, for that matter, even between amorphous polymers of the same family (e.g., see discussion of the alkyl methacrylate polymers that follows) when peak identification may change with changing structure. To add to the confusion of this system, the exact temperature or frequency at which the transition peaks occur depend on whether damping is measured as loss modulus, log decrement or tan δ. Recall, for example, that in Figure 11-14, the loss-modulus peak occurred at a slightly lower temperature than the log decrement peak.

1119 Pendant Group Motion and Secondary Relaxation

To illustrate the role of pendant groups in secondary transitions, we will consider several members of the *n*-alkyl methacrylate polymers, where R

$$-CH_2-\underset{\underset{O}{\overset{\|}{C-OR}}}{\overset{CH_3}{\underset{|}{C}}}-$$

represents the *n*-alkyl group. We chose this particular family of polymers because they exhibit prominent secondary transitions, and the mechanisms of these transitions are identical for materials with similar pendant group structures. In addition to the α- or glass transition, there are four other transitions. These are associated with motion of the —COOR group, the α–methyl group, the R group, and with absorbed water. As illustrated in Figure 11-23 for poly(methyl methacrylate) (PMMA) at 1 cycle/second, the α–transition occurs at about 130 °C and the β-transition—attributed to —COOCH$_3$ group motion—occurs at about 40 °C. Notice that the β-transition is reflected very modestly in the modulus data, but it is quite pronounced in the damping curve. When measured as G'', these transitions occur at 120 and 10 °C. A γ-transition is reported around -173 °C (1 cycle/second) due to α-methyl group motion and a δ-transition around -269 °C (4 °K) (1 cycle/second) due to methyl group motion on the —COOCH$_3$ group.

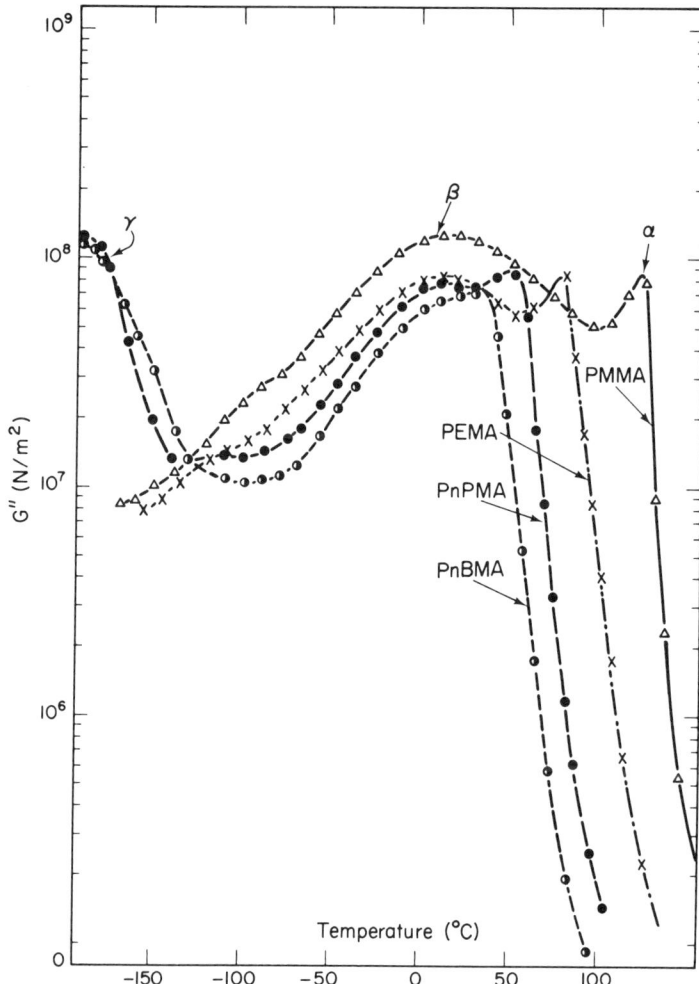

Figure 11-25. Temperature dependence of G″ at 1 c/s for PMMA, poly(ethyl methacrylate) (PEMA), poly(n-propyl methacrylate) (PnPMA) and poly(n-butyl methacrylate) (PnBMA). [J. Heijboer, Physics of Non-Crystalline Solids, (Proc. Inst. Cong. Delft, July, 1964, Int. Union Pure Appl. Chem.) J. A. Prins, ed., 1965.]

Shown in Figure 11-25 are loss-modulus data for four methacrylate polymers at 1 cycle/second. The methyl and ethyl forms show only α- and β-transitions because of the limited temperature range covered; whereas the *n*-propyl and *n*-butyl members also show a γ-transition. In contrast to PMMA, the γ-transition in the ethyl, *n*-propyl, and *n*-butyl members of this family of polymers is attributed to motion of the *n*-alkyl (R) group of the

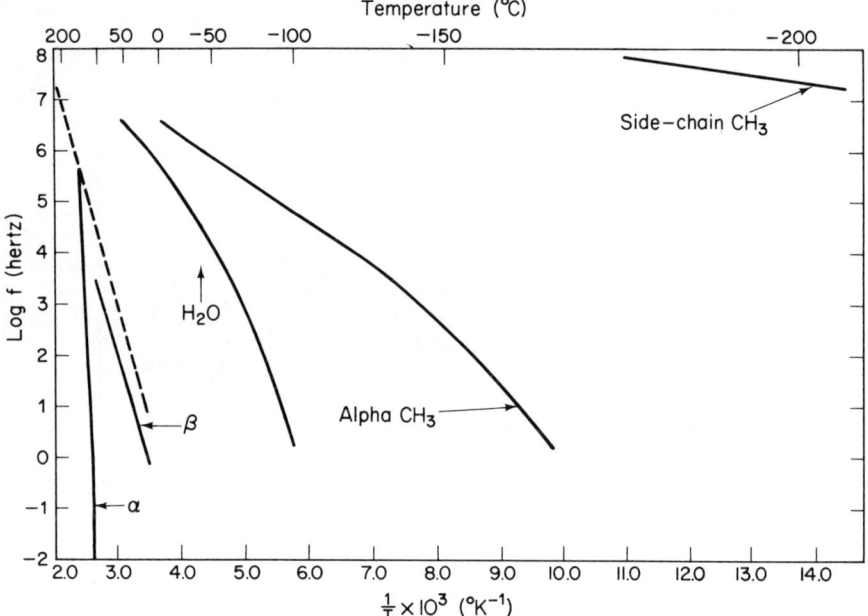

Figure 11–26. Plot of log f vs. (1/T) for the mechanical (solid lines) and dielectric (broken line) loss maxima for PMMA. [After N. G. McCrum, B. E. Read, and G. Williams, *Anelastic and Dielectric Effects in Polymeric Solids*, John Wiley and Sons, New York, 1967.]

—COOR group rather than the α-methyl group. For the higher alkyl methacrylates, the α- and β-transitions merge within certain temperature regions. Finally, Figure 11–26 shows the variation of the log frequency at maximum peak height, log f_{max} (cycles/second) versus $1/T$ for the transitions of PMMA.

1120 In-Chain Motion and Secondary Relaxation

We shall now consider some important polymer systems that display in-chain secondary transitions (and perhaps substituent transitions as well). Referring to Figure 11–24, we see that polyethylene may display three transitions. The α (LD) and α' (HD) transitions are clearly associated with transitions within the crystalline regions. The differences in their temperature values arises because of the differences in their crystalline melting points (about 115 °C for LD and about 135 °C for HD). The best explanation attributes this relaxation to the translational motion of the molecules in the direction of the chain axis during the annealing process which takes place at these temperatures. (See Section 609.) The effect of side-chain branching on secondary relaxation is clearly evident in Figure 11–27. Three samples are shown with

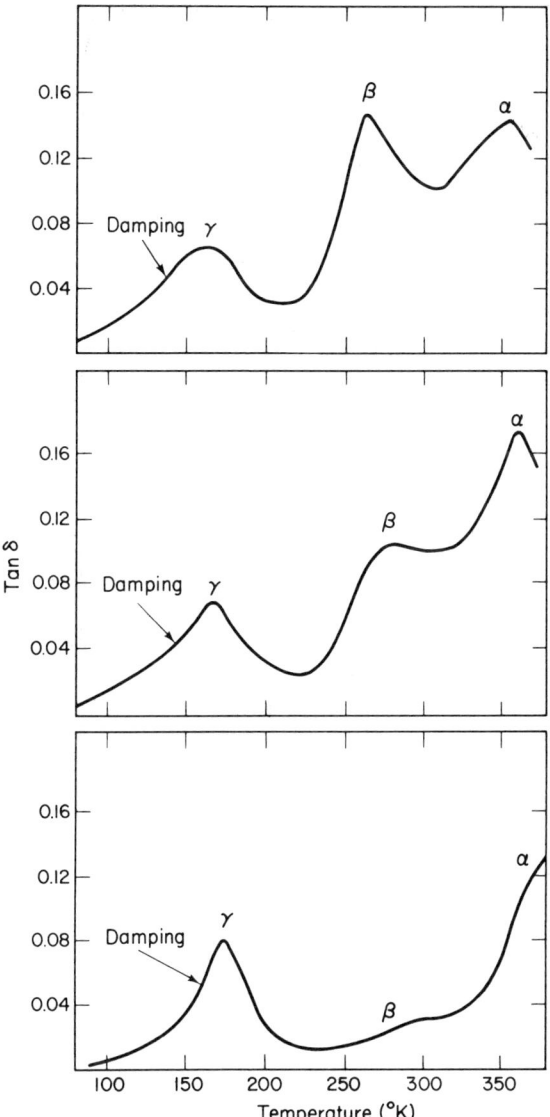

Figure 11-27. Showing the effect of side-branch content on the magnitude of the relaxation in polyethylene. Upper, middle and lower curves for specimens containing 32, 16 and 1 methyl group per 1000 carbon atoms. [D. E. Kline, J. A. Sauer, and A. E. Woodward, J. Polymer Sci., **22**, 455 (1956).]

side methyl contents (and densities) of 32 (0.915), 16 (0.922), and 1 (0.957) per 1000 chain carbon atoms. The β-transition all but disappears in the low-branching content (HD) sample. Notice that the temperature at peak height also decreases with decreasing branch content.

The γ-relaxation, which occurs in the neighborhood of -125 °C (Figure 11–23), is of paramount importance, for other polymers containing at least three to four —CH_2— backbone sequences also display identical transitions in the same temperature region. They are identical in the sense that the same mechanism of molecular motion is involved. The so-called *crankshaft motion*, illustrated in Figure 11–28, requires the simultaneous rotation of the several sequential —CH_2— units about bonds 1 and 7 or

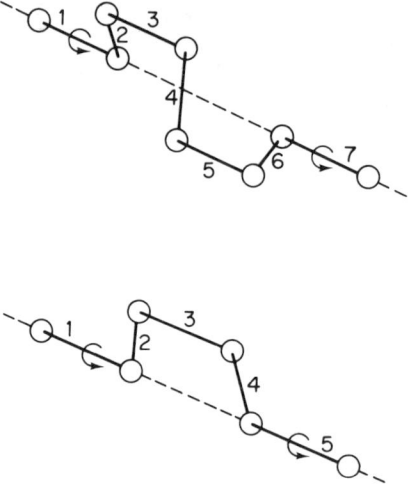

Figure 11–28. Illustration of the crankshaft mechanism.

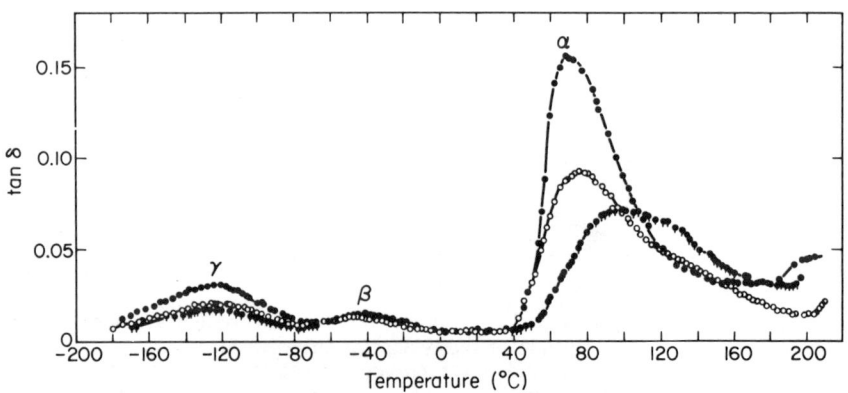

Figure 11–29. Temperature dependence of the mechanical loss tangent, tan δ, at 100 c/s for nylon 6 samples of different crystallinity. (●) sample quenched from the melt to -78 °C (least crystalline); (○) sample crystallized at 200 °C (intermediate degree of crystallinity); (◕) -caprolactam removed (highest crystallinity). [M. Takyanagi, Mem. Fae. Eng. Kushu Univ., **23**, No. 1, 41 (1963).]

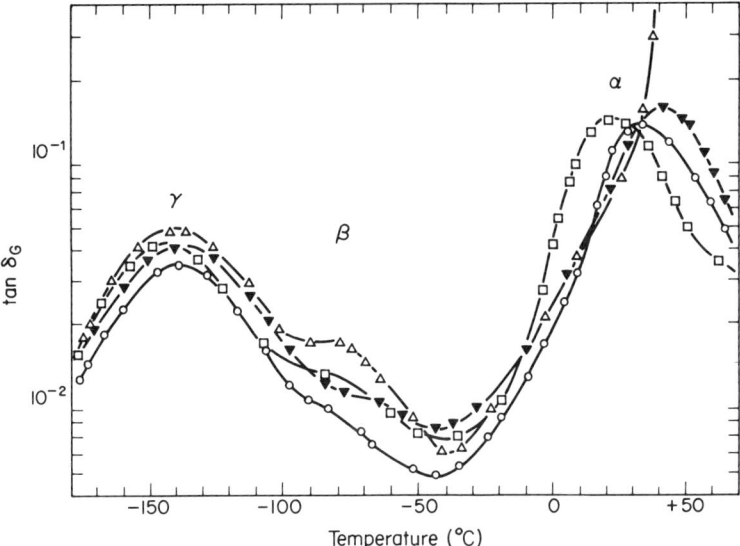

Figure 11-30. Temperature dependence of tan δ_G at 1 c/s for polyurethanes prepared from hexamethylene diisocyanate and, respectively, 1, 4-butanediol (▼), 2, 5-hexanediol (△), 1, 6-hexanediol (○) and 1, 10-decanediol (□). [H. Jacobs and E. Jenckel, Makromol. Chem., **43**, 132 (1961).]

1 and 5 such that the intervening carbon atoms move as a crankshaft. This process requires that bonds 1 and 7 or 1 and 5 be collinear. The chain atoms on either side may remain frozen in. This transition takes place below the temperature usually ascribed to the glass transition. In polyethylene, there is considerable controversy as to the actual value of T_g and as to the exact nature of the glass transition. Values commonly ascribed to T_g for polyethylene range from (1) in the β region just below 0 °C, (2) in the β region at -81 °C, to (3) in the γ region below -100 °C.

What is of primary importance is that the crankshaft mechanism provides a major means of energy absorption. Materials exhibiting these low-temperature, crankshaft mechanisms of relaxation typically exhibit excellent impact resistance. For example, linear polyamides (Figure 11-29) and polyurethanes (Figure 11-30) both exhibit γ-transitions around -120 °C and -140 °C, respectively. For both of these polymers, the β-transition is associated with absorbed water and the α-transition with the glass transition.

Similarly, the polycarbonate shown in Figure 11-31 also exhibits a significant low-temperature damping peak. This polycarbonate has the repeat-unit structure

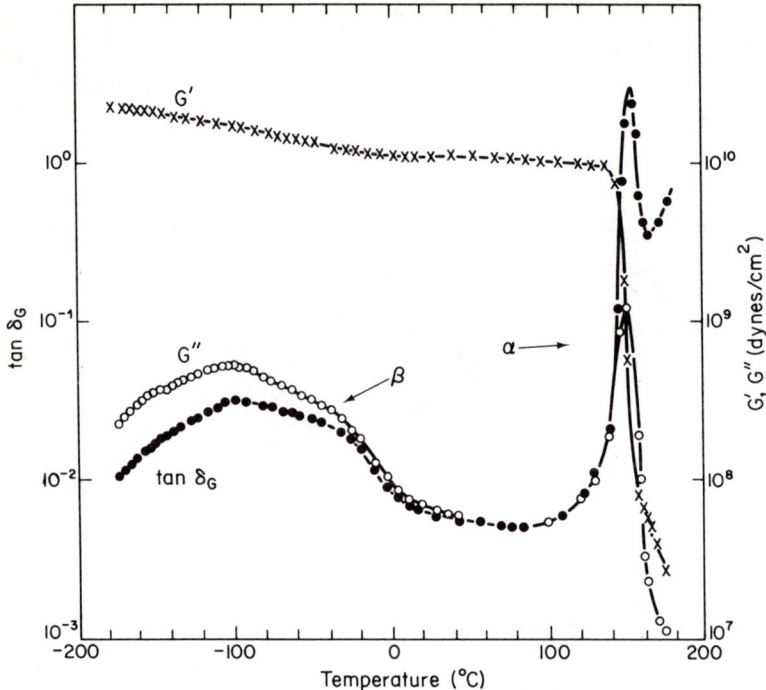

Figure 11-31. G', G" and tan δ_G against temperature at 1 c/s for polydian carbonate. [K. H. Illers and H. Breuer, Kolloid-Z., **176**, 110 (1961).]

and the indicated β peak is ascribed to motion of the carbonate group. The α-transition at 150 °C corresponds to the glass transition. In spite of its regular architecture, the molecular-chain stiffness of this material prevents it from readily crystallizing, and it is usually found in the amorphous state. The polycarbonates, as well as the polyurethanes, are known for their exceptional impact resistance, which is in keeping with the magnitude of the transition peaks, shown in Figures 11-30 (γ-peak) and 11-31 (β-peak).

1121 Relaxation and Retardation Spectra

We now return to pick up the thread in our development of linear viscoelasticity begun in Chapter 10. Consider first the representation of an infinite parallel arrangement of Maxwell elements for a linear amorphous polymer. Recall that the analytic expression for this model for relaxation is given by

$$G(t) = \int_0^\infty G(\lambda) e^{-t/\lambda} \, d\lambda \qquad (10\text{-}26)$$

where $G(\lambda)$ is the so-called relaxation-time distribution or the relaxation spectrum. Since the distribution of relaxation times is so broad, it is more convenient to consider $\ln \lambda$. Hence we introduce the function $H(\ln \lambda)$, where the parenthesis denotes functional dependence, to replace $G(\lambda)$ as

$$G(\lambda) = \frac{H(\ln \lambda)}{\lambda}$$

Then Eq. (10–26) becomes (note the change in limits)

$$G(t) = \int_{-\infty}^{\infty} H(\ln \lambda) e^{-t/\lambda} \, d(\ln \lambda) \qquad (11\text{-}36)$$

and all relaxation times are considered as $\ln \lambda$.

What we desire now is a means to determine $H(\ln \lambda)$ from data obtained as $G(t)$ versus $\ln t$. This is virtually impossible to do directly, and a number of approximate methods have been devised. We shall consider only the simplest, which is adequate for our purposes. Higher-order approximations are discussed in the advanced reference works of Ferry and Tobolsky. Consider a plot for a single Maxwell element under relaxation conditions, as shown in Figure 11–32 with $S/S(0)$ plotted against $\log t/\lambda$. Notice that

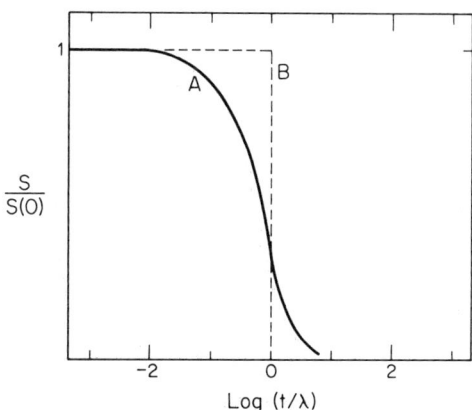

Figure 11–32. Curve A is the stress-relaxation behavior predicted for a single Maxwell model. Curve B is the step function discussed in the text.

most of the decay occurs within one decade of time around the relaxation time—that is, when $t/\lambda = 1$ and $\log t/\lambda = 0$. As a first approximation for determining $H(\ln \lambda)$, $e^{-t/\lambda}$ is replaced by the step function, also shown in Figure 11–37, in which $S/S(0)$ is unity for $t \leq \lambda$ and zero for $t \geq \lambda$. Equation (11–36) then becomes

$$G(t) = \int_{t=\lambda}^{\infty} H(\ln \lambda) \, d(\ln \lambda) \qquad (11\text{-}37)$$

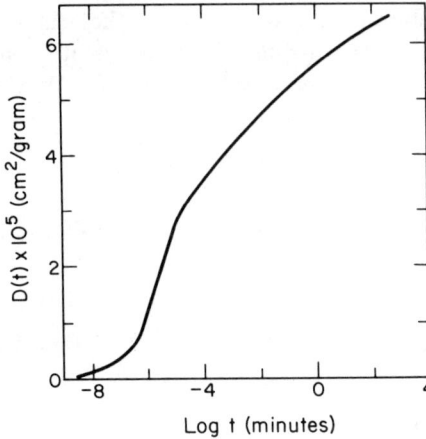

Figure 11–33. The composite creep curve at 25 °C obtained for a cross-linked styrene-butadiene rubber. [V. Kolpe in F. Bueche, *Physical Properties of Polymers*, John Wiley and Sons, New York, 1962.]

Upon differentiation with respect to ln λ, $H(\ln \lambda)$ is found as

$$H(\ln \lambda) = \frac{-dG(t)}{d(\ln \lambda)} \qquad (11\text{--}38)$$

which is simply the slope of the relaxation–ln t curve. As a first approximation, curves of $H(\ln \lambda)$ versus ln λ are found by plotting the slope of $G(t)$–ln t data with ln t = ln λ.

One treats the retardation spectra given by Eq. (10–27) similarly; $J(\lambda)$ is replaced by $L(\ln \lambda)/\lambda$ where $L(\ln \lambda)$ is the new retardation spectrum. A parallel treatment leads to an analogous means of determining this parameter as

$$L(\ln \lambda) = \frac{dJ(t)}{d(\ln \lambda)} \qquad (11\text{--}39)$$

Figures 11–33 through 11–36 illustrate the result of application of the

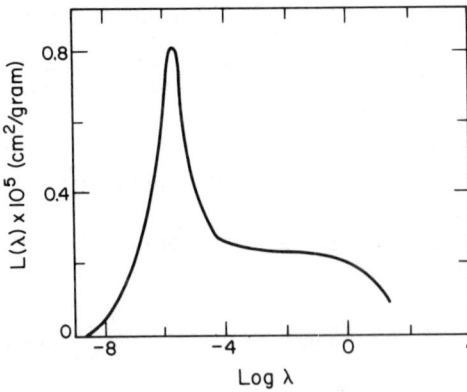

Figure 11–34. The retardation-spectrum obtained from the data of Figure 11–33. [V. Kolpe, op. cit. Figure 11–33.]

Sec. 1121   Relaxation and Retardation Spectra

foregoing analysis to two styrene-butadiene rubbers, one more crosslinked than the other. Compared in Figures 11-34 and 11-36 are the retardation spectra. Notice that the least crosslinked has a larger fraction of material with long response times than the more highly crosslinked. Although, for some purposes, one could characterize the rubber of Figure 11-34 by a single

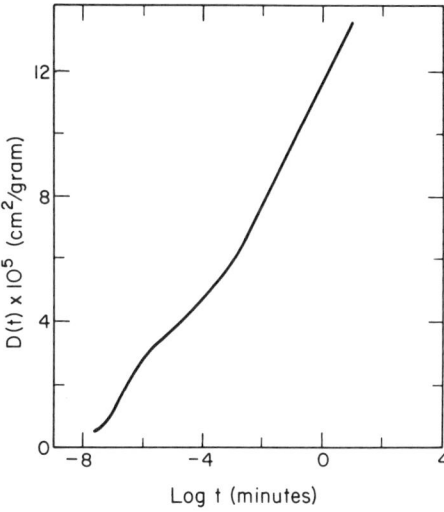

Figure 11-35. The composite creep curve for a rubber similar to the one used for Figure 11-33 but not crosslinked as highly as the material used there. [V. Kolpe, op. cit. Figure 11-33.]

retardation time, the response of the rubber of Figure 11-36 is too complicated for such simplistic representation. In Figure 11-37, the relaxation spectra of the rubber of Figure 11-34 is compared with its retardation spectra.

The significance and utility of retardation spectra like those being considered in Figures 11-34 and 36 can be further explored if we consider, for example, use of either material as a tire rubber. Since a tire rotates at

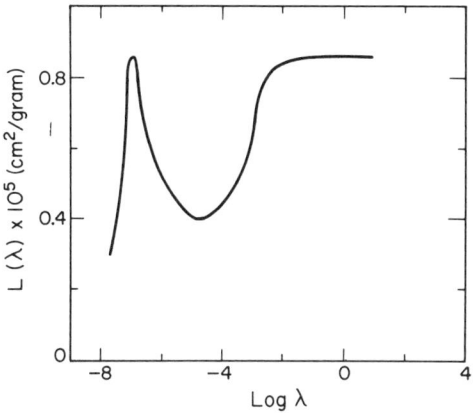

Figure 11-36. The retardation-time spectrum obtained from the data of Figure 11-35. This sample is less crosslinked than the one in Figure 11-34. [V. Kolpe, op. cit. Figure 11-33.]

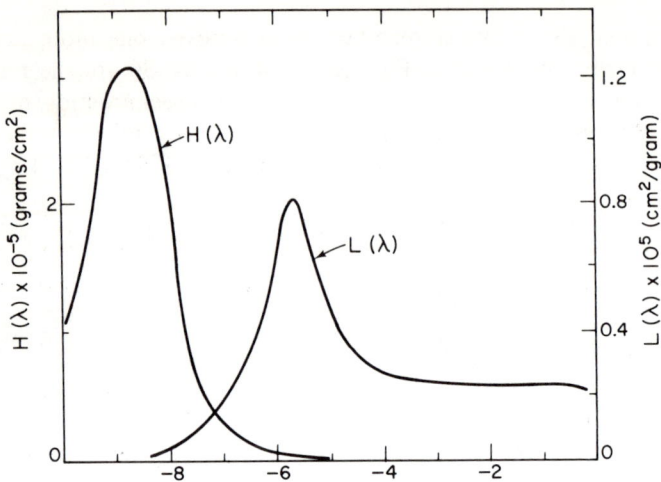

Figure 11-37. The relaxation and retardation spectra obtained for the same vulcanized SBR rubber as shown in Figure 11-33. [V. Kolpe, op. cit. Figure 11-33.]

about 10 times per second, the important retardation times will be those near $\lambda_n = 0.10$ second. The use temperature also ought to be specified, for λ_n depends on temperature. For the present illustration, assume that the specified temperature is 25 °C, the reference temperature. Since the material of Figures 11-35 and 11-36 has a significant fraction of material with response times on the order of 0.1 second, this material—the more lightly crosslinked of the two—will be more subject to heat buildup than the material of Figures 11-33 and 11-34; consequently, it will not function as well in tire applications.

By subjecting the infinite parallel arrangement of Maxwell elements to a sinusoidal strain and separating the responses into component form, one can show that the following relations hold:

$$G'(\omega) = \int_{-\infty}^{\infty} [H(\ln \lambda) \omega^2 \lambda^2 (1 + \omega^2 \lambda^2)^{-1}] \, d(\ln \lambda) \qquad (11\text{-}40)$$

$$G''(\omega) = \int_{-\infty}^{\infty} [H(\ln \lambda) \omega \lambda (1 + \omega^2 \lambda^2)^{-1}] \, d(\ln \lambda) \qquad (11\text{-}41)$$

Similarly, for an infinite series arrangement of Voigt elements subjected to a sinusoidal stress, one can obtain

$$J'(\omega) = \int_{-\infty}^{\infty} [L(\ln \lambda)(1 + \omega^2 \lambda^2)^{-1}] \, d(\ln \lambda) \qquad (11\text{-}42)$$

$$J''(\omega) = \int_{-\infty}^{\infty} [L(\ln \lambda)(\omega\lambda)(1+\omega^2\lambda^2)^{-1}] \, d(\ln \lambda) \qquad (11\text{-}43)$$

According to this analysis, dynamic mechanical properties can be obtained by performing creep and relaxation experiments. Of course, these spectra can also be obtained directly in dynamic mechanical experiments. There is also a connection between $H(\ln \lambda)$ and $L(\ln \lambda)$. Because this derivation is quite complicated, we only present the results here.

$$L(\ln \lambda) = \frac{H(\ln \lambda)}{\int_{-\infty}^{\infty} H(\xi) \, d(\ln \xi)/(1-\xi\lambda^{-1})^2 + \pi^2 H^2(\ln \lambda)} \qquad (11\text{-}43)$$

and

$$H(\ln \lambda) = \frac{L(\ln \lambda)}{\int_{-\infty}^{\infty} L(\xi) \, d(\ln \xi)/(1-\xi\lambda^{-1})^2 + \pi^2 L^2(\ln \lambda)} \qquad (11\text{-}44)$$

Thus the results of a creep, relaxation, or oscillatory experiment can be predicted, provided the results of one is known. However, in order to make these transformations very reliable, closely spaced data are required. Considerable error may find its way into the transformations, for determining the original spectrum requires the measurement of slopes of data that may show considerable scatter.

12
Introduction to Polymer Melt Rheology

Fluids that obey Newton's rheological equation of state, $\tau = \eta \dot{\gamma}$, constitute an important class of industrial fluids. The rheological properties of such fluids are completely described in terms of their densities and viscosities, which are constant at any temperature and pressure. Fluids that consist of simple low-molecular-weight molecules, or solutions thereof, such as water and common organic solvents, generally exhibit Newtonian flow behavior. Materials having a complicated constitution, such as polymer melts, colloids, and suspensions, generally exhibit non-Newtonian flow.

Broadly speaking, one can observe and consider two varieties of *non-Newtonian flow behavior*. In the simplest case of *viscoINelastic fluids*, the departure from Newtonian behavior is strictly viscous, and no elastic, time-dependent effects, such as stress relaxation, are observed. *Viscoelastic fluids*, on the other hand, exhibit both viscous and time-dependent departures from Newtonian behavior, and their flow patterns are generally very complex. The mathematical description of the flow behavior of viscoelastic fluids has been the subject of extensive research for over twenty-five years. Nonetheless, many aspects of this problem still defy adequate description. The main difficulty is associated with the nonlinear phenomena that these fluids exhibit. Complex analytic machinery is required for the treatment of even the most elementary aspects of this problem. Fortunately, certain basic, industrially important polymer flow problems are amenable to analysis within the context of elementary mathematical machinery. Our primary interest in this chapter will be the flow behavior of polymer melts through capillaries. This geometry is the one most frequently encountered in industrial processing, and it is one of the easiest to discuss. A particularly fortuitous circumstance is that we can treat the viscous aspects of viscoelastic flow, independent of the elastic behavior, and in this context we speak of the *strictly viscous* character of viscoelastic fluids.

1201 The Viscous Character of non-Newtonian Fluids

In discussing the viscous character of non-Newtonian fluids, we can no longer simply speak of a viscosity, for its value depends on the flow conditions as specified by the rate of strain. By way of analogy to Newton's law, the viscous character of non-Newtonian fluids can be described by

$$\tau = \eta_a(\dot{\gamma})\dot{\gamma} \tag{12-1}$$

where $\eta_a(\dot{\gamma})$ is defined as the *apparent viscosity*. In a Newtonian fluid, η_a is a constant; in a non-Newtonian fluid, it depends on the value of the local strain rate $\dot{\gamma}$. This approach neglects any elastic effects that may be present. The apparent viscosity is illustrated in Figure 12–1 for two classes of non-Newtonian fluids: *shear thinning*, where η_a decreases with increasing strain

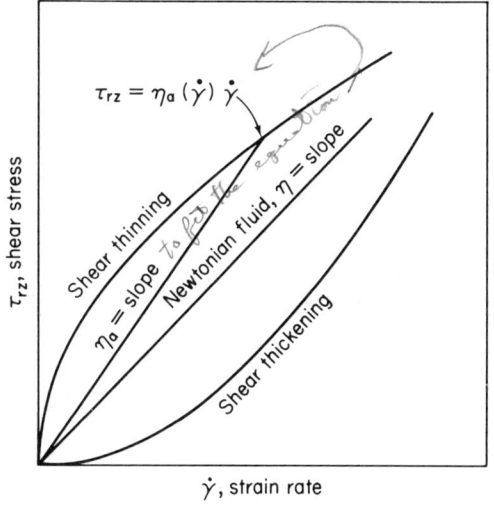

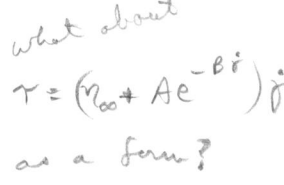

what about
$\tau = (\eta_\infty + Ae^{-\beta \dot{\gamma}})\dot{\gamma}$
as a form?

Figure 12–1. τ_{rz} vs $\dot{\gamma}$ for Newtonian and non-Newtonian fluids.

rate, and *shear thickening*, where η_a increases with increasing strain rate. Polymer solutions and polymer melts are of the shear-thinning variety, and these will be our only concern. Shear thinning is exhibited by those materials whose structure breaks down with increasing shear rate. Polymer melts and solutions may undergo a decrease in apparent viscosity of 3 to 4 orders of magnitude. Shear thickening, a much rarer phenomenon than shear thinning, is observed in systems where a structure is established with the application of stress and where this structure continues to build up with increasing stress.

1202 The Flow Characteristics of Polymer Melts*

The most obvious characteristic of polymer melts is that they are very viscous, with apparent viscosities approaching 10^6 poise at low strain rates. Because of the high viscosities involved, polymer melt flow is always laminar. A plot of apparent shear rate versus apparent shear stress for a polymer melt is illustrated in Figure 12-2. For Newtonian fluids, the relationship is linear,

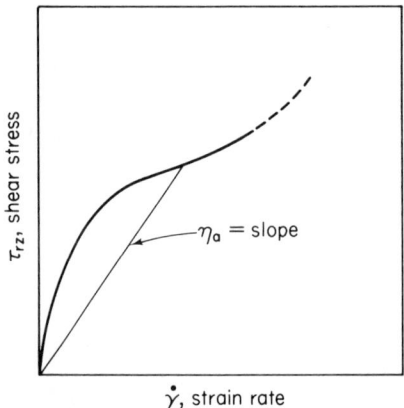

$\dot{\gamma}$, strain rate

Figure 12-2. τ_{rz} vs. $\dot{\gamma}$ for polymer melt.

but for polymer systems it is nonlinear. At low shear stresses—much lower than those used in the polymer processing industries—the relation is nearly linear, and the polymer system resembles a very viscous Newtonian fluid. In this context, the so-called *zero shear-rate viscosity* is used as a characteristic parameter for polymer melts. As the shear rate is increased—into the range ordinarily encountered in ·the polymer processing industries—the

*Except for rather dilute solutions, the properties of polymer melts are shared by polymer solutions as well. The degree to which their properties are shared depends on a variety of factors—principally, the concentrations.

apparent viscosity decreases, until at very high shear rates—beyond the range of usual interest—a linear relation is again observed. This high shear-rate behavior is difficult to attain, and the effect is probably being obscured by viscous heating.

This shear-thinning behavior can be explained on a molecular level. Polymer fluid systems at rest are pictured as an intimately entangled aggregation of long chains, linear or branched. When a stress is applied and flow occurs, the chains must slide past one another. This process requires a certain degree of disentangling. If the stress rate is low, the degree of disentangling within the fluid as a whole is low. As this rate is increased, the degree of disentangling necessarily increases. Since the resistance to flow depends on this degree of entanglement, the resistance to flow, or viscosity, decreases as the rate of strain increases.

The most interesting property of polymer melts is that of elasticity. If a polymer system, either in a state of rest or flow, is subjected to a sudden increase in shear rate, the system responds elastically. For example, if a lump of material is deformed suddenly, and then if the deforming force is immediately removed, the sample will nearly regain its original shape. This property suggests that polymer systems have "memories"—the duration of which depends on their relaxation times. To describe this time-dependent behavior in a picturesque vein, we again refer to a polymer system's fading memory.

This elastic behavior, like shear thinning, can be explained on a molecular scale. In contrast to viscous flow, which involves slip of the molecules past one another, melt elasticity relies on the fact that molecular slip (and thus flow) does not occur within the experimental time scale and that deformations merely strain equilibrium, random conformations. If the deforming force is removed immediately, the strain is relieved by immediate recovery to the original conformations. This state of affairs in an uncrossed linked sample can only be temporary if the deforming force is sustained, for flow will eventually occur. (In elastomers, which are crosslinked networks, flow is prevented and the materials behave in an elastic fashion permanently.) When the time scale of the experiment is low, the molecules do not have time to slip; but as the time scale is increased, the molecules begin to slip and flow occurs. Thus we see how the spectrum of viscoelastic behavior depends at a molecular level on the response time of the polymer system as well as on the process time of the experiment, observation, etc.

The fourth interesting property exhibited by polymer systems is their ability to generate normal stresses. Certain aspects of this phenomenon are well described by mathematical-continuum theories. From a molecular standpoint our understanding is solely qualitative and rudimentary at best. Nonetheless, it is pedagogically useful to undertake such a description at this point.

When a polymer molecule is at rest, its conformation is that of a random coil whose overall shape is spherical. This is a preferred thermodynamic state; if any disturbances distort this shape, elastic forces will seek to restablish the preferred random state. A polymer molecule in a shear field is large compared to the shear gradient. Therefore different parts of a single molecule will experience different shear stresses. This situation is depicted in Figure 12–3 for a single chain in a linear shear field. In contrast, a low-molecular-weight molecule would be small compared to the stress field; and, therefore, any single molecule would experience only a single shear stress. Since a polymer molecule is subjected to variable stresses, it becomes distorted as depicted in Figure 12–3(a) for sufficiently high strain rates, and it tends to become elongated or ellipsoidal-like. The forces seeking to restore the preferred random conformation give rise to the so-called *normal stresses*. To

This requires to be a coupling of z strain to x stress.

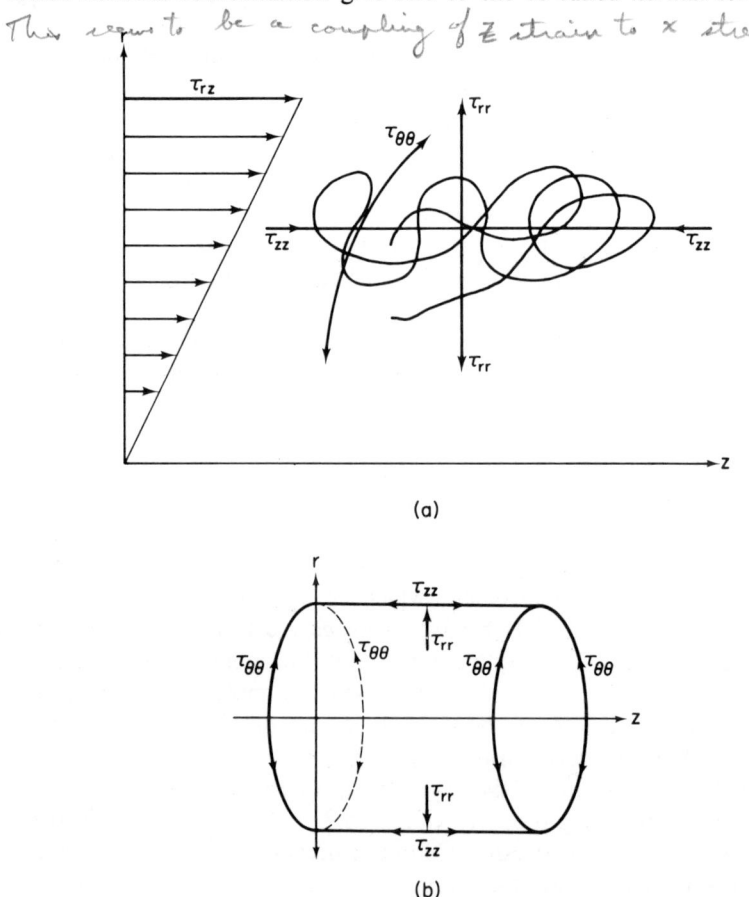

Figure 12–3. Schematic showing normal stresses for (a) a polymer molecule in a cylindrical stress field and (b) sealed, pressurized, hollow cylindrical container.

amplify by analogy, consider a sealed, hollow, cylindrical container that has a positive internal gage pressure. As shown in Figure 12-3(b), three normal stresses will arise: in cylindrical coordinates, these will be σ_{zz}, σ_{rr}, and $\sigma_{\theta\theta}$ (hoop stress). And so in the cylindrical flow of viscoelastic fluids, we must consider the same three normal stresses.

As we have previously stated, the actual situation is not well understood from a molecular point of view, and it is far more complicated than the simple graphical discussion presented here. This point is especially true when one considers an entangled network and when elastic strains are superposed. In many flow situations, normal stress effects must be considered with elastic effects. As we shall observe subsequently, however, normal stresses may occur in the absence of elastic effects.

1203 The Capillary Rheometer

Of the many devices available to study the rheological properties of fluid polymer systems, none is more important to the engineer than the *capillary rheometer*. Many important polymer processes involve the flow of polymer melts through tubes. Thus laboratory studies of the flow of molten polymer through capillaries represent a particularly important and valuable approach to process design, as well as an appropriate starting point for our discussion. A schematic typical capillary rheometer is shown in Figure 12-4. The melt is

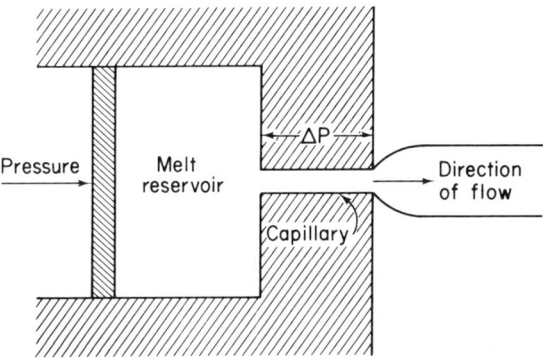

Figure 12-4. Schematic of isothermal capillary rheometer.

forced from a reservoir through a die at constant temperature. The reservoir and die are usually right cylinders aligned along a common axis, but almost any die geometry of interest can be studied. For instance, the die entry may be tapered, or the capillary may be replaced by a slit device to simulate a film extrusion process.

Length to diameter (L/D) ratios for industrial dies may range from 5 to 10 with the diameter ranging from a small fraction of an inch for fiber production to sizes approaching an inch for the production of rodlike articles. Pressures up to 10,000 psi may be employed.

Two modes of operation are employed: constant stress or constant rate. In *constant-stress operation*, the pressure in the reservoir is maintained at a constant value, usually by exposure of the melt to a piston or compressed gas. The rate of melt flow is measured by collecting and weighing the extrudate. In *constant-rate operation*, the piston is driven forward at a constant rate, expelling the melt at a constant volumetric rate, and, therefore, at a constant stress rate. The pressure is measured as that required to move the piston in the reservoir. The instruments used in such experiments are respectively referred to as constant-stress and constant-rate rheometers.

Two important phenomena are associated with the polymer melt as it emerges from the capillary. In *jet swelling*, the polymer expands two to three times its capillary diameter on emerging from the die, because of the recovery of normal stresses and residual elastic energy. *Melt fracture* occurs in a critical high-stress region and manifests itself by the appearance of a twisted and roughened extrudate rather than a smooth extrudate. The importance of these phenomena resides in the fact that they directly determine the size and shape of the product.

We will next give an introductory qualitative description of the phenomena involved in the flow of polymer melts in capillary rheometers, followed by a detailed analysis, which will be quantitative to the extent permitted by our current understanding of this process.

1204 Aspects of Flow Behavior in Capillaries

The extrusion of polymer melts in isothermal capillary rheometers is most conveniently introduced by first considering flow through a very long tube. Discussion will focus on Figure 12-5. The flow character of polymer melts in capillaries is extremely complex. In this section we discuss the phenomena in qualitative summary form. The detailed characteristics are discussed in the following sections. This section should be referred to again for a complete overall view.

As shown in Figure 12-5, the molten viscoelastic polymer is greatly accelerated in flowing from the reservoir into the capillary because of a funnelling effect. This process generates large *elastic strains* in the melt and leads to the rapid dissipation of the driving pressure within an *entrance region*, as the velocity profile rearranges. The *relaxation region* is characterized by a distorted, parabolic, radial velocity profile and the associated normal stresses as well as a diminishing fraction of the entrance-borne elastic energy.

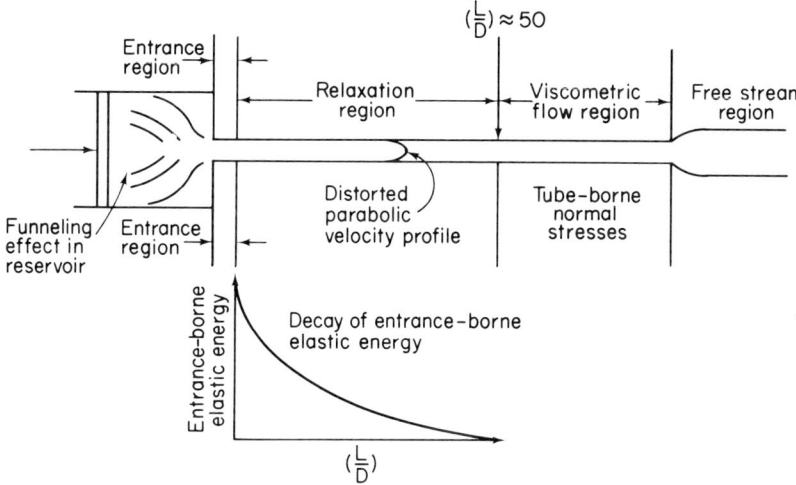

Figure 12-5. Schematic showing regions of flow in a capillary rheometer.

The strictly viscous character of polymer melt flow is successfully treated independently of normal stress effects or elastic-energy dissipation. Such an approach carries the tacit assumption that the establishment and maintenance of the steady-flow velocity profile are independent of these latter effects. Under such circumstances, the entrance-borne elastic energy would have to dissipate without bulk flow in a relaxation mechanism analogous to that encountered in relaxation experiments with viscoelastic solids. If the tube is sufficiently long (greater than 50-tube diameters), the elastic energy will be completely dissipated. This complete dissipation marks the onset of *viscometric flow*, which is characterized by a steady-flow velocity profile and strictly shear-rate-dependent normal stresses. In most engineering applications, however, the tubes are considerably shorter than 50-tube diameters and viscometric flow is never attained. Note that we have a situation with *two entrance regions*: one for establishing the velocity profile and the other for establishing viscometric flow. It may be that the velocity profile undergoes a second rearrangement in anticipation of the outlet. As already noted, the *free stream region* has two very important engineering aspects associated with it: jet swelling in stable flow and melt fracture in unstable flow. The elastic-energy content of the melt is a major determinant in establishing the degree of swelling and the onset of fracture. The normal stresses established by the velocity profile affect jet swelling but only to a moderate degree in the short dies encountered in engineering situations. Our fundamental knowledge of these outlet phenomena and the physical nature of the upstream processes, and hence our ability to design around them in engineering situations, is far from satisfactory.

THE VISCOUS CHARACTER OF POLYMER MELT FLOW

In the next several sections, we shall be concerned with developing analytic expressions that describe the strictly viscous character of polymer melts in capillary flow. Most important, we are going to ignore any effects normal stresses or relaxation may exert on the viscous-flow behavior of polymer melts. Analysis, after the fact, seems to indicate that this assumption is appropriate, at least as a first-order approximation. Many important engineering correlations have been formulated in terms of this analysis. In fact, even manifestations of elastic behavior and normal stress effects can be correlated in terms of the viscous-flow parameters so derived. It is indeed fortuitous that such a separation of viscous effects can be made. Otherwise, design correlations, as well as fundamental rheological studies, might become hopelessly complex.

1205 The Laminar Flow of Newtonian Fluids

To set the stage for subsequent discussion, we must first describe the viscous behavior of Newtonian fluids in laminar capillary flow. To do so, we evoke the following assumptions:
1. The fluid can be treated as a continuum.
2. The tube is sufficiently long so that entrance and exit effects can be neglected.
3. The fluid is incompressible.
4. The fluid does not slip at the wall.
5. The flow is isothermal.

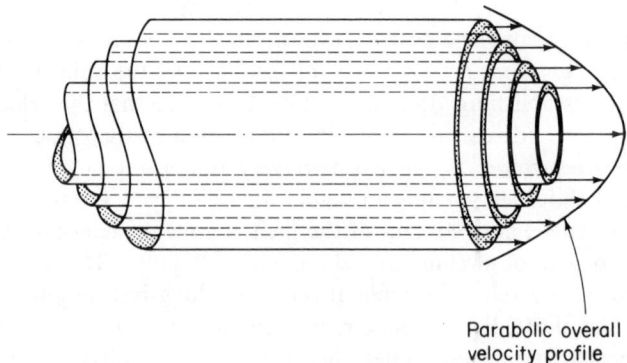

Parabolic overall velocity profile

Figure 12-6. Schematic illustrating laminar flow in a concentric series of lamellar cylinders.

Sec. 1205   The Laminar Flow of Newtonian Fluids

In laminar flow, we picture the fluid as consisting of lamellae of concentric cylinders, as shown in Figure 12-6. The volumetric flow rate of fluid through the tube is the sum of the flow rates of all such lamellae, each having its own velocity. The lamellar layer immediately adjacent to the tube wall has a zero velocity inasmuch as one of its surfaces is formed by the wall. The velocity increases toward a maximum at the center; and at each of the

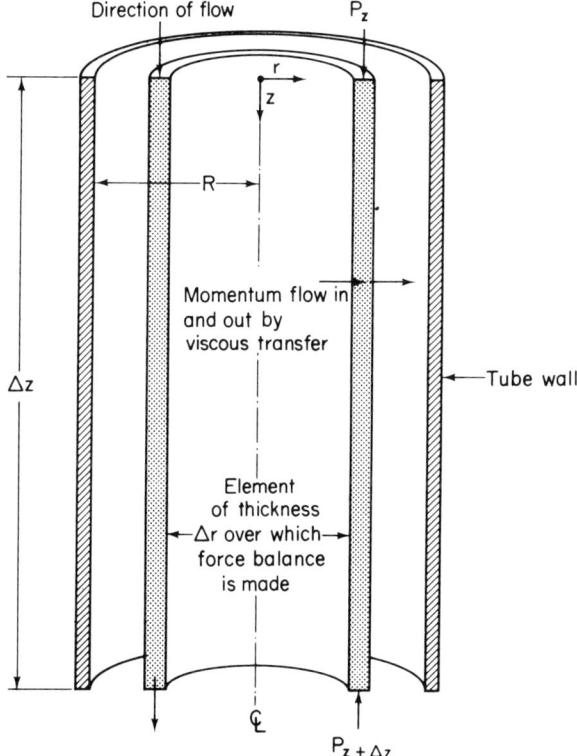

Figure 12-7. Cylindrical element of fluid over which force balance is made. [R. B. Bird, W. E. Stewart and E. N. Lightfoot, *Transport Phenomena*, John Wiley & Sons, New York, 1960.]

annular lamellae there is a radial velocity gradient such that the viscous forces just balance the force exerted by the driving pressure. Consider the situation depicted in Figure 12-7, where we have flow in the z direction of an annular fluid element of length Δz, radius r, and thickness Δr. A force balance on this element yields

$$\Delta z \, 2\pi r \tau_{rz}|_r - \Delta z \, 2\pi r \, \tau_{rz}|_{r+\Delta r} + 2\pi r \, \Delta r \, P|_z - 2\pi r \, \Delta r \, P|_{z+\Delta z} = 0 \quad (12\text{-}2)$$

Dividing by 2π, then rearranging gives

$$\left(\frac{r\tau_{rz}|_{r+\Delta r} - r\tau_{rz}|_r}{\Delta r}\right) = -\left(\frac{P_{z+\Delta z} - P_z}{\Delta z}\right)r \qquad (12\text{-}3)$$

Taking limits, we obtain

$$\frac{d(r\tau_{rz})}{dr} = -\left(\frac{dP}{dz}\right)r \qquad (12\text{-}4)$$

Integrating with respect to r yields

$$\tau_{rz} = -\left(\frac{dP}{dz}\right)\frac{r}{2} + \frac{C}{r} \qquad (12\text{-}5)$$

If τ_{rz} is to remain finite when $r = 0$, then C must be zero. Hence the shear stress τ_{rz} on the surface of a cylindrical element of fluid at radius r is given by

$$\tau_{rz} = -\left(\frac{dP}{dz}\right)\frac{r}{2} \qquad (12\text{-}6)$$

If the overall axial-driving-pressure gradient is linear, then $(dP/dz) = (P_L - P_0)/L$, where $(P_L - P_0)$ is the capillary pressure drop and L is its length. For convenience, let $-\Delta P = (P_L - P_0)$, so that Eq. (12-6) becomes

$$\tau_{rz} = \left(\frac{\Delta P}{L}\right)\frac{r}{2} \qquad (12\text{-}7)$$

and for the shear stress at the wall of the capillary, τ_w, we obtain

$$\tau_w = \frac{\Delta P R}{2L} \qquad (12\text{-}8)$$

where R is the capillary radius. Notice that in deriving Eqs. (12-7) and (12-8), we have not specified the nature of the fluid; and, therefore, they should be valid expressions for all fluids, provided that they satisfy the criteria previously outlined. We see that the shear stress due to viscous forces is linearly dependent on the tube radius (see Figure 12-8) and that it reaches a maximum at the tube wall.

By specifying the fluid as Newtonian, such that $\tau_{rz} = (dv_z/dr)\eta$, we obtain an expression for the velocity gradient, or rate of shear,

$$\dot{\gamma} = \frac{dv_z}{dr} = \left(\frac{\Delta P}{2L\eta}\right)r \qquad (12\text{-}9)$$

Sec. 1205 The Laminar Flow of Newtonian Fluids

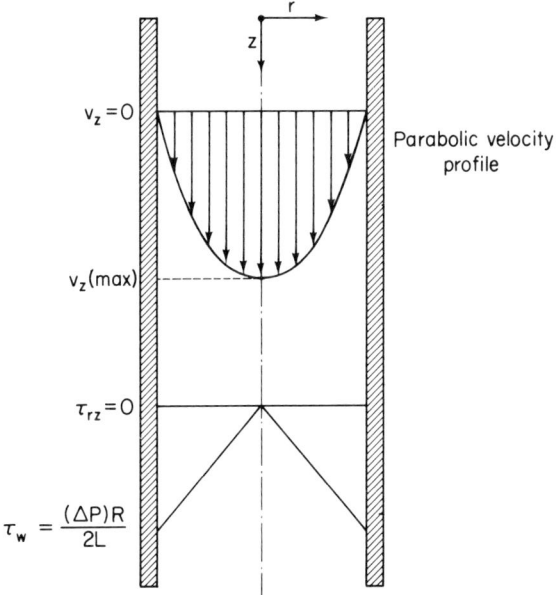

Figure 12-8. Shear stress and velocity profiles in flow in cylindrical tubes. [Bird et al., op. cit. Figure 12-7.]

where both are also linearly dependent on the tube radius. Integration with the boundary condition $v_z = v$ at $r = R$ yields the classical, parabolic, radial velocity-profile for a Newtonian fluid

$$v_z = \frac{\Delta P R^2}{4\eta L}\left[1 - \left(\frac{r}{R}\right)^2\right] \tag{12-10}$$

This relationship is also shown in Figure 12-8.

Since the volumetric flow rates q is given by

$$q = \int_0^R v_z(2\pi r\, dr)$$

we can substitute for v_z and obtain

$$q = \frac{\pi R^4 \Delta P}{8L\eta} \tag{12-11}$$

This is the well-known *Hagen-Poiseuille law* for steady laminar flow in tubes. Equation (12-11) has been amply verified for Newtonian fluids, and it has been used to measure the viscosity of Newtonian fluids. It is informative to note at this point that Eq. (12–10) cannot be directly verified, so that the verification of Eq. (12–11) also serves as a verification of Eq. (12–10). If we solve for η in Eq. (12–11) and substitute in Eq. (12–9), as well as set $r = R$, we obtain an expression for shear rate at the wall,

$$\dot{\gamma}_w = \frac{4q}{\pi R^3} \qquad (12\text{–}12)$$

Even though the Newtonian viscosity term, η, does not appear in Eq. (12–12), it is still only valid for Newtonian fluids.

1206 The Consistency Variables

Very often, for lack of a better approach, the rheological properties of non-Newtonian fluids are related in terms of the preceding relationships for τ_w and $\dot{\gamma}_w$. These variables are renamed the *apparent shear stress* and *apparent shear rate*, respectively. Collectively, they are referred to as the *consistency variables*, and for clarity, they are summarized and represented as

$$S = \text{apparent shear stress} = \frac{\Delta P R}{2L} \; [=] \; \frac{\text{force}}{\text{unit area}} \qquad (12\text{–}13)$$

$$Q = \text{apparent shear rate} = \frac{4q}{\pi R^3} \; [=] \; \text{time}^{-1} \qquad (12\text{–}14)$$

The apparent viscosity, previously introduced, is defined as

$$\eta_a = \text{apparent viscosity} = \frac{S}{Q} = \frac{\pi R^4 \Delta P}{8qL} \qquad (12\text{–}15)$$

The apparent shear stress is the true shear stress at the wall, provided the assumptions previously listed are valid. Development of an expression for the true shear rate at the wall, $\dot{\gamma}_w$, follows in Section 1208. The expression for η_a attempts to give some measure of the viscous nature of the material, but note that even the apparent viscosity will not be a true constant for any particular experiment, since it depends on the tube radius to the fourth power. It has been shown that if experimental results are formulated in terms of consistency variables, the results of all capillary flow experiments can be correlated regardless of dimensions.

Typical experimental relations between S and Q for a polyethylene resin are shown in a log-log plot in Figure 12–9. A curve for a Newtonian

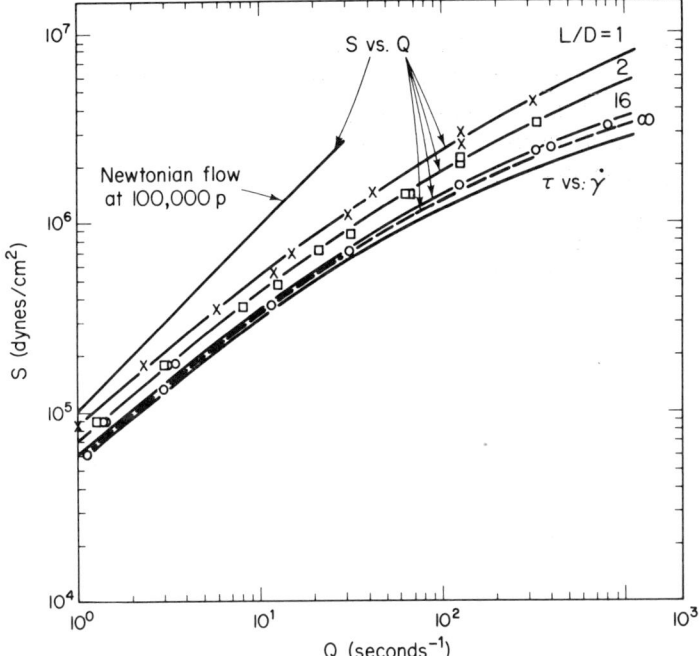

Figure 12-9. Flow of polyethylene (0.953 g./cc. at 23 °C., melt index 1.5) 190 °C. [J. M. Lupton, "Polymer Melt Flow" in *Polymer Processing*, Chem. Engr. Progr. Symp. Ser., **60** No. 49, 17 (1964). J. V. D. Fear, ed.]

fluid with a viscosity of 100,000 poise is also shown for comparison. Such curves are conveniently referred to as *flow curves*. Since entrance and exit effects have not been accounted for, there is a separate curve for each L/D ratio. All the curves merge as L/D gets very large, and these effects become negligible. In ordinary process operations, entrance and exit effects cannot be neglected. Corrections can be made to produce a single curve (Section 1210), but often they are too tedius to make to warrant the trouble. If there is no slip at the wall and geometric similarity is maintained (identical L/D ratios), these curves can be used in engineering design for tube flow—for example, in scale-up.

The consistency variables do not reflect, of course, either the elastic nature of the fluid or the normal stresses that develop.

1207 The Power Law

Plots of log S versus log Q for various polymers are often linear through substantial ranges of operation. We can represent this behavior as

$$S = mQ^n \qquad (12\text{-}16)$$

which is one form of the well-known *power law*, where m is referred to as the flow consistency and n is the flow index. It is the most widely used empirical model for representing melt-flow behavior because of its analytical simplicity as well as its ability to represent the viscous behavior of the system. Nonetheless, the equation, must be used prudently without unjustified extrapolations.

Another form of the power law, sometimes referred to as the Ostwald–deWaele equation, is given by

$$\eta = \eta^0 \left| \frac{\dot{\gamma}}{\dot{\gamma}^0} \right|^{n-1} \tag{12-17}$$

or

$$\eta = \eta^0 \left| \frac{\tau}{\tau^0} \right|^{(n-1)/n} \tag{12-18}$$

where $\dot{\gamma}^0$ and τ^0 represent values of shear stress and shear rate in an arbitrary standard state and η^0 is the apparent viscosity in this state.

Using a form equavalent to Eq. (12–16) $\tau_{rz} = m(dv_z/dr)^n$, substituting in Eq. (12–6), and integrating, we obtain an expression for the velocity profile of a power law fluid

$$v_z = \left(\frac{\Delta P R}{2mL}\right)^{1/n} \left(\frac{Rn}{n+1}\right) \left[\left(\frac{r}{R}\right)^{(n+1)/n} - 1\right] \tag{12-19}$$

Paralleling our Hagen-Poiseuille analysis, we next obtain the volumetric flow rate

$$q = \left(\frac{\Delta P R}{2mL}\right)^{1/n} \left(\frac{\pi R^3 n}{3n+1}\right) \tag{12-20}$$

The average velocity $\bar{v}_z$, defined as $q/\pi R^2$, becomes

$$\bar{v}_z = \left(\frac{\Delta P R}{2mL}\right)^{1/n} \left(\frac{Rn}{3n+1}\right) \tag{12-21}$$

velocity profiles are plotted in Figure 12–10 for various power-law fluids as $v_z/\bar{v}_z$ versus (r/R). The curve labeled $n = 1$ is a parabola characteristic of Newtonian flows. A flattened parabolic profile, for $n < 1$, is characteristic of·polymer melt flow. With $n > 1$, the profiles characteristic of shear thickening fluids are obtained.

As successful as the power law is, it suffers from a number of defects worth nothing.

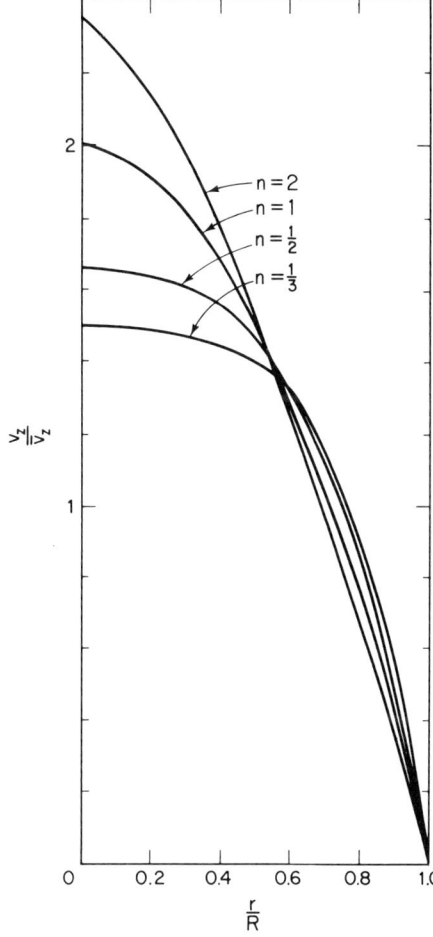

Figure 12-10. Velocity profiles for the isothermal flow of power law fluids through tubes.

1. Since n is less than one for polymer melts, the power law predicts an infinite viscosity or consistency as $Q \to 0$.
2. The parameter m has awkward dimensions that depend on n.
3. The model does not reflect the elastic or normal stress effects.
4. The parameters m and n may depend on the flow geometry.

1208 A General Treatment of Isothermal Viscous Flow in Tubes

We undertake a twofold problem in this section (1) Establish an expression for the true shear rate at the wall, $\dot{\gamma}_w$, which can be evaluated in terms of pressure drops and flow rates. (2) Show that a unique functional relationship exists between S and Q so that such data can be used with assurance in scale up. We assume only that the shear stress is everywhere uniquely related to the shear strain, as $\tau = f(\dot{\gamma})$.

We start with the general expression for volumetric flow rate and integrate by parts to yield

$$q = \pi \left(r^2 v_z \Big|_0^R - \int_0^R r^2 \, dv_z \right) \tag{12-22}$$

To evaluate the first term on the right, we note that at $r = R$, $v_z = 0$ and that the term is zero for $r = 0$; hence the first term vanishes. Equation (12-22) reduces to

$$q = -\pi \int_0^R r^2 \, dv_z = -\pi \int_0^R r^2 \left(\frac{\partial v_z}{\partial r} \right) dr \tag{12-23}$$

Using the relation $\tau = (r/R)\tau_w$, we substitute for r and dr, as well as $\dot{\gamma}$ for $(\partial v_z / \partial r)$, to obtain

$$q = \frac{-\pi R^3}{\tau_w^3} \int_0^{\tau_w} \dot{\gamma} \tau^2 \, d\tau$$

or $\tag{12-24}$

$$\frac{\tau_w^3 q}{\pi R^3} = -\int_0^{\tau_w} \dot{\gamma} \tau^2 \, d\tau$$

Differentiation of both sides with respect to τ_w by the Leibnitz rule yields

$$\frac{1}{\pi R^3} \left(\tau_w^3 \frac{dq}{d\tau_w} + 3\tau_w^2 q \right) = -\dot{\gamma}_w \tau_w^2 \tag{12-25}$$

With $\tau_w = R \Delta P / 2L$, this equation becomes the *Rabonowitsch equation*

$$-\dot{\gamma}_w = \frac{1}{\pi R^3} \left(3q + \Delta P \frac{dq}{d \Delta P} \right) \tag{12-26}$$

Thus, $\dot{\gamma}_w$ can be evaluated from ΔP-q data. However, obtaining $dq/d \Delta P$ from ΔP-q data requires numerical differentiation that is notoriously inaccurate. For this reason, S and Q data are still widely used.

For the second part of our problem, we rearrange Eq. (12–24) to read as

$$-\dot{\gamma}_w = \frac{3}{4}\left(\frac{4q}{\pi R^3}\right) + \frac{\tau_w}{4}\frac{d}{d\tau_w}\left(\frac{4q}{\pi R^3}\right) \qquad (12\text{--}27)$$

and substitute Q for $(4q/\pi R^3)$. The equation becomes

$$-\dot{\gamma}_w = \frac{3}{4}Q + \frac{\tau_w}{4}\frac{dQ}{d\tau_w} \qquad (12\text{--}28)$$

Since $\tau = f(\dot{\gamma})$, we realize that $\tau_w = f(\dot{\gamma}_w)$, and Eq. (12–28) becomes

$$S = f\left[-\frac{3}{4}Q - \frac{S}{4}\frac{dQ}{dS}\right] \qquad (12\text{--}29)$$

which indeed shows that there is a unique functional relationship between the scale-up variables S and Q. This means that if Q is plotted against S, a single curve that is unique for the fluid at a given temperature will result.

Equation (12–26) can also be rewritten as

$$-\dot{\gamma}_w = \frac{3}{4}Q + \frac{1}{4}Q\frac{d\ln Q}{d\ln S} = \frac{3+b}{4}Q \qquad (12\text{--}30)$$

Once b is determined from log Q–log S data, the true shear rate may be readily obtained over a wide range of shear rates if log Q versus log S curves are linear; therefore b will be constant over wide ranges of shear rate. The values of $\dot{\gamma}_w$ are about 1.5 times Q.

1209 The Effect of Temperature

Temperature has an important effect on the rheological properties of polymeric melts, just as it does with polymeric solids. Here we treat the variation of the apparent viscosity with temperature.

The flow behavior of Newtonian fluids as a function of temperature is expressed in terms of the *Arrhenius equation*

$$\eta = AE^{E/RT} \qquad (12\text{--}31)$$

where A is a constant coefficient, E is the activation energy for flow, and R is the gas constant. Plots of log η versus $(1/T)$ are reasonably linear over 100 Farenheit degrees for most Newtonian fluids, including polymeric materials at very low shear rates. Figure 12–11 shows the temperature dependence of the zero shear-rate viscosity for two polyethylene samples and for plasticized poly(vinyl butyral). Note that here the data are represented by straight lines over the entire temperature range of interest.

For non-Newtonian fluids, η_a is a function of the state of shear, as characterized by either S or Q, as well as a function of temperature. Therefore the

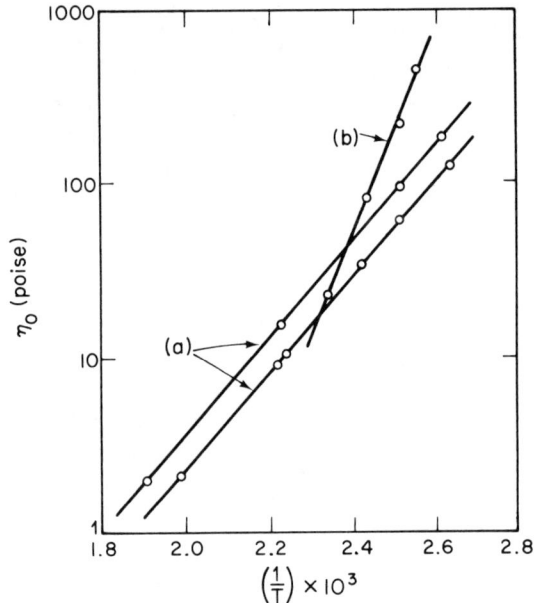

Figure 12–11. Temperature dependence of the zero shear-rate viscosity η_0 of (a) polyethylene and (b) polyvinyl butyral. [W. Philippoff and F. H. Gaskins, J. Polymer Sci., **21**, 205 (1956).]

temperature coefficient of the apparent viscosity must be specified either at constant shear $(\partial \eta_a/\partial T)_S$ or constant shear rate $(\partial \eta_a/\partial T)_Q$. In general, the two will not be equal, and two Arrhenius expressions result

$$(\eta_a)_S = A E^{Es/RT} \tag{12–32}$$

$$(\eta_a)_Q = A E^{E_Q/RT} \tag{12–29}$$

Flow curves for an acrylic resin are shown for various temperatures in Figure 12–12. These curves have been used to cross-plot S versus $(1/T)$ at constant Q to obtain the curves shown in Figure 12–13. At very low shear stresses, the activation energies are the same. Note that E_Q decreases with increasing strain rate, whereas E_S increases with the increasing stress.

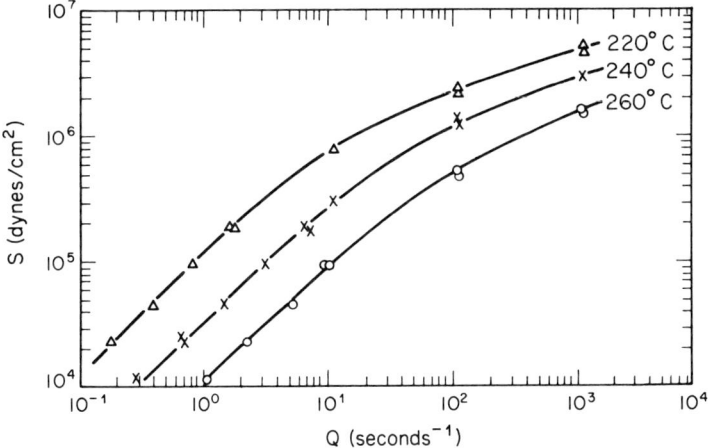

Figure 12-12. Flow of Lucite 40 acrylic resin, L/D = 16. [J. M. Lupton, op. cit. Figure 12-9.]

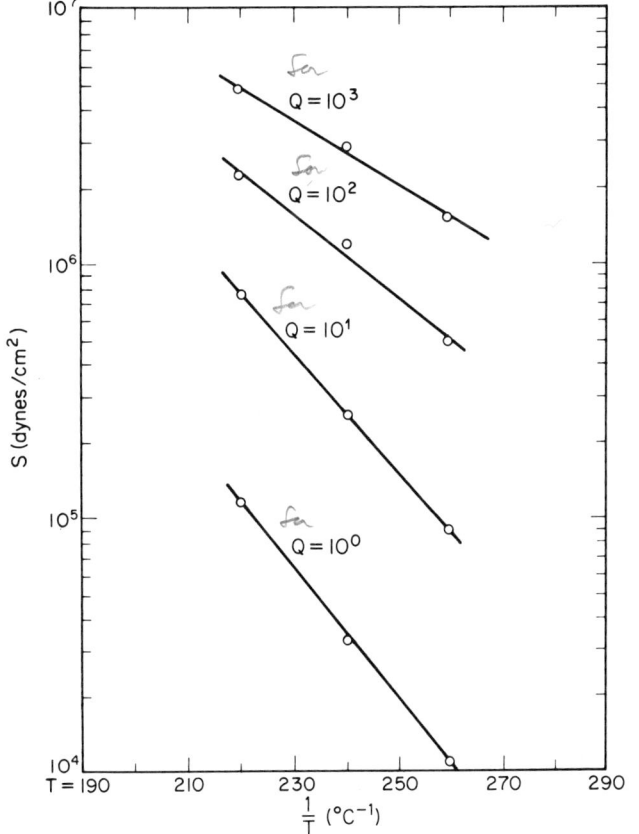

Figure 12-13. Flow of Lucite 40 acrylic resin, L/D = 16. [J. M. Lupton, op. cit. Figure 12-9.]

1210 Entrance and Exit Effects

For polymer melt flow in sufficiently long tubes, one should encounter two entrance lengths: the first accounting for the nonlinear driving pressure drop and for the full development of the viscous flow-velocity profile, and the second for the decay of the entrance-borne elastic energy to establish simple viscometric flow. Best estimates for the length required for the velocity-profile rearrangement are very much less than one-tube diameter, whereas estimates for viscometric flow range from 40 to 60 tube diameters.

In the first entrance region, we are concerned with the loss in driving pressure caused by the abrupt change in flow geometry and consequent velocity-profile rearrangement. In transferring fluid from the reservoir, a funneling effect occurs in the reservoir and the fluid is accelerated as it enters the capillary. The velocity profile existing in the main body of the reservoir is destroyed in this transfer and another profile must be established in the tube. This transfer requires energy and manifests itself by an abrupt decrease in the driving pressure. In Newtonian flow, this energy dissipation is due to viscous losses and to form drag. In viscoelastic flow, the dissipation is due to elastic forces as well. Elastic energy is imparted to a fluid element by the driving pressure. This situation is somewhat analogous to squeezing a rubber ball into a small-diameter tube. The entrance-region energy losses for polymer melts are on the order of hundreds of times greater than those encountered in Newtonian fluids.

E. B. Bagley* was the first to undertake the study of the losses in driving pressure in the entrance region. He postulated that the axial driving pressure dropped suddenly and then assumed a linear form, as shown in Figure 12-14.

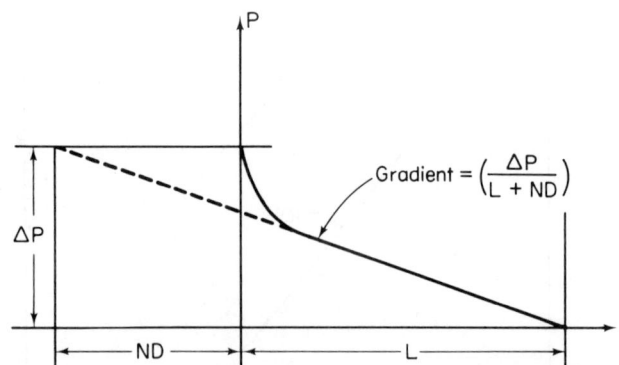

Figure 12-14. Correction of tube length L by addition of length ND in order to calculate the pressure gradient in the steady flow region.

* E. B. Bagley, *J. Appl. Phys.*, **28**, 624 (1957).

Bagely proposed the addition of a fictitious tube length, ND, to the actual length such that the driving-pressure gradient p becomes

$$p = \frac{\Delta P}{L+ND} \qquad (12\text{-}34)$$

The true shear stress at the wall for the steady-flow region can now be written as

$$\tau_w = \frac{D}{4}\left(\frac{\Delta P}{L+ND}\right) \qquad (12\text{-}35)$$

Rearranging Eq. (12–35) yields the equation of a straight line and a means of determining N.

$$\frac{L}{D} = \frac{\Delta P}{4\tau_w} - N \qquad (12\text{-}36)$$

Since τ_w is a unique function of Q (Section 1208), we can write

$$\frac{L}{D} = \frac{\Delta P}{4f(Q)} - N \qquad (12.37)$$

A series of ΔP versus Q flow curves for a series of L/D ratios can be replotted with (L/D) versus ΔP for constant Q. The intercept of the resulting line with $\Delta P = 0$ should yield N.

This procedure is illustrated in Figure 12–15 with data for polypropylene. A straight-line relationship is found for each value of Q, but N is found to depend on Q, generally ranging from 2 to 5. In general, a separate correction must be obtained empirically for each melt, each range of shear rates, and each temperature. In addition, difficulty is encountered in extrapolating with any degree of accuracy. In engineering practice, these difficulties generally negate the advantages gained from such a correction, and flow curves for each L/D ratio are often used as a basis for calculations and design. Recently a group of investigators[*] has shown for polyethylene that N is related to Q by the expression $N = a + b \log Q$. For theoretical work, these corrections must be applied to experimental data to obtain true values of S; Q, of course, is independent of any entrance effects. Figure 12–16 shows the single-flow curve that results when the data of Figure 12–15 are corrected. The fact that full correction is seemingly achieved using overall driving pressure indicates that reservoir and exit corrections are probably negligible.

[*] A. Ram and M. Narkis, *J. Appl. Polymer Sci.*, **10**, 361 (1966).

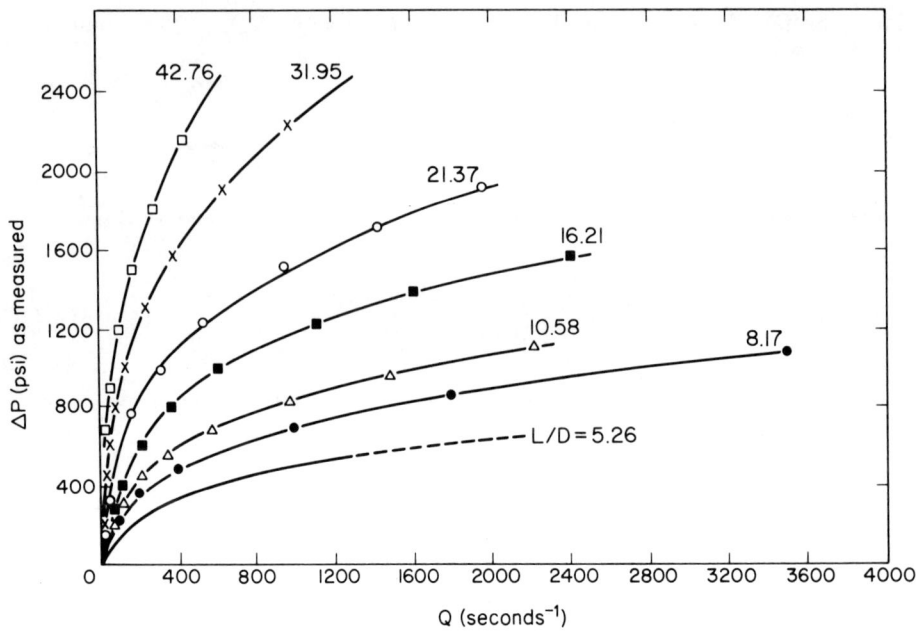

Figure 12–15 (a). ΔP vs Q data for polypropylene, $\overline{M}_v = 320{,}000$.

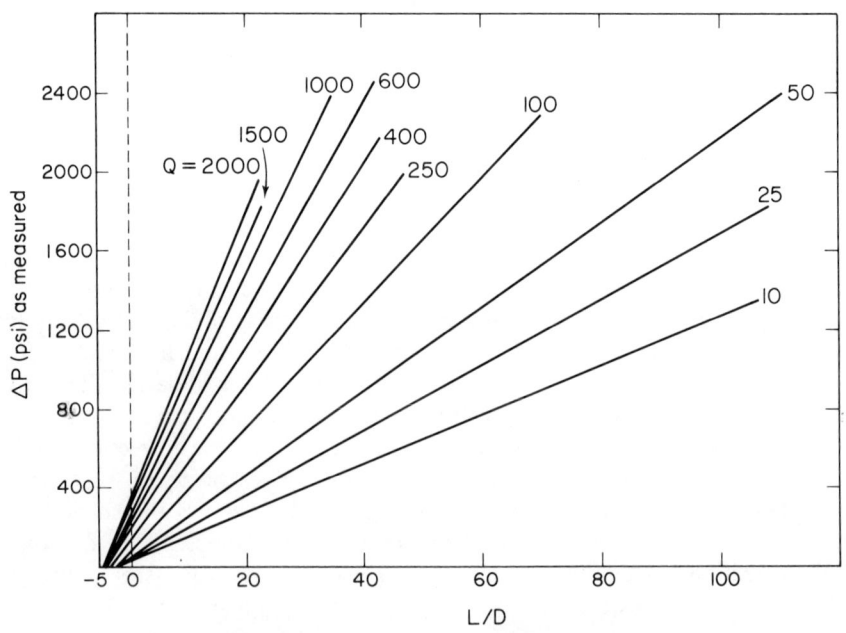

Figure 12–15 (b) "Bagley" plot of data in 12–15 (a). [R. C. Kowalski, Ph.D. Thesis, "Elastic Behavior of Molten Viscoelastic Polymeric Materials", Polytechnic Institute of Brooklyn, 1963.]

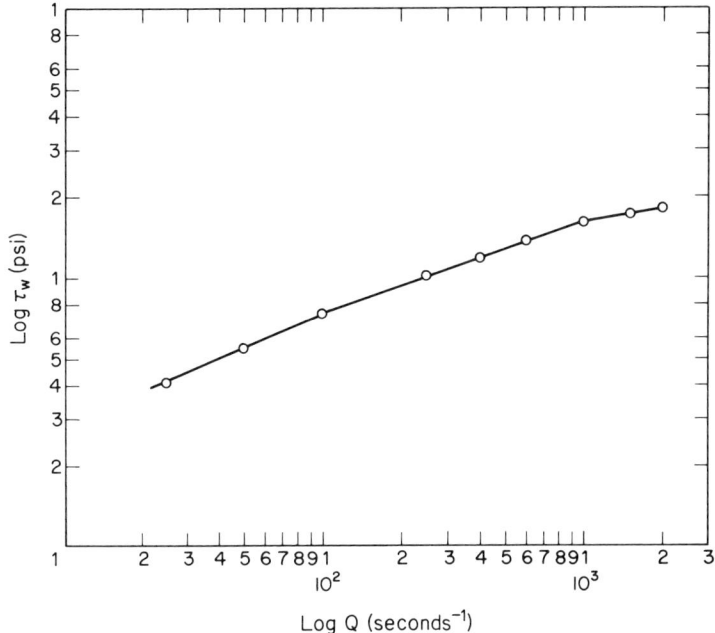

Figure 12–16. Corrected flow curve for polypropylene data of Figure 12–15.

As yet, these effects lack a sound theoretical explanation. Qualitatively, they are attributed to viscous-energy dissipation and elastic-energy storage. Several investigators have elevated the study of entrance corrections to an area of major activity.

J. R. A. Pearson* has suggested a means of making an overall correction by performing two experiments in which polymer is extruded at equal flow rates from the same reservoir through two capillaries of equal radii but different lengths. One would measure the driving pressure at the same piston height for both experiments to obtain ΔP_1 and ΔP_2 for tubes of length L_1 and L_2. The true gradient would then be given by

$$p = \frac{\Delta P_1 - \Delta P_2}{L_1 - L_2} \qquad (12\text{--}38)$$

In practice, this equation would be easier to employ with a constant-rate rheometer than a constant-stress rheometer.

* J. R. A. Pearson, *Mechanical Principles of Polymer Melt Processing* (Elmsford, New York: Pergamon Press, 1966).

1211 The Relaxation Region and Jet Swelling

After the entrance region, it is presumed that the strictly viscous character of polymer melt flow is fully established—that is, the velocity profile is fully developed and stable and the hydrostatic driving-pressure gradient is linear and also stable. In the early part of this flow regime, two stresses are encountered in addition to the one imparted by the hydrostatic driving-pressure. These are the elastic stresses generated by transfer of the melt from the reservoir into the capillary, and the normal stresses generated by the effect of the shear gradient on the conformation of the polymer molecules. The entrance-borne elastic forces in any fluid element begin to decay by some viscous relaxation mechanism as the element passes down the capillary; hence the designation the *relaxation region*. The exact nature of the decay mechanism is not known, but it is generally implicitly assumed that this mechanism does not disturb the velocity profile. If the tube is sufficiently long, the elastic forces will eventually be completely dissipated. In engineering practice, the tubes are not sufficiently long, and the recovery of the residual elastic, as well as the normal stresses, has important implications in jet swelling and melt fracture.

The most dramatic manifestation of elastic-stress relaxation in capillary flow is observed in the decay of jet swelling with die length. Jet swelling results from the desire of the polymer melt to relieve elastic and normal stresses and thereby return to its original dimensions in the reservoir. Die-swell data are normally correlated in terms of the *swell ratio*, defined as

$$\delta = \frac{\text{jet diameter}}{\text{capillary diameter}} \tag{12-39}$$

Curves of the form shown in Figure 12–17 are obtained for constant-strain rate experiments, where δ is simply plotted against t_R, the *residence time* of the polymer in the capillary.* Observe that jet swelling decays to a constant value. This constant level of swell is associated with the recovery of normal stresses. The point at which this level is achieved marks the end of the relaxation region and the onset of viscometric flow, a discussion of which follows. Usually 40- to 60-tube diameters are required to reach this point, so that complete relaxation is never realized in industrial extrusion processes.

Data of the form shown in Fig. 12–17 can usually be fitted to an equation of the form

$$\delta = a + be^{-t_R/c} \tag{12-40}$$

---

* t_R = volume of capillary/q.

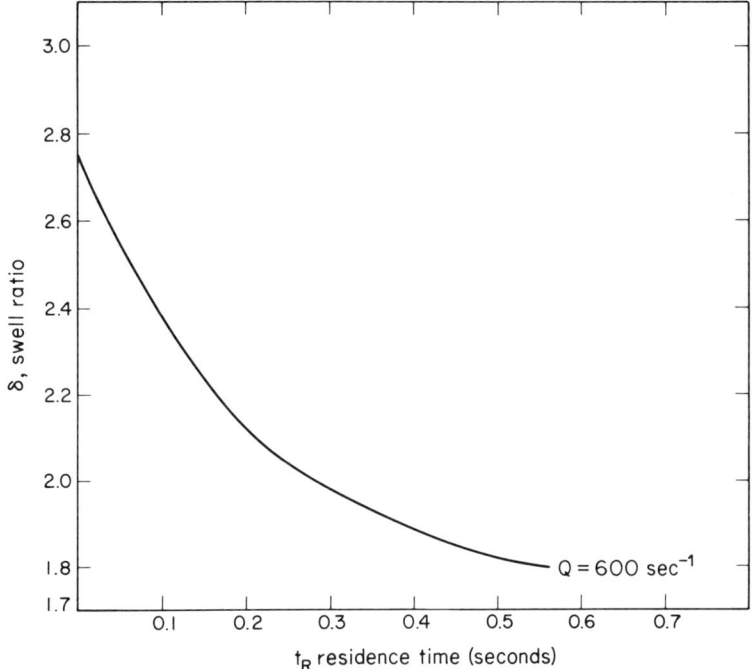

Figure 12-17. Swell ratio data for polypropylene of Figure 12-15 $Q = 600 \sec^{-1}$.

where all the constants a, b, and c are physically interpretable. In accordance with the preceding descriptive material, c may be regarded as a relaxation time; b indicates the amount of elastic energy stored in the entrance region; a is the swell ratio for an infinite residence time after the elastic energy is completely dissipated (it represents the contribution due to the capillary-borne normal stresses).

Finally, note that the behavior just described is another example of how a polymer exhibits a fading memory.

1212 Viscometric Flow

This type of flow is characterized by a steady velocity profile with strictly shear-rate-dependent normal stresses and the absence of radial stresses generated by entrance-borne elastic energy. As previously indicated, the number of hydrodynamical problems for which an exact, or even an approximate, solution exists is rather small, and these problems generally belong to the realm of simple geometries and simple flow situations. The flow in the viscometric region is one of the few problems for which an exact solution

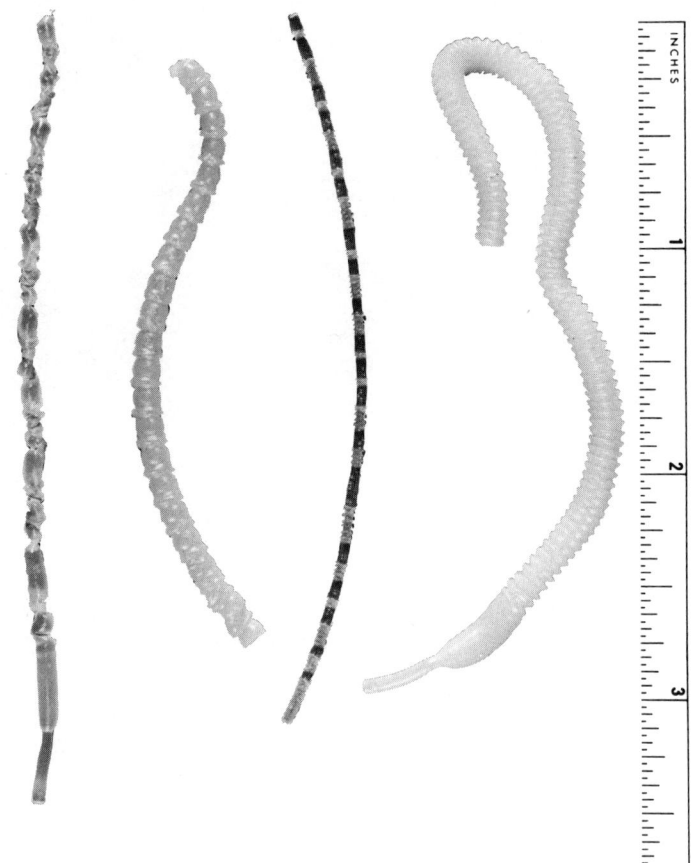

Figure 12–18. Extrudates obtained during unstable flow. From left to right: (1) low-density polyethylene (PEt) [density = 0.923 g/cc at 23°C, melt index = 2.0]. (2) high-density PEt resin (density = 0.95 g/cc at 23°C, melt index = 0.3). (3) FEP–fluorocarbon resin. (4) high-density PEt resin (density = 0.963 g/cc at 23°C, melt index = 0.8). [J. M. Lupton, "Polymer Melt Processing", in *Polymer Processing*, Chem. Engr. Symp. Ser., **60**, No. 49, 17 (1964), J. V. D. Fear, ed.]

exists. We will not develop this solution here, for more complicated mathematics are involved than we intend to handle. Also, this is the region of least engineering importance because industrial processes involve such short die lengths.

1213 Unstable Flow

Up to a critical apparent shear stress S_c, polymer melts extrude smoothly. Above this critical shear stress, about 10^6 dynes/cm^2 for all polymers regardless of process geometry, the extrudate becomes distorted. This distortion may

Sec. 1213 Unstable Flow

(a)

(b)
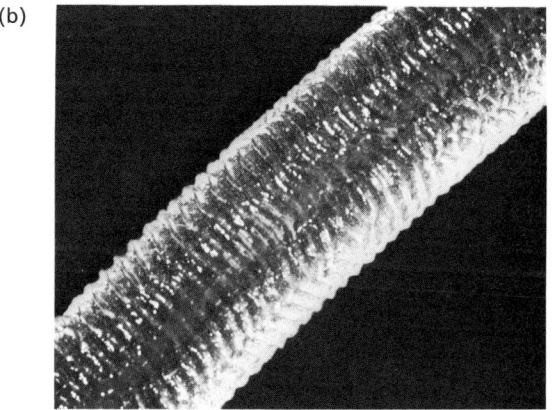

(c)
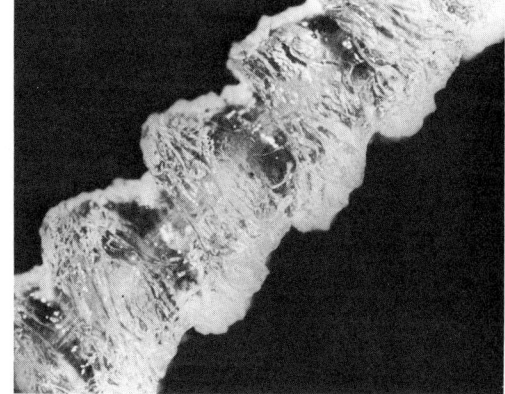

Figure 12-19. Poly(dimethyl siloxane) extrudates obtained during unstable flow.
(a) $Q = 2.5$ sec^{-1}. The flow is stable and the extrudate is smooth.
(b) $Q = 69$ sec^{-1}. The flow is unstable and the extrudate has a shark skin appearance.
(c) $Q = 123$ sec^{-1}. Unstable flow with melt fracture. [J. J. Benbow, R. N. Brown, and E. R. Howells, Coll. Intern. Rheol., June-July, 1960, Paris.]

be a barely visible surface distortion, or it may be a gross distortion. The surface defects have a matte or sharkskin appearance. Mattness is the mildest form of defect and consists of a loss in surface gloss, whereas sharkskin, a more severe form, gives finely spaced, sharp, regular, circumfrential ridges in the extrudate. The gross defects may be jaggedly random at very high stresses; or, at somewhat lower stresses, they may be geometrically regular in a banded bamboolike pattern or in a smooth helical pattern. Some representative samples are shown in Figures 12–18 and 12–19. The grossly distorted

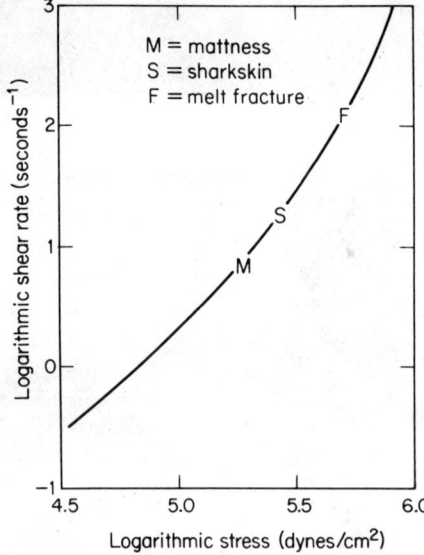

Figure 12–20. Flow curves for material of Figure 12–19. [Benbow et al., op. cit. Figure 12–19.]

state is most often referred to as *melt fracture* and sometimes as *elastic turbulence*. The mild surface distortions occur at slightly lower critical shear stresses than gross distortions; at higher stresses, the surface distortions may be superposed on the gross distortions as shown for the poly(dimethyl siloxane) rubber sample* of Figure 12–19. The surface for this sample was smooth up to an apparent shear rate of 6.3 sec^{-1} Figure 20a when mattness began (not shown). This gave way to sharkskin Figure 20b when the rate was raised to 15.8 sec^{-1}. Gross defects Figure 20c began at a rate of 110 sec^{-1}. (Note that the critical rates cited do not correspond with those of the photographs; they were taken at the rates indicated to illustrate the types of defect.) Figure 12–21 shows the flow curve for the material of Figure 12–20,

---

* Several groups of investigators are using this material in their experiments, for it can be extruded at room temperature at lower stresses than molten polymers. It exhibits all the flow characteristics of molten polymers.

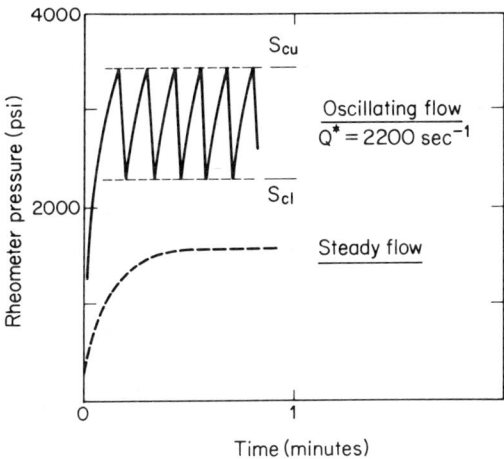

Figure 12-21. Oscillating flow of polyethylene (0.952 g./cc. at 23 °C, melt index = 1) L/D = 16, 190 °C. [J. M. Lupton, op. cit. Figure 12-9.]

and it indicates the points at which the various defects began. Note the narrow shear-stress range in which the total defect range is exhibited.

Another interesting manifestation of unstable polymeric flow is observed in the pressure-time records of the constant-rate rheometer. Figure 12-22 shows typical curves for stable and unstable flow; similar behavior is observed in industrial extruders. At low shear rates, the operation is normal, and the driving pressure moves promptly to a steady state value. Above a critical shear rate, however, the system fails to reach steady state, and it oscillates between two pressure levels. The stresses corresponding to these levels are

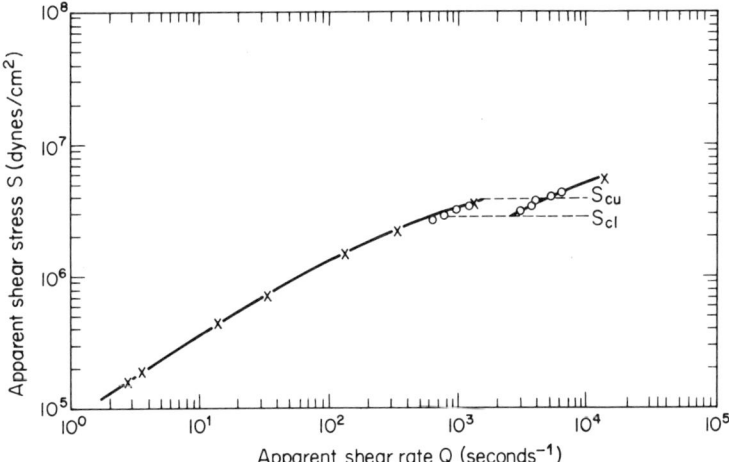

Figure 12-22. Flow of polyethylene (0.952 g./cc. at 23 °C, melt index - 1.0) L/D - 16, 190 °C. [J. M. Lupton, op. cit. Figure 12-9.]

referred to as the *upper critical shear stress*, S_{cu}, and the *lower critical shear stress*, S_{cl}.

In the stress region where such fluctuations are observed, careful measurements have shown, as illustrated in Figure 14–22, that the flow curve is doubly branched, overlapping in the shear stresses but not in the shear rate. The extrudate may again be smooth in the upper branch of the flow curve.

Study of the mechanisms of these defects is important, for they are unacceptable in the process industries and they set upper operating limits on process equipment. These phenomena, like jet swelling, are encountered in all types of industrial-processing equipment. Explanations for the observed behavior is still qualitative. The gross defects are usually associated with an inlet-region mechanism, but some workers have recently suggested that a capillary-region mechanism may also have to be considered. The surface defects are sometimes associated with an exit-region mechanism.

It is interesting to recall in passing that Newtonian fluids also exhibit unstable flow, although it is of a different type: for example, Reynolds turbulence and the breakup of jets as observed with a common garden or fire hose.

1214 Inlet-Borne Instability

A number of investigators have used photographic and tracer techniques to study unstable flow in transparent capillaries. During stable flow, a smooth funneling of the fluid is observed in the reservoir-inlet region, as depicted in Figure 12–23(a). Around the funnel boundary, there are dead spots with internal circulation caused by the shearing action of flow. These dead zones are at a hydrostatic pressure equal to that in the funnel; but geometry prevents

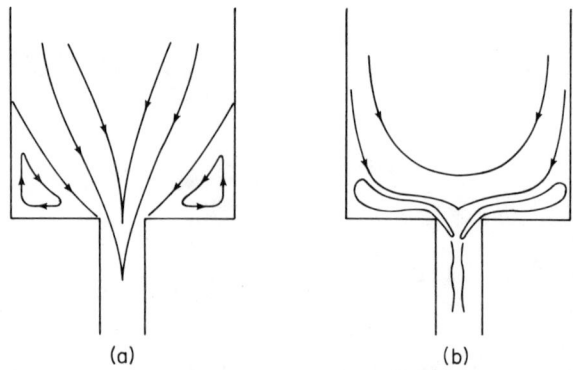

Figure 12–23. Schematic showing mechanism for entrance borne instability.

flow into the capillary and the material moves in an eddy. At a critical shear stress, streamline fracture occurs at the funnel boundary, as shown in Figure 12–23(b). There is a momentary elastic snapback of the funnel, and material from the dead zones surges into the capillary. The regular pressure and the flow patterns in the center of the capillary become reestablished, and the cycle is repeated, causing pulsations in the extrudate.

In unstable flow experiments with a colored central thread lying along the axis, this thread reappears in the extrudate in a series of separate fragments. Also, the frequencies of the inlet phenomena are the same as those observed in the distortions of the extrudate.

Tapered die entries seemingly raise by a factor of ten the shear stress at which melt fracture occurs in the extrudate. Actually, the critical shear stress is the same as with flat entries. Experiments show that fracture in the reservoir occurs at the same critical stress, but the fractures are more frequent and less catastrophic so that the effect is damped out by the time the polymer leaves the capillary. As the stress is increased, melt fracture will again appear in the extrudate.

This phenomenon can be explained on a molecular basis. Recall that molecules in the melt are entangled and high shear causes them to disentangle. At the streamline boundary, there is a region of high shear between the material just inside the funnel boundary and that within the eddy. During ordinary operations, disentangling occurs smoothly; but at high shear rates, the disentangling cannot occur rapidly enough and elastic tensions build up. When the material ruptures near the die inlet, these tensions cause the stream to retract like an extended elastic band cut in the middle, until the elastic energy is dissipated by friction and ordinary flow is temporarily resumed.

The foregoing mechanism is entirely consistent with the experimental evidence of the effect of molecular weight on S_{cl}. It is found that S_{cl} decreases as molecular weight, and therefore as elasticity, increases. (The degree of chain entanglement increases with molecular weight, and thus the elastic nature of the material increases.) In fact, for most linear polymers, the product $S_{cl} \times \bar{M}_w$ is essentially constant throughout a considerable range of shear stresses and molecular weights. The molecular-weight distribution seems to have little effect on S_{cl}.

Short-chain branches also have little effect on S_{cl} but long-chain branches have a considerable effect. For a series of polymers with equal molecular weight, the degree of entanglement should decrease as the branch length increases, since the molecules must become more compact. Hence S_{cl} also decreases with increasing branch length, molecular weight remaining constant. The critical shear stress is practically independent of temperature, but the critical rate is quite temperature dependent.

PART V
Appendices

A

Polymer Nomenclature

For the variety of polymers we will be discussing, nomenclature is a relatively simple matter. Generally, the polymer name is based on the monomer(s) from which it is derived. In the simplest instance, when the monomer consists of only one word as it so often does with homopolymers, we merely add the prefix "poly", e.g., polyethylene or polystyrene. When the monomer is comprised of two words, we bracket the monomer-derived term and then add the prefix "poly", e.g., poly(vinyl benzene) [instead of polystyrene] or poly(vinyl chloride). When two distinct chemical species are required to react to form the repeat unit, as in some step-reaction polymerizations, the nature of the reaction product is indicated, e.g., poly(ethylene terephthalate) and poly(hexamethylene adipamide) are derived from ethylene glycol + terephthalic acid and hexamethylene diamine + adipic acid, respectively.

One class of polymers deserves special mention. The polyamides which contain the interunit linkage –NHCO– are generically known as nylons. A wide variety are derived from linear aliphatic diamines and dicarboxylic acids. Poly(hexamethylene adipamide), cited above, is an example. To designate the specific character of this polymer as a nylon, we count the number of carbon atoms in the diamine portion, and then in the dicarboxylic acid portion. In the present example, these numbers are 6 and 6. Thus, poly(hexamethylene adipamide) is known as Nylon 66 (enunciated as six-six). Poly(hexamethylene sebacamide) is designated Nylon 610 (enunciated as six-ten) and polycaprolactan is designated Nylon 6.

In the case of copolymers it is necessary to indicate whether the copolymer is random, block or graft. For random copolymers the monomers are separated by the suffix –co– as poly(styrene–co–butadiene). For block copolymers the suffix –b– may be used as poly(styrene–b–butadiene) and for graft copolymers the suffix –g– may be used as poly(styrene–g–butadiene).

This very brief discussion should satisfy our needs. For further details consult *Polymer Handbook*, J. Bandrup and E. H. Immergut, ed., Section I: "Nomenclature Rules", John Wiley and Sons, New York, 1965.

B

Problems

Chapter 1

1.1 Write equations analogous to Eq. (1-2) and (1-3) for the number average, $\bar{X}_n$, and weight average, $\bar{X}_w$, degree of polymerization.

1.2 From $\bar{M}_n = \Sigma N_i M_i / \Sigma N_i$, show that $\bar{M}_n = 1/\Sigma(\frac{w_i}{M_i})$ and from $\bar{M}_w = \Sigma W_i M_i / \Sigma W_i$ that $\bar{M}_w = \Sigma w_i M_i$.

1.3 A non-homogeneous polymer system is composed according to the following fractional distribution

Wt. fraction	0.05	0.20	0.35	0.20	0.15	0.05
Mol. Wt. $\times 10^{-5}$	1	2	3	4	6	7

Compute $\bar{M}_n$, $\bar{M}_w$, and $\bar{M}_w/\bar{M}_n$ for this system.

1.4 Compute $\bar{M}_n$ for the following polymer systems where for each $\bar{X}_n = 100$: polystyrene, polycaprolactam, poly(hexamethylene adipamide), and for a polymer made from the following starting materials: adipic acid, sebacic acid, and ethylene glycol in the molar ratios of 1 : 1 : 2.

1.5 For the polymer systems of Problem 1.4, state the average number of repeat units per polymer chain.

1.6 Indicate whether the following monomer systems will produce copolymers or homopolymers. Give the basis for your decisions.
  (a) $H_2N-(CH_2)_5-COOH$
  (b) $H_2N-(CH_2)_6-NH_2 + HOOC-(CH_2)_4-COOH$
  (c) $H_2N-(CH_2)_6-NH_2 + HOOC-(CH_2)_8-COOH$
  (d) $H_2N-(CH_2)_6-NH_2 + HOOC-(CH_2)_4-COOH + HOOC-(CH_2)_8-COOH$
  (e) $H_2N-(CH_2)_5-COOH + H_2N-(CH_2)_9-COOH$

1.7 Consider a polyethylene molecule with a molecular weight of 500,000. Using a C—C bond length of 1.54Å, a C—C bond angle of 109°28' and a cross sectional area of 22Å, calculate its length to diameter ratio in the extended zig-zag chain conformation.

Chapter 2

2.1 Poly(ethylene 1,4-terephthalate) fibers are ordinarily too stiff for direct application. How could we best mitigate this problem with only a slight modification in the chemical nature of the system? Cite examples of other polymer systems where such a modification could be made.

2.2 List the following materials in the probable order of their increasing crystalline melting points and justify your answer. Do not consider molecular length as a factor.

(a) $\{NH-(CH_2)_6-NH-\overset{O}{\overset{\|}{C}}-(CH_2)_8-\overset{O}{\overset{\|}{C}}\}_n$

(b) $\{NH-(CH_2)_6-NH-\overset{O}{\overset{\|}{C}}-NH-(CH_2)_8-NH-\overset{O}{\overset{\|}{C}}\}_n$

(c) $\{NH-(CH_2)_6-NH-\overset{O}{\overset{\|}{C}}-NH-CH_2-\bigcirc-CH_2-NH-\overset{O}{\overset{\|}{C}}\}_n$

(d) $\{NH-(CH_2)_6-NH-\overset{O}{\overset{\|}{C}}-(CH_2)_4-\overset{O}{\overset{\|}{C}}-\}_n$

(e) $\{O-(CH_2)_6-\overset{O}{\overset{\|}{C}}-\}_n$

What factors must we consider before making such a relative comparison? Under what circumstances could we make no judgement?

2.3 Show how neighboring chains of poly (hexamethylene adipamide) might be aligned in sheets through hydrogen bonding.

2.4 Define and cite examples of hydrogen bonding, polarity as well as polar bonding, and dispersion forces.

2.5 Account for the fact that such polymer properties as refracture index, electrical properties, and color are independent of molecular weight.

2.6 Given the two monomers, styrene and butadiene, consider the wide variety of polymers that it is theoretically possible to synthesize. Prepare an outline listing the polymers. Indicate the factors you would consider in preparing these polymers, and discuss your outline in detail.

Chapter 3

3.1 Prove that $p_A = p_B/r$ when $F_A(0)/F_B(0) = r$.

3.2 A mixture of 1.1 moles of ethylene glycol and 1.0 moles of 1,4-phthalic acid are reacted until 100% of all acid functions are converted to ester linkages. The polymer so obtained is then steam stripped at high temperature until 0.099 moles of ethylene glycol are removed. What is $\bar{X}_n$ for the polymer before and after vacuum stripping? Write the chemical equation describing the stripping process. If the stripping is continued until exactly 0.1 moles of ethylene glycol have been removed from the polymer, what weight per cent of the polymer should be expected to have a degree of polymerization of exactly 100 if simultaneous contamination with water vapor during stripping causes 0.001% of the remaining ester linkages to hydrolyze?

3.3 Develop an expression for the degree of polymerization in a polycondensation system of the type ARB which contains trace amounts of the monofunctional reagent R'A. State clearly each step of your development as well as all assumptions or conditions of your derivation. What is highest average value of $\bar{X}_n$ that can possibly be attained?

3.4 Derive an expression for the development of $\bar{X}_n$ as a function of time in bifunctional, acid-catalyzed, step-reaction polymerizations.

3.5 Eqs. (3–16) and (3–18) were derived for a bifunctional system. What rationale must be applied to extend the application of these equations to balanced $RA_2 + R'B_2$ systems?

3.6 Show that a maximum exists in the most probable distribution function $w_x = x(1-p)^2 p^{x-1}$ for any particular value of p. Determine $x_{max} = f(p)$. Show that x_{max} is approximately equal to $\bar{X}_n$ as $p \to 1$.

3.7 The ratio $\bar{X}_w/\bar{X}_n = 1 + p$ is often used as a measure of the breadth of molecular weight distribution in bifunctional step reaction polymerization. Discuss the merit of such a procedure by comparing the relative change in $\bar{X}_w/\bar{X}_n$ as $p \to 1$ with the actual shapes of the curves shown in Figures 3–3 and 3–4.

3.8 Compute $\bar{M}_w$ for the polymers resulting from the following systems if $\bar{X}_n = 10$

(a) $\overset{\overset{\displaystyle NH}{\rule{3cm}{0.4pt}}}{CH_2\text{—}(CH_2)_4\text{—}C=O}$

(b) $HO\text{—}CH_2\text{—}CH_2\text{—}OH + OCN\text{—}(CH_2)_6\text{—}NCO$

3.9 Compute the gel points for the system $RA_3 + R'A_2 + R''B_2$ using the stoichiometric proportions of (a) 2 : 0 : 3, (b) 2 : 1 : 4, (c) 2 : 3 : 6, and (d) 2 : 5 : 8. Also calculate $\bar{X}_n$ at the gel points. Tabulate and compare your results for the various stoichiometries, using the average functionality of each system as a basis.

3.10 It is desired to affect gelation of the following system when 90% of the A functions have reacted. What should the stoichiometric proportions be?

$$RA_3 + R'B_2$$

3.11 For the following condensation system

$$2\ R'A_4 + 3R''A_3 + 50R'''A_2 + 60R''''B_2$$

(a) Find the extent of reaction at gelation, p_A.
(b) Find the number average degree of polymerization at incipient gelation.

3.12 It is desired to design a lightly crosslinked condensation polymer system of the $RA_3 + R'A_2 + R''B_2$ type such that a branch point occurs on the average of once for every 100 units in the chain. Compute the gel point for this system.

3.13 The manufacture of the polyfunctional B-stage resin, $x\ RA_3 + yR'A_2 + zR''B_2$, $\bar{X}_n = 12$, is being considered. Gelation is to be prevented by blocking the reaction with excess $R''B_2$. Determine the stoichiometry which should be used.

3.14 Compute the time to reach the gel point for the following acid catalyzed systems where $C_0 k = 2.24 \times 10^{-4}\ \text{min}^{-1}$.

(a) $2\ RA_3 + R'A_2 + 4R''B_2$
(b) $2\ RA_3 + R'A_2 + 3.5R''B_2$

3.15 Describe in a qualitative way the physical properties of the polymers which would result if the following reactants were polymerized in the molar ratios of (a) 1 : 1 : 0, (b) 3 : 0 : 2, (c) 4 : 1 : 2, (d) 12 : 9 : 2, and (e) 20 : 17 : 2. Use the average functionality of each system as a basis for discussion.

phthalic anhydride + HO—CH$_2$—CH$_2$—OH + HO—CH$_2$—CH—CH$_2$—OH
                                                           |
                                                          OH

3.16 Show that the number distribution of molecular sizes for polymers formed in the system BRA + BR'O, where O is an unreactive end, can be expressed as

$$n_x = (p_A p)^{x-1}(1 - p_A p)$$

p is the ratio of B groups in ARB molecules to the total number of B groups i.e.,

$$p = \frac{F_B(0) \text{ in BRA}}{F_B(0) \text{ in BRA} + F_B(0) \text{ in BR'O}}$$

The following structures form:

I) BRA $-$ (BRA)$_{x-2}$ $-$ BRA

(II) BRA $-$ (BRA)$_{x-1}$ $-$ BRO

3.17 Rationalize the formulation of Eq. (3–30). Derive Eq. (3–13) from (3–30).

Chapter 4

4.1 A head-to-tail configuration is always expected in chain polymerization. Show with a mechanistic scheme why this should be so for the radical polymerization of styrene and the anionic polymerization of vinylidene cyanide.

4.2 Predict the probable order of reactivity in anionic polymerization for the following monomers:

(a) $CH_2=CH-CH_3$; (b) $CH_2=CH-CN$; (c) $NC-CH=CH-CN$;
(d) $CH_2=C-CN$; (e) $CH_2=C-CN$.
　　　　　|　　　　　　　　　|
　　　　CH_3　　　　　　　CN

4.3 Compare graphically both addition polymerization and bifunctional condensation polymerization in terms of dimensionless monomer unit concentrations $M(t)/M(0)$, and number average degree of polymerization as a function of a suitable dimensionless reaction time.

4.4 A steady state free radical styrene polymerization process is being controlled such that the rate of polymerization is constant at 1.79×10^{-3} gm of monomer/min.ml. The initial initiator concentration is 6.6×10^{-6} moles/ml.
 (a) What must be done to maintain the constant rate of polymerization?
 (b) If the rate constant for the first order decomposition of the initiator, k_d, is 3.25×10^{-4} min^{-1}, what is the rate of free radical generation per second per ml? What is $\bar{x}_n$?
 (c) What percent of the original initiator concentration remains after a reaction time of 3 hours?

4.5 Consider the isothermal, solution polymerization of styrene at 60°C in the following formulation:

> 100 gm styrene
> 400 gm benzene
> 0.5 gm benzoyl peroxide

In this case it is expected that benzoyl peroxide will be 100% efficient and have a half-life of 44 hours. At 60°C, $k_p = 145$ l/mole sec and $k_t = 0.130$ l/mole sec. All ingredients have unit density.

Derive the rate expression for this reaction. Calculate the rate of propagation at 50% conversion. How long will it take to reach this conversion?

4.6 For the anionic polymerization of α-methyl styrene in liquid ammonia with potassium amide as catalyst, termination occurs solely by transfer to ammonia. Write the chemical equations for this polymerization and show that

$$v_p = (k_p k_i/k_{tr}) \frac{[KNH_2][M]^2}{[NH_3]}$$

$$\bar{X}_{ni} = (k_p/k_{tr}) \frac{[M]}{[NH_3]}$$

4.7 The cationic polymerization of isobutylene is promoted with a catalyst-cocatalyst combination of BF_3 and H_2O. Write the chemical equations for this polymerization and show that

$$v_p = (k_p k_i/k_t)[H^+A^-][M]^2$$
$$\bar{X}_{ni} = (k_p/k_t)[M]$$

4.8 It has been postulated that in thermal free radical polymerization the initiation process is bimolecular in monomer and the primary active species so produced is

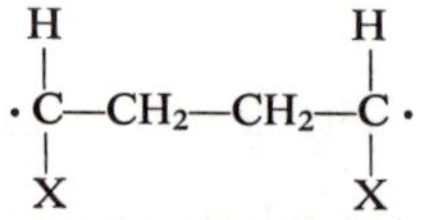

$$\cdot \overset{H}{\underset{X}{C}}-CH_2-CH_2-\overset{H}{\underset{X}{C}} \cdot$$

Derive the steady state rate expression for this polymerization process. State all assumptions. How could this proposed mechanism be verified?

4.9 For the living polymer system show that

$$\bar{X}_n = \frac{2([M(0)] - [M(t)])}{[I]}$$

$$\frac{[M(t)]}{[M(0)]} = \exp(-k_p[I]t)$$

$$\bar{X}_n = \frac{2[M(0)]}{[I]}[1 - \exp(-k_p[I]t)]$$

4.10 Show that chain transfer to polymer, leading to the production of branched molecules, does not affect $\bar{X}_{ni}$.

4.11 The degree of branching which occurs in a chain polymerization process by reaction with polymer can be estimated from chain transfer coefficients of model compounds. Thus, in the radical polymerization of styrene, ethyl benzene may be used, for which $C_{\Phi Et} = 1.08 \times 10^{-4}$ at 80 °C. Show that $v_p/v_{tr} = (1-p)/pC_{\Phi Et}$ where p = fractional conversion. Compare the incremental branching densities at 50 and 90% conversion.

4.12 Using Eq. (4–14) to (4–17) as a basis, discuss the numerous ways in which molecular weight can be controlled in addition polymerization. Discuss the pros and cons of each method, and list them according to feasibility.

4.13 It is desired to produce poly(methyl methacrylate) in bulk at 60°C with $\bar{M}_n = 10^5$. The initiator is 1%(wt) AIBN based on monomer, and n-butyl mercaptan is to be used as a chain transfer agent. For AIBN at 60°C, $k_d = 9.0 \times 10^{-6}$ liters^{-1} and $\epsilon = 0.5$; for n-butyl mercaptan, $C_{SH} = 0.66$; and $(k_p/k_t^{\frac{1}{2}}) = 1.61$. What (wt) percent (based on monomer) of mercaptan is required? What molecular weight would be produced in the absence of the mercaptan?

4.14 Show that Eq. (4–13) exhibits a maximum and determine the conditions under which the maximum corresponds to $\bar{X}_n$.

Chapter 5

5.1 Discuss the product one would obtain if styrene (M_1) and maleic anhydride (M_2) were reacted in the molar ratios of 75/25, 50/50, and 25/75. $r_1 = 0.01$ and $r_2 = 0$ at 60°C.

5.2 In the text it is stated that acrylonitrile (ACN) is copolymerized with vinyl acetate (VAc) or vinyl chloride (VCl) up to about 10% in order to improve certain properties of the polymer product.
   (a) Consider the reasons for establishing a regular distribution of ACN in the copolymer products.
   (b) For the ACN (M_1) – VAc (M_2) combination, $r_1 = 4.05$, $r_2 = 0.061$ at 60°C, and for the ACN (M_1) – VCl (M_2) combination $r_1 = 3.28$, $r_2 = 0.02$ at 60°C. Discuss the difficulties in obtaining a regular distribution in the product and how you would circumvent them.

5.3 Beginning with Eq. (5–10) derive the following expression

$$df_1 = f_1 \left[\frac{M_1 - dM_1}{M_1 - (f_1/F_1)dM_1} - 1 \right]$$

State the feed conditions under which df_1 will be positive or negative. Relate your answer to the numerical example given in Section 504. Discuss the feasibility of copolymerizing under azeotropic conditions.

5.4 For a random copolymer with $r_1 = 1$ and a 50/50 composition, prepare a plot of $N(M_1, n_1)$ vs. n_1.

5.5 It is desired to form a copolymer from CH_2=CHX (M_1) and CH_2=CHY (M_2), containing twice as many X groups as Y groups. The monomers copolymerize ideally, with A_1 adding M_1 twice as fast as M_2. Describe the procedure as well as the feed composition you might use to make this copolymer.

5.6 Reconsider your answer to Problem 2.6.

Chapter 7

7.1 Clearly differentiate between the meaning of the following terms: amorphous, crystalline, crystallizable, and non-crystallizable. Amplify your discussion by considering the application of each of these terms to polyethylene, poly(ethylene terephthalate) and the various tactic forms of polypropylene. Also illustrate your discussion with plots of specific volume vs. temperature wherever appropriate.

7.2 Compare the transition region behavior displayed in Figure 7-3 with that in Figure 7-5.

7.3 Using the discussion centering around Figures 7-5 and 7-12 as a basis, sketch log $E_r(t)$ vs. T curves for a linear polymer at various degrees of crosslinking. Discuss the mechanical behavior one might expect to observe in the various viscoelastic regions.

7.4 Using Figure 7-5 as a basis, sketch log $E_r(t)$ vs. T curves for a semi-crystalline polymer at various degrees of crystallinity as well as for its amorphous counterpart. Discuss the mechanical behavior one would observe in the various viscoelastic regions.

7.5 Consider Figure 7-21 and speculate on the effect light crosslinking would have on the shape of the curves. Discuss the implications of this.

7.6 Explain the shift in T_g observed in Figure 7-8 in terms of molecular mechanisms.

Chapter 8

8.1 In the evaluation of Eq. (8-6) for the polymethylene chain model show that the contribution due to the $j = i + 3$ term is given by $2(\sigma - 3) l^2 \cos^3 \theta$.

8.2 Fill in the details of the development from Eq. (8–20) to Eq. (8–24).

8.3 Verify the results of integration in Eq. (8–30) and (8–31).

8.4 Gaussian statistics are most appropriate in describing polymer chain conformations for polymer chains which are very long and very flexible. Discuss the limitations of Gaussian statistics with decreasing molecular weight and with increasing chain stiffness. How should the shape of the spherical envelope be affected?

8.5 Describe how the analogy between the random flight statistics of a diffusing particle and a polymer chain breaks down.

8.6 Calculate the $\overline{rms}$ dimensions of a linear polymethylene chain with $\bar{X}_n = 10^4$ using Eq. (8–10) and (8–13) with $\beta = 3.5$. What is the standard deviation in h? What fraction of the time could one expect to find h within the limits of the standard deviation?

8.7 Calculate $\bar{h}$ for a hexane molecule—assume free rotation.

8.8 Plot the radial distribution functions for the radius of a polymethylene chain model and for a polystyrene molecule ($\beta = 5.2$), both with $\bar{X}_n = 10^4$.

8.9 If for a polymethylene chain $\bar{h} = 10^3 \text{Å}$, calculate the end-to-end separations which are only 1/10 as probable.

Chapter 9

9.1 Consider a single linear polymer molecule in a good solvent. A good solvent is one in which interaction between the chain segments and solvent molecules is favored, and the overall effect is to expand the random coil beyond its random flight or unperturbed dimensions. The factors seeking to expand the coil size are counteracted by other factors. Discuss the nature of these counteracting factors and how they might be quantitatively characterized.

9.2 Show by a vector representation what is meant by an affine deformation. How does this description fit into our analytic development of the affine deformation of an ideal polymer network?

9.3 Show that the stress in simple shear for an ideal rubber network is given by
$$f = vkT(\alpha - \alpha^{-1})$$

9.4 Natural rubber with $\rho = 0.90$ gm./cc, $\bar{M}_n = 75{,}000$ and $\bar{M}_c = 6000$ is to be stretched from 4 in. to 6 in. What is the force per unit area required? If a single chain of this rubber is subjected to a tensile force of 10^{-5} dyne in the z direction, what will be the average end-to-end distance in the x, y, and z directions?

9.5 Determine the $\bar{M}_c$ values for the rubber samples shown in Figure 9–18.

9.6 Obtain an expression for the heat evolved under conditions of isothermal extension of an ideal rubber.

9.7 Consider the consequences of heating a stressed versus an unstressed elastomer.

Chapter 10

10.1 Discuss the physical significance of the retardation time λ, based on the behavior of a Voigt element, as λ approaches limiting values of zero and infinity.

10.2 For a Voigt element, derive an expression for the strain recovery from an initial strain γ_1, starting at time t_1.

10.3 Verify Eq. (10–11) directly from model analysis.

10.4 Derive Eq. (10–12), and then solve the equation, using appropriate boundary conditions, to obtain Eq. (10–10) and (10–11).

10.5 Show that the general rheological equation for two Maxwell elements in parallel is given by
$$S + (\tau_1 + \tau_2)\dot{S} + (\tau_1\tau_2)\ddot{S} = \ddot{\gamma}(G_1 + G_2)(\tau_1\tau_2) + \dot{\gamma}(G_1\tau_2 + G_2\tau_1)$$

10.6 Derive Eq. (10–19) and (10–27) in detail.

Chapter 11

11.1 Discuss the origin and rationale associated with the correction term (T_0/T) in Eq. (11–1).

11.2 Refer to the shear modulus curve for the 4% HEXA curve in Figure (11–17). Account for its shape at temperatures above T_g.

11.3 Verify Eq. (11–13), (11–17), (11–30), and (11–31).

11.4 Calculate the maximum energy stored per cycle and that dissipated per cycle for a Maxwell element and compare them to Eq. (11–16) and (11–32).

11.5 Show for a Maxwell element with mass that the log decrement $= \pi \tan \delta$. Convert the G'' and G' data of Figure 11–14 into log decrement data and compare with Δ shown in this figure.

11.6 Perform time-temperature superpositions on the data in Figures 11–12 and 11–13 with a reference temperature of 75°C. Determine the shift factor in each case and plot them; compare with a_T calculated from the WLF equation. From the master curves calculate a few values of G' and G'' and compare them with those shown in Figure 11–11. Determine the relation between log f_{max} (in the G'' curve) and $(1/T)$ and plot. Determine the relaxation spectra for the data in Figure 11–11.

11.7 Derive Eq. (11–39).

11.8 Describe the behavior observed in Figure 11–11 in terms of the Deborah number.

Chapter 12

12.1 Determine the power-law constants for the data shown in Figure 12–16.

12.2 Verify Eq. (12–19), (12–20), and (12–21).

12.3 Define the Deborah number in terms of the parameters of Eq. (12–40). Describe the behavior observed in Figure 12–17 in terms of the Deborah number.

12.4 By differentiating Eq. (12–19), show that the apparent strain rate for a power-law fluid is given by $Q = (\Delta PR/2mL)^{1/n}$.

12.5 By using a form of the power law in which the pressure drop is proportional to the nth power of ΔP, show that $\dot{\gamma}_w$ given by the Rabinowitsch equation is identical to Q obtained in Problem 12.4.

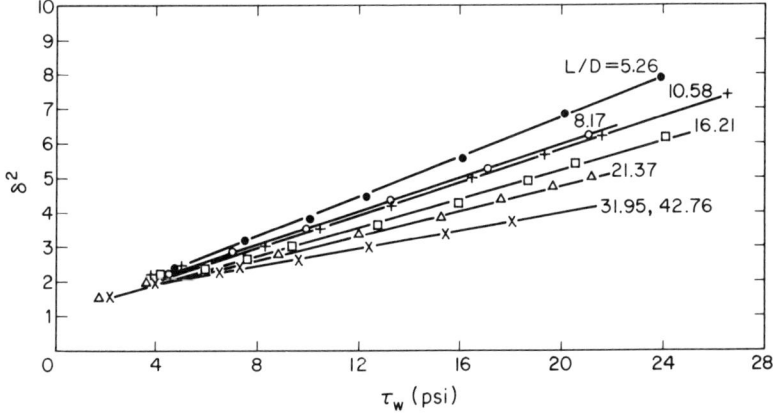

Problem 12–6.

12.6 A uniform rod of 0.765 in. diameter is to be extruded from the material whose flow properties are shown in Figures 12–15 and 12–16. A set of dies 0.30 inches in diameter with the same L/D ratios indicated in the figures is available. The critical shear rate at the processing conditions is 1600 sec^{-1}. The density of the material under processing temperatures is 0.73 while that at room temperature is 0.91. Swell-ratio data are shown in the accompanying figure; these values apply to the jet size at the extruded temperature. Under isotropic expansion or contraction conditions, all dimensions change proportionally by the same amount; and, therefore, (L/D) for the extruded material at room temperature $= (L/D)$ at the extrusion temperature. Which die should you use? Specify the extrusion pressure and the speed with which the rod must be taken up in subsequent processing operations. Calculate the Reynolds number at the critical flow rate. Why was the condition for incipient melt fracture specified in terms of Q rather than S?

C
Bibliography

Part I INTRODUCTION

General References

Bandrup, J. and Immergut, E. H.: *Polymer Handbook*, Interscience, New York, 1965.
Billmeyer, F. W.: *Textbook of Polymer Science*, 2nd ed., Interscience, New York, 1971.
Flory, P. J.: *Principles of Polymer Chemistry*, Cornell Univ. Press, Ithaca, New York, 1953.
Mark, H. (ed): *Encyclopedia of Polymer Science and Technology*, Vols. 1–9 (additional volumes in preparation), Wiley, New York, beginning 1964.
Rodriguez, F.: *Principles of Polymer Systems*, McGraw-Hill, New York, 1970.
Tanford, C.: *Physical Chemistry of Macromolecules*, Wiley, New York, 1961.

Special Topics

L. J. Broutman and R. H. Krode, *Modern Polymer Composites*, Addison Wesley Publishing Co., Reading, Mass., 1967.
Cantow, H. J. (ed.): *Polymer Fractionation*, Academic Press, New York, 1967.
Determann, H.: *Gel Chromotograpy*, Springer-Verlag, New York, 1968.
Frazer, A. H.: *High Temperature Resistant Polymers*, Wiley, New York, 1968.
Kerker, M.: *The Scattering of Light*, Academic Press, New York, 1969.
Kline, G. M. (ed): *Analytic Chemistry of Polymers*, Vols. 1–3, Interscience, New York, 1959–1962.
Mitchell, J. and Billmeyer, F. W. (eds.): *Analysis and Fractionation of Polymers*, Interscience, New York, 1964.

Part II POLYMER SYNTHESIS

General References

Lenz, R. W.: *Organic Chemistry of Synthetic High Polymers*, Interscience, New York, 1967.
Odian, G.: *Principles of Polymerization*, McGraw-Hill, New York, 1970.
Schildknecht, C. E.: *Polymer Processes*, Interscience, New York, 1956.

Special Topics

American Chemical Society: *Polymerization and Polycondensation Processes*, Advances in Chemistry Series, No. 34, Washington, D.C., 1962.
American Chemical Society: *Addition and Condensation Polymerization Processes*, Advances in Chemistry Series, No. 91, Washington, D.C., 1969.
Bovey, F. A., Kolthoff, I. M., Medalia, A. I. and Meehan, E. J.: *Emulsion Polymerization*, Interscience, New York, 1955.
Burlant, W. J., Hoffman, A. S., *Block and Graft Polymers*, Van Nostrand Rheinhold, New York, 1960.
Ceresa, R. J., *Block and Graft Copolymers*, Butterworths, Washington, 1962.
Fettes, E. M. (ed.), *Chemical Reactions of Polymers*, Interscience, New York, 1964.
Ham, G. E. (ed.): *Copolymerization*. Interscience, New York, 1964.
Ham, G. E. (ed.): *Vinyl Polymerization*, Interscience, New York, 1967.
Morgan, P. W.: *Condensation Polymers*, Interscience, New York, 1965.
Smets, G. and Hart, R.: *Advances in Polymer Science*, 2, 173, 1960, "Block and Graft Copolymers."

Part III PHYSICS OF THE SOLID STATE

General References

Bueche, F.: *Physical Properties of Polymers*, Interscience, New York, 1962.
Flory, P. J.: *Principles of Polymer Chemistry*, Cornell Univ. Press, Ithaca, New York, 1953.
Miller, M. L.: *The Structure of Polymers*, Van Nostrand Rheinhold, New York, 1966.

Special Topics

Bovey, F. A.: *Polymer Conformation and Configuration*, Academic Press, New York, 1969.
Dusek, K. and Prins, W.: *Advances in Polymer Science*, 6, 1–102, 1969, "Structure and Elasticity of Non-Crystalline Polymer Networks."
Flory, P. J.: *Statistical Mechanics of Chain Molecules*, Wiley, New York, 1969.
Geil, P. H.: *Polymer Single Crystals*, Interscience, New York, 1963.

Mandelkern, L.: *Crystallization of Polymers*, McGraw-Hill, New York, 1964.
Morawetz, H.: *Macromolecules in Solution*, Interscience, New York, 1965.
Treloar, L. R. G.: *The Physics of Rubber Elasticity*, 2nd ed., Oxford, New York, 1958.

Part IV POLYMER RHEOLOGY

General References

Eirich, F. R. (ed.): *Rheology*, Vols. 1–4, Academic Press, New York, 1956–1967.
Ferry, J. D. *Viscoelastic Properties of Polymers*, 2nd ed., Wiley, New York, 1970.
Frederickson, A. G.: *Principles and Applications of Rheology*, Prentice-Hall, Englewood Cliffs, N. J., 1964.
Tobolsky, A. V.: *Properties and Structure of Polymers*, Wiley, New York, 1960.

Special Topics

Bueche, F., *Physical Properties of Polymers*, Interscience, New York, 1962.
Coleman, B. D., Markovitz, H. and Noll, W., *Viscometric Flows of Non-Newtonian Fluids*, Springer-Verlag, New York, 1966.
McCrum, N. G., Read, B. E. and Williams, G. : *Anelastic and Dielectric Effects in Polymeric Solids*, Wiley, New York, 1967.
McKelvey, J. M.: *Polymer Processing*, Wiley, New York, 1962.
Middleman, S.: *The Flow of High Polymers*, Interscience, New York, 1968.
Nielsen, L. E.: *Mechanical Properties of Polymers*, Van Nostrand Rheinhold, New York, 1962.
Pearson, J. R. A.: *Mechanical Principles of Polymer Melt Processing*, Pergamon, New York, 1966.
Skelland, A. H. P.: *Non-Newtonian Flow and Heat Transfer*, Wiley, New York, 1967.
Tadmoor, Z. and Klein, I., *Engineering Principles of Plasticating Extrusion*, Van Nostrand Reinhold Co., New York, 1970.
Van Wazer, J. R., Lyons, J. W., Kim, K. Y. and Colwell, R. E.: *Viscosity and Flow Measurements*, Interscience, New York, 1963.

Index

A

active center, 10, 80
addition copolymerization, 128–159
addition polymerization: see polymerization, chain reaction
amorphous polymers, 200, 202, 314: see molecular packing, conformation, and transitional phenomena
annealing polymer crystals, 188
apparent shear rate, 352
apparent shear stress, 352
apparent viscosity, 341, 352
atactic, 31
autoacceleration, 100

B

bifunctional and bi-bifunctional systems, 11, 46
blends, 13: see polymer blends
blocked polymerization, 55
Boltzmann principle, 288
bonding forces, intermolecular 8, 19–22, 207
Boyer-Beamen rule, 206
branched polymer(s), 5, 11
branching mechanisms, 11, 94
branching probability, 63–65
branch point, 62
branch unit, 11, 61

C

capillary rheometer, 345
  flow in capillaries, 346
catalysts, 81
chain dimensions, 227–236
  average, 227
    general ideal expression, 235
    restricted rotation, 233
chain flexibility: see steric factors
chain folding, 42, 174
chain packing: see molecular packing
chain reaction polymerization: see polymerization, chain reaction
chain section, 62
chain transfer, 81, 103, 155
  agent, 83, 104
  constant, 103
chemical modification, 14
cohesive energy density, 21
combination, 87
compatibility, 14, 129
  effect on transitional phenomena, 222

complex variables, 305–307
composites, 13
condensation polymerization: see polymerization, step-reaction
configuration, 6, 23
  crystalline polymers, 165–172
  head-to-tail, 30
conformation, 6, 39–42, 165–172, 225–247
  crystalline polymers, 165–172, 175, 184
  helical, 39
  local vs. long range, 39
  random, 41, 225–247
  zigzag, 39
consistency variables, 352
copolymer composition, 134, 137–141
  distribution of sequence lengths, 133
  heterogeneous vs. homogeneous, 135
copolymer equation, 129
  integration, 137
copolymerization, 12
  addition, 128–159
    alternating, 129, 132
    azeotropic, 136
    block, 150, 152
    graft, 150, 155
    ideal, 131
    random, 129, 133
  step-reaction, 77
copolymerization chemistry, 143–149
  α-olefins, 148
  free radical, 144, 152, 155
  ionic, 147, 152, 159
copolymerization kinetics, 129
  rate, 141
copolymers, 13
  damping behavior, 322
  properties, 128
  transitional phenomena, 218–224
core-shell morphology, 123

corresponding temperatures, principle of, 200
crankshaft motion, 332
creep experiment, 280, 294
critical branching probability, 63
crosslinked polymers, 4, 48
crosslink formation, 11, 60–74: see vulcanization
crosslinking, 36
  degree of crosslinking, 37
  effect on properties, 36, 199, 209, 320
  in elastomers, 199, 249
  transitional phenomena, 209
crystal growth, 173
  spherulitic, 182
crystalline melting transition (temperature), 10, 29 193–196: see transitional phenomena
crystalline polymers, 163–192, 199, 202: see molecular packing, morphology and order, and transitional phenomena
crystallographic repeat unit (distance), 166
crystal structure, 164–172
  representative, 167–172

D

damped motion, 312
damping peaks, 314, 320–334
  amorphous polymer, 314
  copolymerization, 322
  crosslinking, 320
  multiple relaxation peaks, 325
  plasticization, 321
  secondary (relaxation) transitions, 325
Deborah number, 287
deformation (mechanical), 276

degradation, 15
degree of polymerization, 16: see molecular weight
dendritic, 180
depolymerization, 15
disproportionation, 87
dissipation factor, 306
drawing and orientation, 190–192
draw ratio, 191
dynamic mechanical testing, 313–334: see oscillating experiments
time-temperature relations, 314
dynamic modulus, 306

E

elastic (Hookean) behavior, 278, 303
elasticity (mechanical), 302
elastic modulus, 298–301, 313
elastomers, 202, 248–272: see rubber elasticity
emulsion polymerization, 115–127
kinetics, 120–127
core-shell morphology, 123
molecular weight development, 126
Smith-Ewart theory, 121
end-to-end distance, 228
distribution function, 241–246
radial distribution function, 245
engineering-use temperatures, 201
entrance effects, 360
entrance region, 346, 360
equal reactivity principle, 51, 98
exit effects, 360

F

fading memory, 280, 364
fibers and fibrous polymers, 190, 202: see molecular packing

first-order transitions, 193
five regions of viscoelastic behavior, 199
flexible chains: see steric factors
flow curves, 353
correction, 360
fold period, 175
fold plane, 42, 174
freely orienting chain, 230
random flight, 237–247
freely rotating chain model, 231
fringed micell model, 163
functionality, 10
functional groups, 46

G

G-values, 158
gel, 61
gelation, 61
theory, 60–74
gel effect, 100
gel point, 61
observations, 65
glass transition (temperature), 10, 196–198: see transitional phenomena
glassy, region, 199
growth twin, 180

H

Hagen-Poiseuille law, 351
head-to-tail configuration, 30, 82
hedrites, 180
helical conformation, 39
Hookean behavior, 278
compared to rubber elasticity, 259
hydrogen bonding: see bonding forces

I

ideal rubber, 259
impact resistance, 129, 157, 326
in-chain motion, 330
inhibitor, 83
initiation, 81
initiators, 81, 85
interchange reactions, 48
interfacial polycondensation, 74
interlamallae ties, 185
intermolecular bonding, 8, 19–22
   transitional phenomena, 207–210
intermolecular forces of attraction, 19: see bonding forces
interunit linkage, 46
ionic bonding, 21
irradiation, 157
isotactic, 31

J

jet swelling, 346, 364

K

kinetic chain length, 103
kinetics
   chain reaction polymerization, 96–110
      steady-state, 96
      thermal dependencies, 104
   emulsion polymerization, 120–127
   free radical polymerization, 97–102
   step-reaction, 52

L

laminar flow, 348
leather-like polymers, 202
linear polymer(s), 4
linear viscoelastic behavior: see viscoelastic topics
living polymers, 91, 107
   block copolymerization, 152
logarithmic decrement, 313
loss modulus, 306
loss tangent, 306

M

master curve, 301
Maxwell model, 281, 310
   generalized, 290
   Maxwell-Weichart, 290
   with mass, 311
   with Voigt in series, 294
melt fracture, 346: see unstable flow
melting point, equilibrium crystalline, 194: see transitional phenomena
memory: see fading memory
molding powder, 79
molecular architecture, 8, 23–37
molecular motion: see segmental motion
molecular packing, 8, 29, 37–42, 161
   crystalline polymers, 165–172
      oriented, 190
   random (amorphous) systems, 225–227
molecular weight, 16
   averages, 16
   effect on properties, 24–27
      transitional phenomena, 216
   in polymerization
      chain-reaction, 102–105
      monodispersed, 91
      step-reaction, 55–60, 67
      instantaneous number-average, 102
molecular weight distribution, 16, 27, 83

molecular weight distribution (*cont.*):
  in polymerization
    chain-reaction, 83, 105–110
    step-reaction, 57–60, 68–74
monomer(s) (units), 3–4
  vinyl, 85
monomer reactivity ratios, 131
morphology and order, 42, 163–192
  annealing, 188
  crystal structure, 164
  effect of pressure, 187
  melt crystallized, 182–187
    interlamellar ties, 185
    spherulites, 182
  polymer single crystals, 172–181
    complex morphologies, 178
    hollow pyramids, 176
most probable distributions, 57–60
Mooney-Rivlin equation, 269

N

necking, 191
network flow, 26
Newtonian flow, 340, 342
  laminar flow, 348–352
Newton's law, 279
nomenclature, 375
non-Newtonian flow, 340
  viscous character, 341
normal stresses, 344
nylon development, 27

O

organometallic polymers, 7
orientation and drawing, 190
oscillatory experiments, 293, 298, 303–334: also see dynamic mechanical testing
  model analysis, 303–313

P

packing: see molecular packing
pendant groups: see repeat unit
pendant group motion, 328
phase angle, 301
photochemical activation, 155
physics of the solid state, 161
plasticization, 210
  damping peaks, 321
Poisson distribution, 110
Poisson ratio, 278, 298
polar bonding: see bonding forces
polarity, effect on polymerization, 82, 146
polyblends, 14: see polymer blends
polyfunctional system, 11, 46, 60
polymer blends, 13, 222, 324
  compatible, 219
  transitional phenomena, 218–224
polymerization, 10–11, 43–160
  chain reaction, 45, 80–127
    kinetic processes, 81
    substituent effect, 84
  distinguishing features, 45
  ring scission, 49
  step-reaction, 46–79
    crosslinking, 60–74
    high-temperature, 50
    low-temperature, 50, 74
    mechanisms, 50
    stoichiometry, 55, 67
polymerization chemistry
  anionic, 89
  cationic, 87
  coordination, 92
  copolymerization, 143–149
  free-radical, 85
  step-reaction, 46
  Ziegler-Natta, 92
polymerization kinetics: see kinetics
polymer production — chain reaction, 110–127

polymer production — chain reaction (*cont.*):
  bulk, 112
  emulsion, 115–120
    kinetics, 120
  homogeneous and heterogeneous, 110
  mass, 112
  solution, 112
  suspension, 113
polymer production — step reaction, 78
  low-temperature, 74
polymer melts; see rheology
polymer(s), 3
  chemical modification, 14
  classification, 44
  crosslinked, 4: see crosslinked polymers
  linear vs. branched, 4–5
  special properties, 9
    examples, 10
  stereoregular, 30
  synthesis: see polymerization
  thermal stability, 15, 29, 34
polymethylene chain model, 231
power law, 353
prepolymer preparation, 57
probability concepts, 239
propagation, 81

Q

Q-e scheme, 149

R

Rabinowitsch equation, 356
radial distribution function, 245
radioactive activation, 356
radius of gyration, 228
randomly coiled chain, 225

random conformation, 41, 225–247
random flight analysis, 237–247
rate of strain, 278
reactivity ratios, 131
recurrence regularity, 30
redox decompositions, 86
reduced temperature, 201
regulator, 104
relaxation (experiments), 279, 296
  in elastomers, 256–260
relaxation modulus, 198
relaxation peaks, 320
  multiple, 325–334
relaxation region, 346, 364
relaxation spectra, 290, 334–339
relaxation time, 282: see response time, relaxation spectra, and retardation spectra
repeat unit(s), 4
  structure and properties, 32–36, 212–216, 314–334
    backbone composition, 32–34, 212, 330
    pendant groups, 34–37, 214, 328
resilience, 302
resonance, 82, 144
response time, 282, 286
restricted rotation, 233
retardation spectra, 290, 334–339
retarded elastic behavior, 284
retarder, 83
rheology, 273–371: also see viscoelastic topics
  polymer melt, 340–371
    flow characteristics, 342
    jet swelling, 346, 364
    temperature, 357
    unstable flow, 366–371
    viscous character, 348–359
ring scission polymerization, 49
root-mean-square dimensions, 228
rubber elasticity, 248–272

rubber elasticity (*cont.*):
  networks
    chain sections, 266
    ideal, 261
    nonideal, 265
    structure, 271
  statistical theory, 260–267
    testing the stress-strain relation, 267–271
  thermodynamics, 253–260
    equilibrium stress-elongation, 256–260
    ideal rubber, 259
rubber plateau region, 199

S

secondary (relaxation) transitions, 205, 320
  in-chain motion, 330
  pendant group, 328
second-order transition (temperature), 205
segment density distribution function, 246
segmental motion, 203, 249, 330
sequence length distribution, 133
shear modulus and compliance, 278
shear rate: see strain rate
shear thickening and thinning, 342
shift factor, 301
simple extension, 277
simple shear, 277
single polymer crystals, 172–181
size distribution: see molecular weight distribution
Smith-Ewart theory, 121
sol, 61
solubility parameter, 22
spherulites, 182
spiral growth, 179
step-function experiment, 293

step-reaction polymerization: see polymerization, step-reaction
stereoregular (stereospecific) polymers, 30
  formation, 82, 92
steric factors, 29–36
  in polymerization, 82, 144
  transitional phenomena, 212–216
stiffness: also see steric factors
  mechanical, 302
stoichiometry, 55, 67
strain, 277
  rate, 278
    apparent, 352
    true, 355
strength, 302
stress-relaxation experiments: see relaxation experiments
stress-strain experiments, 296, 302
  in elastomers, 250
structural regularity, 29: see steric factors
substituent effect in addition polymerization, 84
superposition principle, time-temperature, 299
swell ratio, 364
syndiotactic, 31

T

tacticity, 31
termination, 81
thermal stability, 15, 29, 34
thermoplastic vs. thermosetting, 36
time-scale of experiment, 204, 287, 298
  correlated with molecular motion, 204
time-temperature equivalence, 299, 314–320

time-temperature superposition
principle, 299
application in dynamic mechanical
testing, 314–320
toughness, 302: also see impact
resistance
transfer: see chain transfer
transfer constant, 103
transitional phenomena, 10, 193–224:
see damping peaks
effect of composition, 205–224
compatibility, 222
copolymers and polyblends,
218–224
crosslinking, 209
intermolecular bonding, 207
molecular weight, 216
plasticization, 216
steric factors, 212
mechanical properties, 198–202
five regions of viscoelastic
behavior, 199
molecular motion, 202–205
multiple relaxation (transition)
peaks, 325
in-chain motion, 330
pendant group motion, 328
transition region, 199
transitions
first-order, 193
glass, 196
secondary relaxation, 314, 320–334
second-order, 205
transition temperatures
engineering-use, 201
equilibrium crystalline, 194
glass, 10, 196–198
Trommsdorff effect, 100

U

unit cell, 165

unperturbed chain dimensions, 237
unstable flow, 366–371
inlet-borne, 370

V

van der Waals forces: see bonding
forces
viscoelastic behavior of polymers,
9, 275, 293–339: also see
rheology
linear and nonlinear, 276, 293–339
polymeric solids, 293–339
viscoelastic behavior, five regions of,
199
viscoelastic fluids, 340
viscoelastic models, 281–292
general linear, 288–292
in oscillating experiments, 303–313
simple linear, 281–287
viscoINelastic fluids, 340
viscometric flow, 347, 365
viscosity, 278
viscous behavior, 278, 303
viscous flow of polymers, 348–359
effect of temperature, 357
non-Newtonian, 341
pure (Newtonian), 278
general treatment, 355
Voigt model, 283, 307
generalized, 292
with Maxwell in series, 284
vulcanization, 15

W

WLF equation, 301

X

X-ray diffraction, 163, 190, 249

Y

Young's modulus, 198

Z

zero shear-rate viscosity, 342
Ziegler-Natta catalysts, 92
zigzag conformation, 5, 39